GEOPHYSICAL AND ASTROPHYSICAL CONVECTION

THE FLUID MECHANICS OF ASTROPHYSICS AND GEOPHYSICS

A series edited by ANDREW SOWARD,
University of Exeter, UK and
MICHAEL GHIL,
University of California, Los Angeles, USA
Founding Editor: PAUL ROBERTS,
University of California, Los Angeles, USA

This book is part of a series. The publisher will accept continuation orders which may be cancelled at any time and which provide for automatic billing and shipping of each title in the series upon publication. Please write for details.

GEOPHYSICAL AND ASTROPHYSICAL CONVECTION

Edited by

PETER A. FOX
*National Center for Atmospheric Research
Boulder, Colorado, USA*

and

ROBERT M. KERR
*National Center for Atmospheric Research
Boulder, Colorado, USA*

Contributions from a workshop sponsored by the Geophysical
Turbulence Program at the National Center for Atmospheric Research,
October 1995

CRC Press
Taylor & Francis Group
Boca Raton London New York

CRC Press is an imprint of the
Taylor & Francis Group, an **informa** business

CRC Press
Taylor & Francis Group
6000 Broken Sound Parkway NW, Suite 300
Boca Raton, FL 33487-2742

First issued in paperback 2019

© 2000 by Taylor & Francis Group, LLC
CRC Press is an imprint of Taylor & Francis Group, an Informa business

No claim to original U.S. Government works

ISBN-13: 978-90-5699-258-3 (hbk)
ISBN-13: 978-0-367-39829-3 (pbk)

British Library Cataloguing in Publication Data

A catalogue record for this book is available from the British Library.

ISSN: 0260-4353

Visit the Taylor & Francis Web site at
http://www.taylorandfrancis.com

and the CRC Press Web site at
http://www.crcpress.com

Contents

List of Figures

ix

List of Tables

Preface

Since 1993, the Geophysical Turbulence Program at the National Center for Atmospheric Research together with the National Science Foundation has sponsored interdisciplinary workshops each summer in Boulder, CO. The general theme of the workshops relates to turbulence research but with a wide community focus.

The material in this volume arose out of a workshop held in October of 1995. The purpose of the workshop was to bring together researchers and students working on both astrophysical and geophysical systems, in which convection was an important effect. Contributions to the meeting and to this volume included those from both experimental and observational views, as well as theory and computational efforts.

Each of the chapters herein represents an effort to capture the state of understanding of convection in a variety of contexts, including systems influence by external physical effects (such as rotation, radiation, etc.). Even though the initial contributions were solicited in 1995, many have been updated during the preparation of this volume.

Each of the chapters in this volume has been peer reviewed, edited and proofread. All remaining errors are the responsibility of the editors.

Acknowledgements

We thank Barbara Hansford and Chin-Hoh Moeng for assistance in organizing this meeting. We also thank Amy Bauer for assistance in the preparation of the manuscripts. We wish to acknowledge the support and patience of the staff at Gordon and Breach Science Publishers during the preparation of this volume. All of the manuscripts in this volume were peer reviewed and we wish to express our appreciation to those who performed these reviews. Support of the Geophysical Turbulence Program, NCAR and NSF is appreciated. NCAR is sponsored by the National Science Foundation.

Chapter 1

ATMOSPHERIC CONVECTION WITH ANALOGIES IN ASTROPHYSICS AND THE LABORATORY

ROBERT M. KERR

Geophysical Turbulence Program
National Center for Atmospheric Research
P.O. Box 3000, Boulder, CO 80307-3000

An overview of atmospheric convection is provided and a discussion of some analogies between atmospheric and solar convection. This includes the basic concepts of turbulent scaling, dry and moist atmospheric convection, and the role of large-scale organization in atmospheric convection. A hierarchy of length scales ranging from small-scale convective boundary layer turbulence to large-scale precipitating mesoscale convective systems will be emphasized. New approaches to modeling turbulence and convection are discussed.

1.1 Introduction

In the absence of turbulent fluid motion, heat and energy transfer are relatively steady and are driven by the thermal diffusivity and radiative properties of the fluid; for example, in laminar flows and the dense interior of the sun. For a turbulent fluid, the emphasis of this volume, the heat transfer is chaotic and is usually treated in terms of statistical quantities such as the turbulent heat flux. On larger scales, evidenced by solar granulation and atmospheric convective patterns, it has long been recognized that convective flows are organized. What was not appreciated until more recently is that this 'large-scale organization' persists into the turbulent scales and perhaps controls important scaling properties. Organization of turbulent scales implies coupling between disparate scales, which is usually ignored, but appears to be universal.

As an example of one of the different types of convection represented in this volume, let us consider a slice through a high Rayleigh number direct numerical simulation of laboratory convection (Kerr, 1996) in Figure 1.1. In this slice, taken near the top boundary, narrow, cool flows moving away and down from the surface are darkest, while arrows indicate horizontal flows converging onto the cellular walls. This demonstrates a tendency for surface convergence patterns to generate relatively narrow vertical motions, often marking the boundaries of convective cells. In the sun, the most obvious sign of cellular patterns is solar granulation,

Figure 1.1: Velocities from a $Ra = 2 \times 10^7$, Prandtl number = 0.7, aspect ratio 6 direct simulation of Rayleigh-Bénard convection. The domain runs from $z = -1$ to 1 in the vertical and is 12×12 in the horizontal. Horizontal velocities are visualized using arrows indicating strength and direction at level $z = .61$, near the upper boundary, overlaying grayscale contours of vertical velocity at level $z = .32$.

the possible origins of which are discussed in the contribution by Rast. In the atmosphere, a striking realization of cellular patterns, discussed in the contribution by LeMone, appears in radar images of reflectivity from insects in convective updrafts. The different patterns in these images are associated with the type of shear. Without shear, the patterns are cellular. With shear, convection is organized into streamwise rolls.

This influence of larger-scale shear on convection, discussed in LeMone's contribution, does not have a clear analogy on the sun, and brings up an important distinction between what is known about solar convection and atmospheric convection. This is that solar convection, as seen on the surface and where magnetic fields are not strong, is homogeneous, while in the atmosphere there is a hierarchy of scales for convection, all of which while coupled, are also governed by independent forcing and dynamical mechanisms. It is this hierarchy of scales in the atmosphere that will be the basis for the organization of this contribution. While it might be difficult to decouple the dynamics of different scales in atmospheric convection, separating convective phenomena by scales – small, nearly laminar scales, intermediate turbulent scales, and organized large scales – is the traditional means of dividing convection.

The physics of convection and the simulations to be discussed in this volume, both geophysical and astrophysical, can be divided into three classes. For simulations, going from the least to most parameterized, these are direct numerical simulations DNS (Bartello, Brandenburg, Brummell, Grabowski, Werne, Yuen), large-eddy simulations LES (Canuto et al., Cabot, Marshall), and cloud-resolving models (Bretherton, Rast), where the simulations of solar granulation by Rast have been put in the cloud-resolving category. Often, but not always, the three different classes cover different length scales, going from smallest to largest. The definition and usefulness of each class is discussed further below.

To begin, the common parameterizations, modeling techniques, and scaling properties that will be discussed in this this volume will be outlined following the hierarchy of scales and the degree of parameterization. The properties of fundemental turbulence will be discussed first, then the properties of mixed layers and dry convection. Next, some introduction to the basic concepts governing moist convection and clouds in larger-scale complexes will be given. This is in recognition of fundamental differences between moist and dry convection such as the effects of latent heating and evaporation. Along the way, some of the jargon and concepts that are specific to atmospheric convection, but the astrophysical community is probably not familiar with, will be explained. After the basic concepts are introduced, how these different regimes are interconnected into a hierarchy of scales will be discussed.. Finally, new modeling techniques for turbulence will be discussed to complement the contribution by Canuto et. al. The discussion of the concepts of dry and moist convection will be based in part on the conference presentation by Emanuel and his book (Emanuel, 1994). For a discussion of the basics of solar convection see Gilman's contribution.

1.2   Reynolds number and modeling

Historically, turbulent flows were first considered from a statistical point of view, where the objective was to predict bulk statistical properties such as the heat flux. This approach has led to a number of scaling laws such as Prandtl scaling (Prandtl, 1932), Deardorff scaling (Deardorff, 1970), and Monin-Oboukov scaling (Monin and Oboukhov, 1953), where statistical properties depend upon the bulk properties of the fluid motion, such as its shear and heat flux. The principle is to apply standard turbulent parameterizations for boundary layers and mixing between turbulent and non-turbulent fluid in order to model turbulent mixing and scaling in the region between the surface and the penetrative mixed layer at the top of the planetary boundary layer.

The equations to be solved are the Navier-Stokes equation for velocity $\mathbf{u}$

$$\frac{\partial \mathbf{u}}{\partial t} + (\mathbf{u} \cdot \nabla)\mathbf{u} = -\frac{1}{\rho}\nabla P + (\nabla \cdot \nu \nabla)\mathbf{u} - g\hat{z} \quad , \tag{1.1}$$

and scalar equations of the general form

$$\frac{\partial c}{\partial t} + (\mathbf{u} \cdot \nabla)c = \kappa \nabla^2 c \quad , \tag{1.2}$$

for temperature T, moisture, density ρ, or various combinations of these discussed in section 1.4. In these equations, bold-face indicates vectors, P is the pressure, ν is viscosity, g is the gravitational acceleration, c is the scalar, and κ is the diffusivity. Each of these equations has a nonlinear $\mathbf{u} \cdot \nabla$ advection term and a ∇^2 dissipative term. Pressure is a function of T and ρ in general, although for geophysical flows the incompressible or anelastic approximations $\nabla \cdot \rho(z)\mathbf{u} = 0$ are often made and P is found using a Poisson solver.

Let us define some properties that scaling laws should relate, how these properties dictate what can be studied with models, then discuss the basic assumptions that lead to the classical scaling laws for convection. What one often wants to predict are fluxes in terms of the mean temperature gradient, shear, and dimensionless numbers describing the turbulence intensity. In a neutral fluid, the Reynolds number

$$Re = \frac{UL}{\nu} \tag{1.3}$$

describes the turbulence intensity, where L is a length scale of the turbulent kinetic energy, U is a velocity on this length scale, and ν is viscosity. It can be viewed as the ratio of the nonlinear to viscous terms in (1.1) or as related to the ratio of two length scales, L and a small-scale length for dissipation and viscous effects, which is usually the Kolmogorov length scale $\eta = (\nu^3/\epsilon)^{1/4}$, where ϵ is the rate of dissipation of kinetic energy. If these scales are sufficiently separated, the energy spectrum as a function of wavenumber $E(k)$ will have an inertial subrange, a wavenumber range where fluxes of energy and other conserved quantities determine the scaling properties. For fully-developed three-dimensional turbulence, the most recent experimental review by Sreenivasan (1995) confirms the classical

$$E(k) = \mathrm{Ko}\epsilon^{2/3}k^{-5/3} \tag{1.4}$$

spectrum of Kolmogorov with a universal value for Ko, the Kolmogorov constant. There are several variants upon the Reynolds number depending on which length scales are chosen. When L is the scale of the turbulent domain, one obtains the large-scale Reynolds number $R_L \sim (L/\eta)^{4/3}$. Another is the Taylor microscale Reynolds number

$$R_\lambda = U_{rms}\lambda/\nu \quad , \tag{1.5}$$

which uses the average fluctuating velocity U_{rms} and the Taylor microscale λ defined by the dissipation rate ϵ, $\lambda \sim U_{rms}/(\epsilon/\nu)^{1/2}$. Use of this length scale, which is typically in the middle of the -5/3 inertial subrange, implies $R_L \sim (R_\lambda)^2$.

When the exact equations of motion (1.1,1.2) with real Newtonian (∇^2) viscosities and diffusivities are used and all scales of the simulation are resolved, one is doing direct numerical simulation or DNS. It is R_λ that determines the range of scales and how many grid points are necessary to properly resolve a direct numerical simulation or DNS where the dissipative scales are not parameterized. Just as R_λ indicates how many grid points are required to properly resolve a flow, given the number of grid points available, an estimate for the largest R_λ feasible can also be made. This principle can be applied to large-eddy simulations or LES, where the primitive equations of motion are used for the largest scales and the smallest turbulent scales are parameterized. For LES, the effective numerical Reynolds numbers are the order of the number of grid points squared and this Reynolds number must be taken into account when in deciding whether small-scale effects in LES are related to the large-scale forcing, or are associated with the effective finite Reynolds number. The largest meshes discussed in these proceedings are 256^3, which implies a maximum effective Reynolds number of the order of $200^2 = 40,000$. Let us consider how these limitations determine what of value can be simulated, and what must be parameterized, in simulations of any scale in the atmosphere, laboratory, or sun, using the three classes of models considered here, DNS, LES, and cloud-resolving.

The value of DNS is that it is the only reliable tool if viscous, diffusive or resistive effects are expected to play a fundamental role in the phenomenon being studied. Some drawbacks with DNS is that while true viscous boundary layers can be simulated efficiently for simplified boundary conditions, this is at the cost of numerical complexity and a restriction to these boundary conditions. In addition, since much of the resolution is committed to these purposes, either one studies domains that are many orders of magnitude smaller than true atmospheric motions, or the Reynolds numbers is far below atmospheric values. As discussed in the contribution by Yuen, the only geophysical flow where the Reynolds number is low enough for DNS to be directly applicable is in the earth's interior.

The value of LES is that since the smallest turbulent scales are parameterized, in particular the surface layers, realistic atmospheric forcing can be applied to the turbulent scales and small-scale atmospheric motions. What range of scales a particular LES covers is determined by the types of small-scale parameterizations used and the type of resolved scale motion to be represented. When only the smallest scales are parameterized, it is customary to augment the dissipation, replacing ν with an eddy viscosity $\nu_e(\mathbf{x})$ in (1.1) that is designed to represent the dissipative effects of the subgrid turbulent scales upon the resolved scales. While this simple approach has been found to be adequate for parameterizing the subgrid scales in variety of situations, including convective and shear flows with and without boundaries, new approaches to subgrid modeling are discussed in section 1.6 and in the contribution by Canuto et al.. Only the subgrid scale stresses and shears are parameterized in LES, bulk shears are still used to force an LES and intermediate scale shears are fully simulated. Besides being a tool for finding turbulent parameterizations that can be used in simulations of larger-scale phenomena (Ayotte et al., 1995), LES is a useful comparison for boundary layer observations. Cloud-resolving models (CRMs) are a class of LES that might use turbulent as well as other parameterizations of small-scale physical processes to simulate larger-scale cloud systems that span intermediate or mesoscale regions.

As a variety of LES, CRMs use the primitive equations for the determining the velocity, which due to increases in computational power is currently feasible in three-dimensions over thousand kilometer, mesoscale domains with horizontal mesh sizes the order of kilometers. Previously, due to coarse resolution, mesoscale models were hydrostatic. That is, while horizontal velocities were determined prognostically, the vertical velocity was determined diagnostically from a hydrostatic relationship between pressure and gravitational forces. This is an appropriate modeling technique when the horizontal resolution is much larger than the vertical resolution. It remains the primary means for determining the velocity in global models for climate and numerical weather prediction. In the hierarchy being presented, CRMs have roles in determining the structure of local meteorological events such as storms, understanding convective organization, and giving insight into parameterizations of properties such as the bulk effects of deep convection upon global scales (Moncrieff 1995; Grabowski et al. 1996).

1.3 Dry convective scaling

While the Reynolds number is a useful concept for determining the intensity of the turbulence and setting the scales for what can be simulated, in convection it is more traditional to use the Rayleigh number,

$$Ra = \frac{\alpha \sigma g d^3 \Delta T}{\nu^2},$$ (1.6)

where α is the coefficient of thermal expansion, $\sigma = Pr = \nu/\kappa$ is the Prandtl number, κ is the thermal diffusivity, d is the height, and ΔT is the temperature difference over d. Given the square of viscosity in the denominator, typically $Ra \sim R_L^2 \sim R_\lambda^4$. Assuming that the maximum R_λ is given by the mesh size, for three-dimensional direct simulations to be discussed here of the order of 100^3, the maximum Rayleigh number of DNS is the order of $Ra \sim 10^8$. Atmospheric Rayleigh numbers on the other hand are typically the order of $Ra \sim 10^{16}$, solar values can be as high as 10^{28} and astrophysical Rayleigh numbers can be even higher. Nonetheless, if turbulent scaling relations are independent of Reynolds and Rayleigh numbers once a certain value is reached and if there are no intermediate forcing scales, then studies of convective scaling at the highest Rayleigh numbers accessible to current direct simulations should have some relevance to these large geophysical and astrophysical values.

Another dimensionless number relating convection and shear is the Richardson number, one definition of which using the velocity gradient is

$$Ri = N^2 / (\frac{\partial u}{\partial z})^2$$ (1.7)

where $N^2 = g\alpha(dT/dz)$ is the Brunt-Väisälä frequency for gravity waves. Again, like the Reynolds number there are several other varieties of the Richardson number that depend on how the competitive effects of buoyancy and shear are represented. When Ri is strongly positive, the atmosphere is stabilized by stratification, which will only be touched upon briefly in this volume. When Ri is negative, the atmosphere is convectively unstable. However, Ri positive and small, less than $1/4$, is a necessary but not sufficient condition for a stratified flow to be unstable to shear. A related quantity is the *Convective Available Potential Energy* CAPE, which can be used in the numerator of the Richardson number. CAPE is discussed further in the section 1.4 for its role in moist convection.

Rotational effects are important whenever the Rossby number $Ro = v/Lf < 1$, where $f = 2\Omega \sin\phi$ is the Coriolis term with Ω the rotation of the system and ϕ the latitude. For the earth $f = 1.47 \times 10^{-4} \sin\phi$. These effects will be felt over the Rossby radius of deformation $L_R = (gH)^{1/2}/f$, which represents the length scale over which deformation of a layer of cellular motion height H can be balanced by rotation. For oceans, depending on whether it is a shallow sea where $H \sim 40$ meters or the deep ocean where $H \sim 4$ kilometers, L_R can vary between 200 and 2000 kilometers. When L_R is small, Coriolis interactions between mixed fluid in convective vortices or chimneys and the surrounding stratified fluid can form fronts in the ocean, as discussed for the case of convective chimneys in the contribution by Marshall. Simulations of rotating convection in the contribution by Werne have demonstrated that chimneys, strong vertical vortices containing most of the convection, can develop through interactions between convection and rotation. For the atmosphere, if H is taken to be the height of the tropopause, the resulting $L_R \approx 3000$km is too large to be consistent with the size of observed mesoscale convective systems. The theoretical reason (Gill, 1982, p. 207) for this is that over these distances the constant f-plane approximation, that is $\phi =$constant, is invalid and therefore gravitational instabilities on a smaller scale become important in the atmosphere. Atmospheric internal waves are one source of a smaller length scale for gravitational effects that would balance the tendency for the Coriolis term to deform the height of the cell. Therefore, in the atmosphere it is more appropriate to compare of the frequency of gravity waves N to f, giving a size for mesoscale convective systems of $L_R = NH/f \sim 1000$ kilometers. This is the scale on which dry, subsiding air is influenced by rotation. This does not necessarily apply to smaller-scale rotating convective disturbances, where the horizontal scales are usually set by the mean horizontal velocity, the evaporation rate and the time it takes for precipitation to fall out.

When there are boundaries, the classical scaling laws of convection, following Prandtl (1932), result from assuming that there is a single thin layer that matches the properties of surface with the bulk of the fluid, in this case the thermal properties. In the case of Rayleigh-Bénard convection (Malkus, 1954) this leads to the prediction that the normalized heat flux or Nusselt number should depend on the Rayleigh number as

$$Nu = < w\theta > /\kappa \frac{d\Theta}{dz} \sim Ra^{1/3} \quad , \tag{1.8}$$

where $d\Theta/dz$ is the mean temperature gradient.

The assumption that heat flux as a function of Rayleigh number can be modeled with only one boundary layer length implies that there are no intermediate scale features in the velocity field that would modify the heat flux relationships. By intermediate scale, one might mean smaller than large-scale properties such as the mean flow, but larger than the dissipation scales. This implicit assumption has been called into question after a series of experiments under Libchaber, partially summarized by Castaing et al. (1989), where the primary conclusion was that the heat flux law should be modified to read $Nu \sim Ra^{2/7}$. The new 2/7 regime was termed 'hard' turbulence, as opposed to the classical 1/3 power law for Nusselt number. As will be discussed more in the contributions by Adrian, Zaleski, Belmonte/Libchaber, and Werne, 2/7 rather than 1/3 seems ubiquitous in experiments and simulations in the laboratory regime. Figure 1.1, a horizontal slice from a simulation with a 2/7 regime, demonstrates an important property of flow that gives 2/7. This is that the flow is not a homogeneous mixture of turbulent eddies, but instead shows larger-scale organization. This is further illustrated by a cartoon of a vertical slice in Figure 1.2. The important point is that broad downdrafts when they hit a surface generate surface shears, which in turn collide to generate narrow updrafts. These spread some, but remain sufficiently organized across the convection cell so that when they hit the opposite boundary, new surface shears are generated and the process is renewed. The development of these intermediate-scale shears creates the second length scale that is required in order for the classical 1/3 law to be broken. Whether this regime has much to do with atmospheric convection has been properly questioned since atmospheric convection is usually evolving, penetrating, and often lacks smooth boundaries. As a result, surface shears generated by convection are weaker and

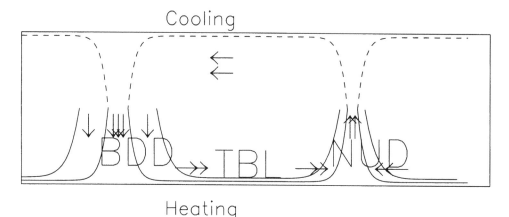

Figure 1.2: Fluid motion in Rayleigh-Bénard convection in the hard turbulence regime is characterized by broad downdrafts (BDD), thin surface shear layers (TBL) and narrow updrafts (NUD). In classical Rayleigh-Bénard convection the large scale flows in the upper half (dashed lines), along the rigid upper surface, mirror those on the lower surface. In atmospheric marine stratocumulus and solar convection at the photosphere, the region cooled by phase transitions and radiation is not a rigid surface and there is asymmetry in the vertical flow.

one of the key components of explaining the 2/7 regime, the importance of surface shears, is more difficult to justify.

The lesson that might be learned from hard convective turbulence is not the particular scaling law, but the more general principle that organization on scales larger than the scales of individual plumes can change the entire dynamics. For example, the strong convergence that produces narrow updrafts could be compared with how surface convergence is a precondition for generating deep convection in the atmosphere.

Other important questions that remain for the ideal Rayleigh-Bénard problem are the dependence of Nusselt number on Prandtl number and whether there are new transitions at Rayleigh numbers beyond the current experiments, either to 1/3 or something new. The high Rayleigh number experiments that give 2/7 are all for gases, $Pr = 0.7$. In the oceans, for temperature and salinity, $Pr = 7$ and 700 respectively, while for a radiative fluid like the sun, $Pr \approx 0$. The assumptions about the relative positions of the thermal and momentum boundary layers that are used by the current models for 2/7, would not necessarily apply in these Prandtl number regimes. The only theoretical attempt to address the Prandtl number issue, by Shraiman and Siggia (1992), does not agree with the preliminary study of low Rayleigh number numerical simulations in Kerr et al. (1995), as discussed in the contribution by Zaleski.

Returning to the classical assumptions, these have been developed further for free convection in the atmosphere by Priestley (1954) and Deardorff (1970) to give scaling with height and a velocity scale, respectively. To get the scaling of velocity fluctuations in Rayleigh-Bénard convection, if one takes the heat capacity $\rho c_p = 1$ and starts from the requirement that the energy input, the heat flux $Q = <w\theta>$, must be balanced by the kinetic energy dissipation $\epsilon_d = \int dV \nu (\nabla u)^2$, then

$$(Nu - 1)Ra = <(\nabla u)^2 > d^4/\kappa^2. \tag{1.9}$$

If one then makes the usual assumption for turbulence that the kinetic energy dissipation depends only on the large velocity and length scales, that is $\epsilon_d = U^3/L$, where L is taken to be the height of the box d, then it follows that for $Nu \sim Ra^{1/3}$ that $U \sim Ra^{4/9}$. Replacing L with z for the atmosphere one then predicts that the vertical velocity should scale as

$$w \sim (Qz)^{1/3} \quad , \tag{1.10}$$

from which it follows that if Q is independent of height, temperature fluctuations should go as

$$\theta \sim Q^{2/3} z^{-1/3} \quad . \tag{1.11}$$

However, as noted above, atmospheric flows can rarely be characterized as simple convective flows, since there usually is large-scale shear. For purely sheared flows, boundary layer scaling should obey the classical laws of aerodynamics, in particular the log law near the surface

$$\bar{u} \sim |M|^{1/2} \log(z/z_0) \quad , \tag{1.12}$$

where $M = \overline{u'w'}$ is the turbulent momentum flux. A convenient way to represent the competition between the effects of shear and convection is to introduce a length determined from the two energy inputs, momentum stress and heat flux, the Monin-Oboukov length scale

$$l_{MO} \sim -M^{3/2} Q^{-1} \quad . \tag{1.13}$$

In Monin-Oboukov scaling, the significance of this length is that below it shear layer scaling is used, that is the log law, and above it free convective scaling is used. Monin-Oboukov scaling is found to work remarkably well locally in the atmosphere and is the basis of most parameterizations of the planetary boundary layer. Monin-Oboukov scaling is derived from the assumption that the shear and convective regions do not directly influence each other, which, based upon our experience with Rayleigh-Bénard convection, is not always true.

1.4   Precipitating Convection

While the smaller-scale processes that are modeled in the atmosphere can primarily be modeled by dry dynamics, motions that are broader in the horizontal and deeper in the vertical require consideration of phase changes involving moisture and precipitation. Due to these phase changes and the competing effects of density, temperature and moisture, thermodynamic effects are often best represented not by the primitive variables, but combinations thereof that better represent cumulative effects. A few of these terms will be defined here because they are repeatedly referred to in the contributions. These and other thermodynamic quantities are defined in more detail in standard texts such as Emanuel (1994) and Gill (1982). The most commonly used replacement for temperature is the potential temperature, which for a dry atmosphere is

$$\theta = T \left(\frac{p_0}{p}\right)^{R_d/c_p} \quad , \tag{1.14}$$

where R_d is the gas constant for dry air and c_p is the dry heat capacity at constant pressure. θ can also be written in terms of the dry specific entropy $s_D = c_p \ln(T/\overline{T}) - R_d \ln(p/\overline{p})$ as

$$\theta = \overline{T} \exp(s_d/c_p) \quad . \tag{1.15}$$

R and c_p are often replaced by effective values that include the effects of moisture, which can be introduced through modifications of the entropy. For adiabatic displacements of unsaturated air, θ is conserved. That is, as a parcel of dry air changes height and pressure, which by the ideal gas law changes the temperature, θ will not change. Therefore, in order to track buoyancy it is usually easier to work with θ and its variants than the fundamental variables of temperature, moisture, and pressure.

A quantity that does not change along trajectories in a moist atmosphere is the equivalent potential temperature θ_e, which can be defined in terms of the total specific entropy

$$s = s_d + r s_v + r_l s_l \tag{1.16}$$

as

$$\theta_e = \overline{T} \exp \left(\frac{s}{c_p + (r + r_l)c_l} \right) \quad , \tag{1.17}$$

where the subscripts d, v, and l refer to contributions from the dry, water vapor, and liquid water components respectively. r and r_l are the vapor and liquid water mixing ratios, and c_l is the liquid water heat capacity. Equivalent potential temperature is what the temperature would be if all the moisture were condensed out of a parcel that has been displaced to a reference pressure. This is the quantity most relevant for stability studies, where sometimes, the weight of condensed water is also included. However, neither the potential temperature, the equivalent potential temperature, nor other temperature equivalents determine buoyancy directly. The buoyancy of a parcel depends on the difference between the temperature equivalent of the parcel 'p' and the value for the background or ambient air 'a' which the parcel must compete against in order to rise.

The strength of deep convection depends on the potential energy contained in surface layers before parcels begin to rise above these layers. The quantity that has been constructed to represent these effects in the parcel being considered is the *Convective Available Potential Energy* or CAPE. CAPE is an integral quantity over depth up to the level of neutral buoyancy LNB of a parcel at height z_n, which is typically at or below the tropopause. For a given reference height L

$$\text{CAPE} = \int_L^{z_n} B \, dz = \int_L^{z_n} g \frac{(\theta_{\rho p} - \theta_{\rho a})}{\theta_{\rho a}} \, dz = \int_{p_n}^{p_L} R_d (T_{\rho p} - T_{\rho a}) \, d \ln p \tag{1.18}$$

where B is the buoyancy, T_ρ is the density temperature, which includes the buoyancy effects of moist, cloudy air, and the second relation comes from the ideal gas law and the thermodynamics of a moist adiabatic process. It is this form that allows one to calculate CAPE from thermodynamic stability charts as an area between the parcel adiabat and the ambient adiabat, such as in the contribution by LeMone. The principal processes that generate CAPE are surface fluxes and large-scale advection.

The presence of CAPE within the boundary layers, which are between 600 and 2000 meters deep, is not destabilizing by itself and some mechanism for overcoming the convective inhibition energy, or CIN, is needed. How boundary layers can break through capping layers or potential barriers, release CAPE, and develop into deep convection is discussed in the contribution by LeMone. In addition to local instability mechanisms (Clark et al., 1986), another mechanism results from the balancing of large-scale convergence, schematized in Figure 1.3. This is that in the region of convergence, where the inflow from opposite directions ultimately meets, upward motion can overcome the large-scale subsidence and CIN. Once vertical motion has started, the ascent to lower pressure and the release of latent heat, lifts the air even higher, resulting in deep convection. Squall lines that form along lines of convergence in an MCS are responsible for some of the most violent weather at mid-latitudes, such as thunderstorms.

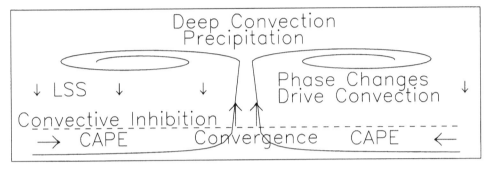

Figure 1.3: Deep convection results when the *Convective Available Potential Energy* CAPE in the boundary layer, generated by the moist, warm ocean surface, is forced upwards by convergence. Until convergence pushes the air up, the boundary layer is held in place by a potential barrier of convective inhibition and large-scale subsidence LSS. Once the air in the boundary layer is forced high enough, there are phase changes and the release of latent heat, providing buoyancy that pushes the air even higher. The conditional stability of the boundary layer, measured by CIN, is one of the key features of moist convection.

Once one of these mechanisms has ignited deep convection, additional organization and shears are needed in order to maintain a mesoscale convective system. For example, there must be a balance between energy inflow and dissipation mechanisms in convection and precipitation, a meteorological equivalent of the concept of fluctuation-dissipation mechanisms in nonlinear physics. This balance can be used to parameterize convective effects in global models using concepts such as the quasi-equilibrium assumption of Arakawa and Schubert (1974). However, the detailed mechanisms for maintaining deep convection require local shear patterns that drop precipitation away from updrafts, suck in warm, moist air from outside the immediate neighborhood of the storm, move the location of deep convection over new sources, and so allow the convective engine to be continually restoked. Some mechanisms for how this can be accomplished and their relationship to larger-scale organization are given below.

For hurricanes, Charney and Eliassen (1964) showed that the release of latent heat in the eye wall of the hurricane drives strong updrafts that are balanced by strong, convergent flow into the hurricane along the surface. In the more typical situation over the oceans, nearly continuous deep convection and precipitation along a squall line can be maintained by shear that develops perpendicular to the squall line and moves the squall line so that it is always over new energy sources. An analogous situation over land where warm, moist surface inflow on one side of the squall line rises over a low-level cold pool on the other side of the squall line, as discussed by Rotunno et al. (1988), is diagrammed in Figure 1.4 to show the differences with dry Rayleigh-Bénard convection in Figure 1.2 and the effects of large-scale convergence leading to deep, moist convection in Figure 1.3. For tropical cumulonimbus, it has been shown by Moncrieff and Miller (1976) that the propagation velocity can be related to the CAPE between the level of warm inflow L and the upper outflow z_n. A feature of convective storms over land is that they often develop only intermittently along the squall line due to variations in the local forcing. An example is rotating supercell storms which, as discussed in the contribution by Lilly, are the source of the strongest tornadoes. Supercell storms derive their energy by tracking along the squall line instead of moving the squall line over new sources. These and other precipitating convective systems are discussed in more detail in texts such as Houze (1993). An underlying theme in this volume is that despite distinct differences in the particular geometries, a number of local, precipitating convective systems depend on forcing and dissipative mechanisms connected by organized fluid motions within the storms themselves.

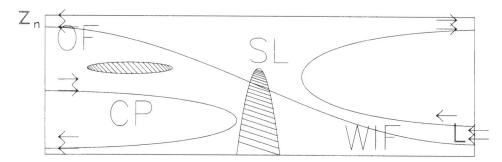

Figure 1.4: Fluid motions near a squall line (SL) that maintain precipitation. The entire system is moving to the right. Relative to the squall line, there is warm inflow (WIF) at low levels L from the right, which is lifted over a cold pool (CP) and leaves the system at high levels z_n on the left. Hashed areas indicate convective precipitation that occurs along the squall line due to condensation of the warm, moist inflow and stratiform precipitation in the outflow.

Complimenting the organization within storms, evaporatively-driven downdraft outflows or density currents originating from regions of precipitation, such as those indicated in Figures 1.3 and 1.4, can provide a mechanism for maintaining convection through larger-scale organization. That is, falling precipitation, a source of buoyancy reversal, can produce strong cold downdrafts that, as they hit the surface, generate outflows that spread out and produce surface shear. The outflow from one local system can become a cold inflow into another, such as the inflow on the left of Figure 1.4 that is the source of the cold pool. The cold inflow, by colliding with the larger-scale warm, moist flow represented by the inflow on the right of Figure 1.4, can then produce new lines of convergence and new convection. This is discussed further in section 1.5. Field experiments such as the (Global Atmospheric Research Programme) Atlantic Tropical Experiment, GATE, reviewed by Houze and Betts (1981), support this view. An astrophysical analogy might be how compression waves from supernovae in gas clouds are believed to trigger the formation of new stars, which in these clouds tend to be massive and end their lives as supernovae, which in turn trigger a new cycle of formation.

1.5   Hierarchy of scales

Let us now consider how the different physical regimes of convection that have been outlined fit together into a hierarchy of scales. The separation of convective regimes was arbitary, but has value and necessity based upon the different physical processes and types of simulations that can be applied. Recall the three classes of simulations, direct numerical simulation DNS, large-eddy simulation LES and cloud resolving models. A simulation within each class can be justified only if the true, unresolved small-scales oscillate sufficiently rapidly to allow averaging and parameterization, and if there would be minimal feedback from the largest simulated scales into the larger dynamics that provides the large-scale forcing in the simulation. How this can be approached mathematically and numerically for the case of stable, rotating flows is discussed in the contribution by Bartello. Complete decoupling of dynamical regimes is rarely realized and as computing power increases and the ranges available to each modeling method overlap, the importance of coupling between widely varying scales is becoming recognized.

Let us consider some of the simplifications that are used to study convection on different scales. To study how shear changes convective patterns from cells to rolls, convection with imposed shear can be studied in isolation without considering the possibility that the origin of the shears themselves might derive from small-scale convection. This is the justification for the LES parameterization study discussed by Ayotte et al. (1996) and models such as those discussed in the contribution by Canuto et al., where locally homogeneous and steady conditions on the small scales are assumed. One place where homogeneous and steady conditions can be established is laboratory Rayleigh-Bénard experiments, as described in the contributions by Adrian and by Belmonte/Libchaber.

Despite strong time dependences, an atmospheric situation where the assumptions of steady growth and horizontal homogeneity are applicable is to non-precipitating convection, from which condensation and large-scale organization can develop and profoundly later the nature of the convection. For example, during the morning, air heated by surface warming can intrude steadily into stratified, stable layers formed by nighttime cooling near the ground, such that boundary layer processes and small-scale turbulence, discussed in section 1.2, have roles. Another simplification that applies to this situation is that one can replace the primitive buoyancy variables, temperature and moisture, with one of the variety of buoyancy equivalents discussed in section 1.4, so that simulations and theory for dry convection can be applied. This is one of the points of the entrainment study discussed in the contribution by Grabowski.

What determines the demise of the period of steady growth, and when larger length scales become important asin Figures 1.3 and 1.4, is when convection starts to interact with warmer air at or below the tropopause, whose height varies from 6 to 16 kilometers between the poles and the equator. The tropopause is the bottom of a deep layer of stable stratification heated by absorption of ultraviolet radiation by the ozone layer. Below it, in the troposphere, there can be additional stable layers above the surface, but for the most part conditions favor convection. Above the tropopause, convective growth is inhibited by stable stratification, although overshooting from the convective layers below occurs. As an example of how convection can be inhibited below the tropopause, let us consider stratocumulus formation. In this case, phase changes occur when dry, anti-cyclonic subsiding air aloft is entrained by moist rising air, leading to cooling and negative buoyancy. The effectiveness of the entrainment mechanism is limited by the scale height for moisture, which from the Clausius-Clapeyron equation for saturation as a function of temperature, is about 3 to 5 kilometers. Radiative cooling of the resulting cloud layers also plays a role. Stratocumulus formation resembles laboratory Rayleigh-Bénard convection in Figure 1.2 in the sense that there is convective forcing from both the top and bottom.

This analogy to Rayleigh-Bénard convection has been carried further in our most recent understanding of stratocumulus formation. The cooling that occurs when moist air meets dry subsiding air can occur relatively uniformly over a large area, forming horizontally homogeneous cloud layers. It has long been recognized (Squires, 1958; Lilly, 1968), that the entrainment is due to turbulence of unknown origin. For example, eddy diffusivities and thermodynamic relationships can be used to estimate heights for boundary layers that are qualitatively correct. However, it is now understood that these models fail to properly explain the turbulence generation mechanism needed to maintain cloud top entrainment and do not properly account for the observed uniformity in the boundary layer height in the ITCZ, the *Intertropical Convergence Zone*. Recent fine-resolution calculations by Grabowski (1993), discussed in his contribution, have shown that the concepts of large-scale organization in convective flows with a structure similar to that in Figure 1.2 must be invoked in order to maintain the necessary levels of shear and turbulence needed to maintain the mixing at the top. The traditional concepts of cloud top mixing (Squires,1958; Lilly, 1968), utilizing eddy diffusivities to mix the moist air below with the dry air aloft, are inadequate because the turbulence necessary to maintain the eddy diffusivities is transported to the ground by downdrafts, as discussed in the contribution by Bretherton. While the traditional approaches, and simple physics that uses the scale height of moisture (Malkus, 1958), can give a qualitative picture, Schubert et al. (1995) have suggested that coupling with larger scale flows and generalization of the geostrophic balance of Coriolis and pressure forces in the atmosphere might be necessary to explain features such as the uniformity in boundary layer height in the ITCZ.

In classical Rayleigh-Bénard convection, downward flow from a rigid, cooled plate has the same character as the upward motion as in Figure 1.2. In the atmosphere, downward motion is largely associated with return flow from the upward convective plumes. This is true for growing convective layers, overshooting convection into the tropopause in Figure 1.3, and marine stratocumulus. Because of the different structure and the asymmetry between upward and downward convective motion in the atmosphere, simple eddy diffusive arguments for mixing need to be modified, as demonstrated by the LES calculations of Moeng and Wyngaard (1984). One suggestion is to modify eddy diffusivity arguments taking the skewness of the vertical velocity into account (Wyngaard and Weil, 1991), treating the updrafts and downdrafts with different eddy diffusivity formulations. Another approach is to mix widely separated layers directly, as suggested by the transilient matrix method discussed in the contribution by Stull.

The above two mechanisms for limiting convective depth in the atmosphere, namely interaction with warm stable layers and phase changes, have solar analogues discussed in the contribution by Gilman. The analogue to the effects of stratification upon overshooting convection occurs at the base of the solar convection zone, where it is believed that downward plumes will cross the overshoot layer that represents the boundary between the outer convective zone and the strongly stratified radiative zone in the solar interior. Simulations discussed in the contribution by Brandenburg designed to represent small volumes within the solar convective zone have shown that strong, localized plumes in downdrafts are capable of generating magnetic fields through dynamo action. The analogue to phase changes limiting convection occurs on the sun's surface, where granular convection is homogeneously distributed and is limited in the vertical by cooling from recombination and radiation, as discussed in the contribution by Rast.

Compared to the size of the sun, granulation is an inherently small-scale process. Evidence for supergranulation, with mesogranulation inbetween, is discussed in the contribution by Gilman, although no mechanisms for producing these structures are known. Because large-scale patterns appear to be secondary in solar granulation, local simulations of compressible convection like those in the contribution by Brummell are easier to justify for solar convection than for the atmosphere.

This is not to suggest that multiple scales in solar convection do not play a role. Instead, homogeneity in part implies that the essential dynamics should obey scaling laws over a wide range of length scales such that only a small portion of that regime needs to be simulated in order to reproduce the types of structures and scaling that are observed. This is one of the important lessons from the fundamental turbulence community that is being stressed in this volume. The principle is that simulations using the order of 256^3 points (see the contribution by Werne) coupled with scaling theories (see the contribution by Zaleski) can be used to determine the structure and scaling observed in the experiments discussed in the contribution by Belmonte/Libchaber.

In the absence of additional capping layers such as dry enviromental air, convection originating in the atmospheric boundary layer can develop into deeply penetrating plumes and overshoot into the stable layer above the tropopause, forming 'deep' convection. Boundary layer convection will penetrate up to the tropopause either by steadily eroding the stable layers with time, or it can be forced upwards by strong convergence near the surface to the point where the release of latent heat will make it more buoyant. The importance of convergence is discussed further in section 1.3 using Figure 1.3. 'Deep' convection can be viewed as moist convection that has penetrated far enough, and therefore cooled enough, for there to be phase changes and precipitation. Deep convection is invariably associated with precipitation and plays a role in precipitating systems ranging in size from afternoon thundershowers to hurricanes. One of the fundamental aspects of deep convection is the occurrence of flow organization of a type quite different from the horizontal homogeneity that is associated with stratocumulus and laboratory convection. In this regard, the vertical shear of the horizontal mean flow and coupling to larger-scale dynamics have fundamental roles in both convective initiation and maintenance. Among the largest organized systems are the so-called mesoscale convective systems (MCS), which occur at both tropical and mid-latitudes and are a source of severe weather. Important dynamical scales within an MCS range between on the order of a kilometer, the scale of boundary layer eddies and gravity waves at the top of the surface inversion layer, and the synoptic scale of 1000 kilometers, determined by the Rossby radius of deformation L_R. While these are the largest atmospheric structures to be considered in this volume, questions concerning how mesoscale systems interact with organized flows and other processes in the atmosphere, up to global scales, are also being addressed (Moncrieff, 1995).

Compared to an atmospheric MCS, in solar convection one might speculate that the range of scales one must consider is from smallest scales that can be seen, roughly the size of the faintest lanes between granules, about 0.1 Megameters, to the depth of the convection zone, the order of 200 Megameters. In both the atmospheric MCS and solar cases the range of dynamically important horizontal scales is the order of 1000. This is the range of scales that must be simulated if all of the dominant, coupled dynamical mechanisms are to be represented. In the atmosphere, this is certainly within the range accessible to two-dimensional simulations similar to Krueger (1988) and depending on how much vertical resolution is necessary, is on the margin of the current range of three-dimensional simulations.

For example, a two-dimensional cloud-resolving model over an ocean with scales between 1 kilometer and 900 kilometers by Grabowski et al. (1996) has demonstrated how, by simulating a wide range of scales, new dynamical effects related to coupling between different regimes can be realized. An important difference between what is known about atmospheric convection and solar convection is that there are no known large-scale dynamical systems like an MCS on the sun. However, it is interesting to suppose that given the depth of the solar convection zone, approximately 1/3 the radius of the sun, there might be giant cells that form in response to rotational, magnetic and convective forces, as discussed in the contribution by Gilman.

For the earth's atmosphere, the roles of the tropopause and the scale height for moisture have already been mentioned. One might ask whether the scale heights of density and temperature have roles in either describing or determining convective phenomena in the terrestrial atmosphere. In the sun, due to the enormous variation in height of density and temperature within the convection zone, their scale heights are often used to define the depth of convective layers (see Gilman). This is not the case in the atmosphere, where convection typically occurs well within the scale heights of density and temperature. This does not imply they do not have roles. The scale height of density is uniform across the globe at about 7 kilometers, and is represented in models by using pressure instead of height as the vertical variable. The scale height of temperature is typically much higher, about 30 kilometers and is so far above the tropopause that it has little effect upon convection.

1.6   Improving LES

The general principle behind large-eddy simulation is to use the statistical properties of the small subgrid scales to parameterize their effect upon the large resolved scales. The Smagorinsky subgrid eddy viscosity is one of the simpler and most effective formulations for addressing this, using only the magnitude of the resolved scale stress tensor multiplied by a coefficient to determine the eddy viscosity. This is usually done by adding a subgrid transport to (1.1) of the form

$$\partial \tau_{ij} / \partial x_j \tag{1.19}$$

where the subgrid scale Reynolds stress is

$$\tau_{ij} = \overline{u_i u_j - \overline{u}_i \, \overline{u}_j} \quad , \tag{1.20}$$

and the overbars indicate averages over the subgrid scales. In the original Smagorinsky (1963) formulation τ_{ij} is modelled using

$$\tau_{ij} - (\delta_{ij}/3)\tau_{kk} = -2C_s\Delta^2(e_{ij}e_{ij})^{1/2}e_{ij}, \tag{1.21}$$

where Δ is the mesh size, e_{ij} is the strain and often using with a variable coefficient C_s (Schumann, 1975; Sullivan et al., 1994) ranging between 0.1 and 0.2. However, the choice of C_s is *ad hoc* and the primary benefit from use of this model is allowing enough fluid motion to keep the flow turbulent while also controlling unexpected excursions in fluid velocities that would make a simulation unstable or unphysical. This section will discuss other approaches, based more directly upon the structure and statistical nature of the small scales. The goal is to develop methods that will useful for a wide variety of flows with a minimal number of adjustable constants and parameterizations. However, despite the increasingly sophisticated models discussed, it should be understood that even the simpliest models, combined with reasonable assumptions about the properties of turbulence, usually predict adequate qualitative behavior.

The most direct approach to improving upon the Smagorinsky model is to make the coefficient C_s dependent upon additional properties of the resolved scales. A value of $C_s = 0.2$ can be estimated for flows away from boundaries by simple considerations based on how energy must cascade to the small scales (Lilly, 1966). In the simple *ad hoc* parameterizations, C_s is reduced for shear flows near boundaries to compensate for the observation in simulations that $C_s = 0.2$ produces too much dissipation near boundaries, suppressing instabilities and turbulence. For example (Sullivan et al., 1994),

$$C_s = 0.2 \frac{|\frac{du'}{dz}|}{|\frac{d<u>}{dz} + \frac{du'}{dz}|} \quad , \tag{1.22}$$

where $|\frac{du'}{dz}|$ represents the fluctuating strain and $|\frac{d<u>}{dz}|$ the average strain or shear. But there is a caveat. Namely, however simple these parameterizations are at present, detailed verification was necessary to determine that this is an adequate solution. Modeling in this fashion implies that for every new situation, detailed parameterizations studies would be necessary to determine the appropriate models and coefficients. Therefore, it would be more desirable to have a subgrid formulation that adjusts more readily to a variety of physics. In order to accomplish this goal and to predict more than just the amount of eddy viscosity needed to suppress instabilities, more information is needed about the types of subgrid structures, their statistics, and how they interact with the resolved scales.

One approach to resolving this issue is to use some of our current understanding about the statistical nature of turbulence. For example, second-order closures (Herring et al., 1982) embody much of the classical statistical character of the turbulent energy cascade. Related to this are various renormalization schemes (Yakhot and Orszag, 1987). This approach underlies the contribution by Canuto et al. herein and the work of Lesieur and Chollet discussed in detail by Lesieur (1990). This approach assumes that the energy cascade isotropizes the small scales and is always dissipative. Both of these assumptions are questionable. Let us consider the validity of these assumptions in terms of small-scale coherent structures and 'backscatter'.

The available experimental and computational evidence on small-scale structures in turbulent flows is that, contrary to the usual assumptions of isotropy and homogeniety, they are strongly anisotropic and depend upon the type of turbulent flow. Different types of subgrid structures with different statistical properties could require different approaches to subgrid modeling. Evidence from direct numerical simulations in a uniform shear (Rogers and Moin, 1987) and convection (Kerr, 1996) show that subgrid scale structures are strongly aligned with large-scale flow properties. In Rayleigh-Bénard convection, this alignment combined with persistent shears, could be strong enough to cause the observed differences between the observed and predicted scaling laws, as discussed in section 1.3. Examples of the effect of shear on atmospheric flows can be found in the contribution by LeMone. Therefore any method for modeling the subgrid scales that either explicitly or implicitly assumes isotropic or homogeneous subgrid scales could be inherently flawed, which could be the case with the Smagorinsky model and those based on statistical closures.

An objective might be to use some additional properties of the resolved scales to predict non-isotropic and inhomogeneous properties of the subgrid scales. This is the principle behind the 'backscatter' approach. The concept behind 'backscatter' is that not only does small-scale turbulence act to dissipate energy, but it can also feed energy back into the larger scales. Let us consider the spectral energy budget,

$$\frac{d}{dt}E(k) = T(k) - 2\nu k^2 E(k) \quad . \tag{1.23}$$

The right hand side is composed of two terms, a nonlinear contribution, the energy transfer spectrum $T(k)$, which is cubic in the velocity, and a strictly dissipative term. It has long been understood that $T(k)$ is composed of two parts, a forward dissipative part and a backscatter term. For example, in two dimensions Kraichnan (1976) introduced the concept of replacing $T(k)$ with a negative eddy viscosity in spectral space. In isotropic three-dimensional turbulence, $T(k)$ averaged over the domain should be dissipative or forward. However, while $C_s \approx 0.2$ on average, the sum of the local forward and backscatter terms will not be constant and might even be negative. The recent use of backscatter is to try to predict when and where in physical space, not wavenumber space, backscatter or production of energy by the subgrid scales will occur. This has been done in two ways, stochastically and dynamically.

The initial use of this concept was stochastic. That is, an artifical, subgrid stochastic forcing term was added to the equations in addition to the usual Smagorinsky dissipation, with the magnitude of the stochastic forcing in principle representing the forcing expected from the amount of backscatter in the energy transfer spectrum. It was found that this extra forcing tended to induce the proper instability mechanisms in shear flows (Leith, 1990; Mason and Thomson, 1992) where the standard Smagorinsky formalism with a constant coefficient would not. It was then suggested that the extra forcing could be determined dynamically, based on truncations of the LES equations (Germano et al., 1991), and simply represented by a variable C_s that could be negative in places. The original formulation led to unstable behavior (Sullivan, private communication), because a negative C_s in Smagorinsky (1.21) adds energy at the small scales based on the locations of steep gradients, thereby accentuating those gradients and simply increasing the energy input. This has been corrected by using self-consistent formulations (Ghosal et al., 1995). However, the extra degree of work necessary brings into question whether one has actually gained anything over simply increasing the resolution of the LES simulation.

The latest suggestion is that, based on *a priori* analysis of DNS, the physical space subgrid forward and backscatter components can be identified with different subgrid components of the Lamb vector $u \times \omega$, and that these can be modeled separately (Kerr et al., 1996). A distinct relationship to, but not correlation with, subgrid vortical structures was also noted in this work, which suggests that before anything like a true backscatter formulation can be made, that a better understanding of the subgrid structure of turbulence will be necessary. The concept has been applied using a double filtering scheme in turbulent channel flow and been shown to significantly improve low resolution LES results when compared to high resolution DNS (Saiki and Domaradzki, 1997).

1.7   Conclusion

A summary of the basic principles of atmospheric convection, from the size of mesoscale convective systems down to turbulent scales has been presented. While coupling of wide-ranging scales in a mesoscale convective system might be expected, the discussions of laboratory convection in this volume are designed to demonstrate that even on turbulent scales there can be large-scale organization that can significantly change the dynamics and scaling laws. Could changes in the dynamics related to organization on turbulent scales spill over into atmospheric scales covered by cloud-resolving models? A final example of how coupling between scales could be important is the poorly understood case for how convection is initiated in regions of weak convergence and small synoptic shears over warm tropical oceans. For this case, while large-scale simulations by Miller et al. (1992) using standard surface parameterizations for moderately high surface shear seem to inject the right amount of heat into the atmosphere, in light winds the heat flux has to be arbitrarily increased as if shears from subgrid-scale turbulence of the type discussed in this contribution are influencing the dynamics.

It is hoped that from here, and the references listed, that the reader will be prepared for the more technical contributions to this volume. In some cases, such as dry turbulent convection, the basic principles were laid down as far back as Prandtl (1932), but are still being refined today. However in other cases such as moist convection, the basic scaling laws are only being developed. Theory and modeling of such phenomena as supercell thunderstorms and squall lines that are discussed in this volume has led to a comparatively rich understanding of these phenomena. Beyond these scaling relationships, there is also a need to improve our understanding of the statistical properties of cloud ensembles that are needed for parameterizations in global models. For this, further work on how ensembles of convective systems discussed in this volume interact with large-scale atmospheric circulations interact is needed.

Acknowledgements

Discussions with contributors to this volume and participants in the meeting are appreciated. Input during preparation by M. LeMone, M. Moncrieff, and P. Gilman is appreciated. NCAR is supported by the National Science Foundation.

Bibliography

Ayotte, K.W., Sullivan, P.S., Anders, A., Doney, S.C., Holtslag, A.A.M., Large, W.G., McWilliams, J.C., Moeng, C.-H., Otte, M.J., Tribbia, J.J., and Wyngaard, J.C., "An evaluation of neutral and convective planetary boundary-layer parameterizations relative to large eddy simulations,", *Boun. Lay. Meteor.* **79**, 131–175 (1996).

Arakawa, A. and Schubert, W.H., "Interaction of a cumulus cloud ensemble with the large-scale enviroment Part I," *J. Atmos. Sci.* **31**, 674–701 (1974).

Caflisch, R.E., and Papanicolaou, G.C., eds. *Singularities in Fluids, Plasmas and Optics*. Proceedings of the NATO-ARW, Heraklion, Greece. Kluwer Academic Publishers (1993).

Castaing, B., Gunaratne, G., Heslot, F., Kadanoff, L., Libchaber, A., Thomae, S., Wu, X.Z., Zaleski, S. and Zanetti, G. "Scaling of hard thermal turbulence in Rayleigh-Bénard convection," *J. Fluid Mech.* **204**, 1–30 (1989).

Charney, J.G and Eliassen, A., "On the growth of the hurricane depression," *J. Atmos. Sci.* **21**, 68–75 (1964).

Clark, T.L, Hauf, T. and Kuettner, J.P., "Convective-forced internal gravity waves: results from two-dimensional experiment"s *Quart. J. Roy Met. Soc.* **112**, 899–926 (1986).

Deardorff, J. W., "Convective velocity and temperature scales for the unstable planetary boundary layer and for Rayleigh-Bénard convection," *J. Atmos. Sci.* **27**, 1211–1213 (1970).

Emanuel, K.A., *Atmospheric Convection.* Oxford University Press (1994).

Germano, M., Piomelli, U., Moin, P., and Cabot, W., "A dynamic subgrid-scale eddy viscosity model," *Phys. Fluids A* **3**, 1760–1765 (1991).

Ghosal, S., Lund, T.S., Moin, P., and Akselvoll, K., "A dynamic localization model for large-eddy simulation of turbulent flows," *J. Fluid Mech.* **286**, 229–255.

Gill, A., *Atmosphere-Ocean Dynamics* Academic Press (1982).

Grabowski, W.W., Moncrieff, M.W., and Kiehl, J.T., "Long-term behaviour of precipitating tropical cloud systems: A numerical study," *Q.J.Roy.Met.Soc.* **122** 1019–1042 (1996).

Grabowski. W.W. , "Cumulus entrainment, fine-scale mixing and buoyancy reversal", *Q.J.Roy.Met.Soc.* **119**, 935–956 (1993).

Houze, R.A., *Cloud Dynamics* Academic Press (1993).

Houze, R.A. and Betts, A.K., "Convection in GATE," *Rev. Geophys. Space Phys.* **19** 541–576 (1981).

Herring, J.R. and McWilliams, J.C., *Lecture Notes on Turbulence.* Lecture notes from the NCAR-GTP Summer School, June, 1987, World Scientific (1989).

Kerr, R.M., "Rayleigh number scaling in numerical convection," *J. Fluid Mech.* **310**, 139–179 (1996).

Kerr, R. M, Herring, J. R. and Brandenburg, A., "Large-scale structure in Rayleigh-Bénard convection with impenetrable side-walls," *Chaos, Solitons & Fractals,* **5** 2047–2053 (1995).

Kerr, R.M., Domaradzki, J.A., and Barbier, G., "Small-scale properties of nonlinear interactions and subgrid-scale energy transfer in isotropic turbulence," *Phys. Fluids* **8** 197–208 (1996).

Kraichnan, R.H., "Eddy viscosity in two and three dimensions,", *J. Atmos. Sci.* **33**, 1521 (1976).

Kreuger, S.K., "Numerical simulations of tropical cumulus clouds and their interaction with the subcloud layer," *J. Atmos. Sci.,* **45**, 2221–2250 (1988).

Herring, J.R., Schertzer, D., Lesieur, M., Newman, G.R., Chollet, J.P., and Larcheveque, M., "A comparative assessment of spectral closures as applied to passive scalar diffusion." *J. Fluid Mech.* **124**, 411–437 (1982).

Leith, C.E., "Stochastic backscatter in a subgrid-scale mode: plane shear mixing layer," *Phys. Fluids* **2**, 297 (1990).

Lesieur, M., *Turbulence in Fluids: Stochastic and Numerical Modelling,* 2nd rev. ed., Kluwer Academic Publishers (1990)

Lilly, D.K., "On the application of the eddy-viscosity concept in the inertial subrange of turbulence," Manuscript No. 123, National Center for Atmospheric Research, Boulder, CO, 19 pp. (1966).

Lilly, D.K., "Models of cloud-topped mixed layers under a strong inversion," *Q. J. Roy. Met. Soc.* **94**, 292–309 (1968).

Malkus, J.S., "On the structure of the trade wind moist layer," *Papers Phys. Ocean. Meteor.,* **13**, WHOI, 47pp (1958).

Malkus, W.V.R., "Discrete transitions in turbulent convection," *Proc. Roy. Soc. A* **225**, 185–195 (1954); "The heat transport and specturm of thermal turbulence," *Proc. Roy. Soc. A* **225**, 196–212 (1954).

Mason, P.J. and Thomson, D.J., "Stochastic backscatter in large-eddy simulations of boundary layers," *J. Fluid Mech.* **242**, 51 (1992).

Miller, M.J. Beljaars, A. and Palmer, T.N., "The sensitivity of the ECMWF model to the parameterization of evaporation from the tropical oceans," *J. Climate* **5**, 418–434 (1992).

Moeng, C.-H. and Wyngaard, J.C., "Statistics of conservative scalars in the convective boundary layer," *J. Atmos. Sci.* **41**, 3161–3169 (1984).

Moffatt, H.K. and Tsinober, A., eds. *Topological fluid mechanics.* Proceedings of the IUTAM Symposium, Cambridge, August 13-18, 1989, Cambridge University Press (1990)

Moffatt, H.K, Zaslavsky, G.M., Tabor, M., and Comte, P., eds. *Topological aspects of the dynamics of fluids and plasmas.* Proceedings of the NATO-ARW workshop at the Institute for Theoretical Physics, University of California at Santa Barbara. Kluwer Academic Publishers (1992).

Moncrieff, M.W., "Mesoscale convection from a large-scale perspective," *Atmos. Res.* **35**, 87–112 (1995).

Moncrieff, M.W. and Miller, M.J., "The dynamics and simulation of tropical cumulonimbus and squall lines," *Quart. J. R. Met. Soc.* **102**, 373–394.

Monin, A.S. and Oboukhov, A.M., "Dimensionless characteristics of turbulence in the atmospheric surface layer," *Doklady AN SSSR* **93**, 223–225 (1953).

Prandtl, L., "Meteorologische Anwendungen der Strömungslehre," *Beitr. Phys. fr. Atmos.* **19**, 188–202 (1932).

Priestley, C.H.B., "Convection from a large horizontal surface," *Austr. J. Phys.* **7**, 176–201 (1954).

Rogers, M. M. and Moin, P. "The structure of the vorticity field in homogeneous turbulent flows," *J. Fluid Mech.* **176**, 33–66 (1987).

Rotunno, R., Klemp, J.B. and Weisman, M.L., "A theory for strong, long-lived squall thunderstorms," *J. Atmos. Sci.* **45** 463–485 (1988).

Saiki, E.M. and Domaradzki, J.A., "A subgrid-scale model based on the estimation of unresolved scales of turbulence," Preprint (1997).

Schumann, U., "Subgrid scale model for finite difference simulations of turbulent flows in plane channels and annuli," *J. Comp. Phys.* **18** 376–404.

Shraiman, B.I. and Siggia, E.D., "Heat transport in high-Rayleigh-number convection," *Phys. Rev. A* **42**, 3650–3653 (1990).

Smagorinsky, J., "General circulation experiments with the primitive equations", *Mon. Weath. Rev.* **93**, 99 (1963).

Squires, P., "Penetrative downdraughts in cumuli," *Tellus* **10**, 381–389 (1958).

Sreenivasan, K.R., "On the universality of the Kolmogorov constant," *Phys. Fluids* **7**, 2778–2784 (1995).

Sullivan, P.P., McWilliams, J.C., and Moeng, C.-H., "A subgrid-scale model for large-eddy simulation of planetary boundary layer flows," *Boun. Layer Met.* **71** 247–276 (1994).

Wyngaard, J.C. and Weil, J.C., "Transport asymmetry in skewed turbulence," *Phys. Fluids A* **3**, 155–162 (1990).

Yakhot, V., and Orszag, S.A., " Renormalization group analysis of turbulence." *Phys. Rev. Lett.* **57** (1986).

Chapter 2

SOLAR AND STELLAR CONVECTION: A PERSPECTIVE FOR GEOPHYSICAL FLUID DYNAMICISTS

PETER A. GILMAN

High Altitude Observatory
National Center for Atmospheric Research
P.O. Box 3000, Boulder, CO 80307-3000

Convection occurs in the outermost 30% of the suns radius, filling the volume except where magnetic fields are strong. Main sequence stars hotter and more massive than the sun have thinner convection zones, while cooler less massive stars have deeper zones. Granulation, with scale 10^3 km compared to convection zone depth 2×10^5 km, is the most vigorous, with typical velocities of 1-2 km/sec. Supergranulation, with scale 3×10^4 km also exists, with largely horizontal velocities of .3–.5 km/sec. Convective motions up to global scale may also exist, but have not been convincingly observed. The sun also rotates differentially, with the poles rotating 30% slower than the equator at the solar surface, and experiences poleward meridional flow of about 20 m/sec. Helioseismology results show that the surface differential rotation profile is largely unchanged down through the convection zone, but reduces to solid rotation in a shear layer near the base.

The solar convection zone is characterized by a thin highly superadiabatic boundary layer at the top, in which the thermodynamic and radiative effects of partial ionization play a dominant role; a thick, near-adiabatic bottom boundary layer which may have a hierarchy of different dynamical regimes; and a well mixed central region. Granulation and supergranulation near the top are little influenced by rotation, but any larger scales of motion at depth would be, and must be to maintain the differential rotation.

Simulations of solar convection at true solar Rayleigh numbers are impossible, so virtually all studies have relied on high viscosity analogs. Simulations of small scale convection have successfully reproduced observed radiative signatures of granulation, and highlighted the role of asymmetries between broad updrafts and concentrated downdrafts, as well as the degree to which convective eddies can extend intact over multiple scale heights. Global convection models have reproduced the observed latitudinal differential rotation, but not the absence of a radial gradient in the convection zone. They also indicate that broad equatorial acceleration such as the sun has requires a deep convecting layer.

The most important unsolved problems regarding solar convection appear to be to find the origin of supergranulation; understand how convective motions occurring deep in the zone maintain the differential rotation; and understand the intricacies of the dynamics of the base of the convection zone and the shear layer immediately below it.

2.1   Introduction

The purpose of this survey is to introduce the subject of solar and stellar convection to the rather larger community of geophysical fluid dynamicists, so that they might learn something from it, but more importantly, so that some might become interested and contribute. Now is a particularly opportune time to pursue problems in this area, since the techniques of helioseismology are revealing new information about the interior of the sun.

Convection in stars and planets has much physics in common, but also important differences. And convection is a common, almost ubiquitous phenomenon in both, contributing crucially to energy and momentum balances. One potentially important difference between the physics of convection in the sun and stars, and in planetary atmospheres and the ocean, is the role played by electric currents and magnetic fields. We will not focus on this question here. In general the existence of stellar convection is not determined by magnetic fields, though their presence can modify it substantially. But at least in the visible layers of the sun, dynamically significant magnetic fields are sparse, occupying no more than one percent of the area, so it is sensible to focus largely on the nonmagnetic problem.

It is now well established that in the sun convection occurs in a shell between about .7 of the solar radius and the outer visible surface, or photosphere. More precisely, helioseismological analysis of solar acoustic frequencies now indicates that the bottom of the convection zone, defined as the level where the radial temperature gradient becomes radiatively controlled and therefore substantially subadiabatic, is at .713 of the solar radius (Christensen-Dalsgaard *et al.*, 1985). By this definition, a mixed layer with virtually adiabatic gradient, such as might be produced by overshooting, is included within the convection zone. (Helioseismology is not sensitive enough to distinguish between slightly superadiabatic and slightly subadiabatic stratification.)

Beneath the solar convection zone, the sun is believed to be in radiative equilibrium all the way into the core. 99% of hydrogen burning occurs inside .26 of the radius. Above the solar convection zone and photosphere are the chromosphere and corona, both highly inhomogeneous media, largely dominated in structure by solar magnetic fields. The corona in turn is the source of the solar wind, the high speed ionized plasma that, accompanied by magnetic fields, streams past the earth and other planets, out to the interstellar medium. Thermal convection as conventionally defined is not found in these regions, so they will not be mentioned further here.

The sun is a typical middle aged star, so we should expect a large fraction of so-called main sequence stars to have convection zones. That they do is seen in Doppler broadening of common spectral lines. From conventional stellar structure theory–essentially hydrostatics and energy balance–it is common wisdom that stars of similar mass, radius, and surface temperature to the sun have convection zones of similar depth, while hotter, more massive, larger stars on the main sequence have thinner zones, and cooler, less massive, smaller stars have deeper zones, all the way to fully convective stars. Physically, the convection zone thickness is determined primarily by the temperature at which atomic hydrogen ionizes, in particular to the H^- ion.

2.2   Solar motions

There are two distinct scales of motion on the sun that appear to be forms of convection: the so-called granulation and supergranulation. An image of granulation, seen in white light, is shown in Figure 2.1. It consists of a pattern of bright, irregular polygons, separated by a network of dark lanes. It is well established that there is rising motion in the bright areas, and sinking motion in the dark. Velocity amplitudes vary, but are at least 1 km/sec rms, and there is evidence of local areas, particularly downdrafts, where the velocity exceeds the local sound speed of about 6 km/sec. Granulation is very small in scale compared to the dimensions of the sun: spacing between bright centers is about 1.4×10^3 km, compared to an equator-pole distance of about 1.1×10^6 km. Granulation covers the solar surface except where magnetic fields are strong, such as in sunspots. The thermal contrast between bright and dark areas corresponds to a temperature difference of about 125K. Measured velocities in granules are primarily vertical. Lifetime of individual granules varies from less than 5 min, to about 12 min, depending on the means of breakup.

A "Doppler image" of supergranulation for the whole sun is shown in Figure 2.2. This image represents the Doppler shift of a particular spectral line due to the velocity pattern. Light regions here represent flow toward the observer, dark areas flow away. In contrast to granulation velocities, supergranulation velocities are predominantly horizontal on the sun. This can be inferred from Figure 2.2, since the center of the image shows much less variation in contrast compared to the outer parts. A Doppler shift is produced only by the line-of-sight velocity, so near the center of the solar image only radial velocities will contribute to such a shift. The particular pattern of brightness seen in Figure 2.2 also tells us that the motion field consists of horizontal divergence from centers, obviously typical of a convection pattern. Horizontal velocities are estimated to be 3–500 m/sec. Spacing between centers of divergence is about 30×10^3 km, or twenty times greater than for granulation, but still quite small compared to the equator-pole distance. Supergranule lifetimes are also much greater–1–2 days. In contrast to granules, no clear evidence of hot centers and cool boundaries has been seen, although supergranulation boundaries tend to be collection points for strong magnetic fields, which themselves produce thermal signatures.

It should be apparent from Figures 2.1 and 2.2 that a significant difference between atmospheric and oceanic convection, and solar and stellar convection, is that the former is much more patchy, reflecting localized sources of excitation, such as topography and locally differential heating. There is apparently no analog on sun to vigorous locally confined convection such as a thunderstorm.

On the global scale, two types of persistent motions have clearly been measured on the sun: differential rotation and meridional circulation. Representative profiles of both are seen in Figure 2.3, taken from Hathaway *et al.* (1996). The differential rotation, shown as a linear velocity, is measured relative to the angular velocity at about 30° latitude. In angular measure, the total rotation decreases monotonically from equator to the poles by about 30%. The differential rotation is an apparently permanent circulation feature on the sun, whose profile is centered virtually on the equator. Small differences in the latitude gradient in north and south hemispheres are observed, as are small departures from the broad profile–the so called torsional oscillations (Howard and LaBonte, 1980), which seem to be associated with the sunspot cycle. Amplitude of this feature is about 10 m/sec; the velocity departures migrate toward the equator from high latitudes at about the same rate that the sunspot cycle evolves. Meridional circulation shown in Figure 2.3 represents an average poleward flow in each hemisphere that peaks at about 20 m/sec at about 45° latitude. As with

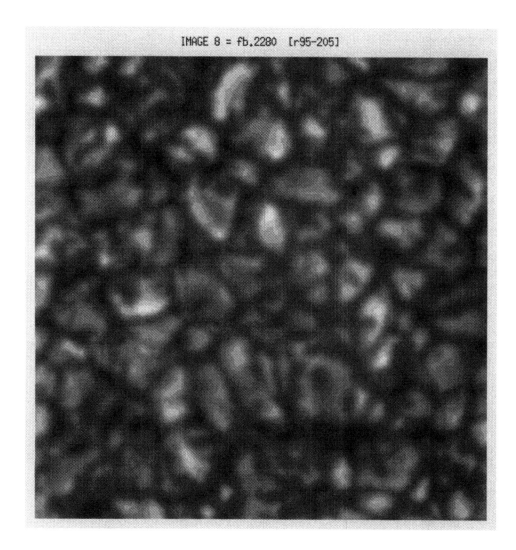

IMAGE 8 = fb.2280 [r95-205]

Figure 2.1: High resolution photograph of solar granulation, showing dark lanes and bright nearly polygonal centers. Observed June 5,1993 by P.Brandt, G. Simon, and G. Scharmer, Swedish Vacuum Solar Telescope, La Palma, Spain.

Figure 2.2: Full disk dopplergram from the MDI instrument on the SOHO satellite, depicting supergranulation. Differential rotation has been subtracted out, and solar acoustic oscillations have been averaged down.

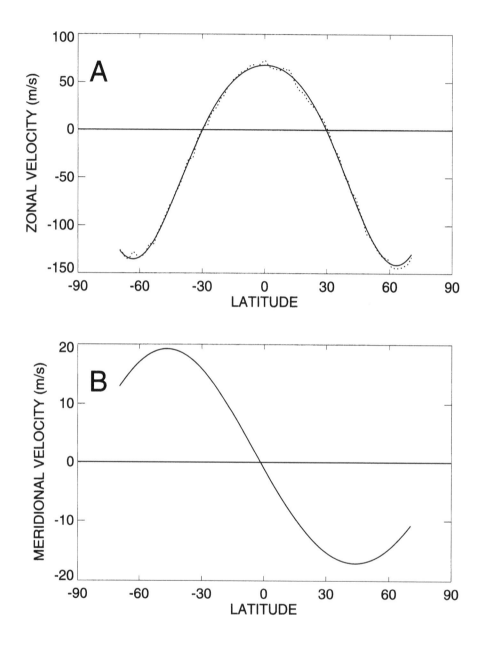

Figure 2.3: a) Differential rotation linear velocity, from Global Oscillations Network Group (GONG) data, first published in Hathaway *et al.* (1996). b) Mean meridional flow for the same time period. Reprinted with permission from SCIENCE, 272,1307. Copyright 1996, American Association for the Advancement of Science.

supergranulation, these global circulation patterns are measured from Doppler shifts.

In addition to the motion fields described above,there is much weaker evidence for two other, intermediate, scales of motion, both of which could be convective in nature. One is the so-called mesogranulation, (November *et al.*, 1981) originally seen as a motion of perhaps 60 m/sec intermediate in horizontal scale between granulation and supergranulation. More recent work by November (1989, 1994) identifies mesogranulation instead as the concentrated vertical flow component of supergranulation. The other is the so-called giant cells, larger in scale than supergranulation, but smaller than meridional circulation and differential rotation, which have been even more elusive. The latest measurements, shown in Hathaway *et al.* (1996), suggest a horizontal velocity of less than 10 m/sec for all spherical wavenumbers less than 10. Global scale convection is thought to exist because the convection zone depth is a substantial fraction of the solar radius, and at the base of the convection zone the scale heights for density and pressure are of order 0.1 of the radius. But even if it does, it is not clear how much amplitude would be seen at the solar surface, since it must compete with the indigenous and very vigorous smaller scale granulation and supergranulation.

All of the velocities described above are measured at the solar surface. But helioseismology (see 31 May 1996 issue of *Science* for representative new results and citations of earlier work), which uses the well measured frequencies of acoustic modes of the sun to infer properties of the solar interior, is yielding new information about solar rotation variations with depth through and even below the convection zone. The general picture that is emerging is that the surface differential rotation seen in Figure 2.3 extends, with some small departures, through most of the convection zone. Below that, the rotation rates at different latitudes converge rather rapidly toward a common value, the surface rate of about 30°. The latest estimates (Kosovichev, 1996) place the upper boundary of this shear layer at about .74 of the solar radius, with a thickness of sightly less than .1. Helioseismology also holds promise of yielding information about other large scale motions at depth, such as meridional circulation and global convection, but that is in the future.

2.3 Global features of convection zone

2.3.1 Structure with radius

Estimates of the run of time average thermodynamic variables in the solar convection zone are usually made by solving equations that express hydrostatic and energy balance in the radial direction for an assumed composition of hydrogen, helium, and heavier elements. Where the radiative temperature gradient exceeds the adiabatic, convection is assumed to occur and the degree of superadiabaticity is estimated using mixing length concepts. These model calculations must match to the present solar mass, radius, and luminosity. For a more complete current description of this type of calculation, see Christensen-Dalsgaard *et al.* (1996). A summary result of such a calculation, provided to the author by J. Christensen-Dalsgaard, is shown in Figure 2.4. This diagram is a variant of one first developed by Gough and Weiss (1976).

In Figure 2.4, the curves depict the fractional superadiabatic gradient obtained from a mixing length form for the convective flux; the fraction of total energy carried by convection; and the temperature. Each of these is plotted against the log of the depth, so the upper boundary layer of the convection zone is substantially spread out. For comparison, the hydrostatic pressure is shown at the top. Different formulations of mixing length theory, as well as more sophisticated theories involving nonlocal effects, penetration, and turbulent

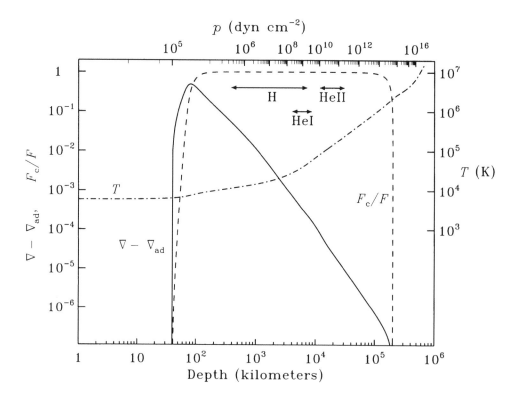

Figure 2.4: Structure of the solar convection zone as output of a standard solar model (Christensen-Dalsgaard *et al.*, 1996), showing run of temperature T, fractional departure from the adiabatic gradient, $\nabla - \nabla_{ad}$, and fraction of energy carried outward by convection, F_c/F. Zones of 10%–90% ionization of hydrogen and helium are also depicted. Based on similar diagram for earlier solar model seen in Gough and Weiss (1976).

pressure change this structure very little except where the fractional departures from the adiabatic gradient are substantial. From Figure 2.4, this occurs only in the uppermost few hundred kilometers of the zone.

We see in Figure 2.4 that convection takes over from radiation at the top in a very thin layer, substantially less than one pressure scale height (150 km) thick. The transition back to a radiative temperature gradient at the base of the convection zone is also seen to occur in about one scale height, but this represents a much greater depth, perhaps 5×10^4 km. Thus the physical scale of the bottom boundary layer is much larger than for the top layer.

Within the convection zone, the departure from the adiabatic gradient peaks at a value of order unity within the first 50 km. Below that, the average gradient falls off steadily by a factor of about 10^6 to the base of the zone. This large decline reflects primarily the great increase in efficiency of convection due to the 10^6 increase in the density of the sun over the depth of the zone. Below about 100 km depth, the modest changes in slope of the superadiabatic gradient, as well as the much larger change in slope of the temperature at 3×10^3 km, are due to the increasing fraction of hydrogen, and to a lesser degree, helium, that are ionized. The double-arrowed lines in the figure show the ranges over which ionization increases from 10% to 90%.

2.3.2   Influence of rotation

Without doubt, the substantial differential rotation of the sun must arise from two facts: the sun rotates, and it has a convecting layer. The meridional circulation may also owe its existence to these two facts. The question then is, 'What kind of interaction occurs, and how large is it?' The amount of influence rotation has on the convection is obviously a key factor. Given a solar rotation period of about 25 days (siderial), granulation with eddy turnover times of several minutes is influenced virtually not at all. Supergranulation, with turnover times of one or two days, will be influenced a small amount. But it is clearly still larger scales of convection, such as the giant cells if they exist, that would be profoundly influenced by rotation. From mixing length type arguments, the turnover time in the lower half of the convection zone ought to be a few weeks to one or two months. This would be enough time for rotation to have a substantial influence. But we do not now know from observations what the dominant scale of convective motion actually is at these depths. We can say that in order for the influence of rotation on the convection to not be large in the lower half of the convection zone requires that mixing length theory greatly overestimate the dominant scale of convection at these depths. We return to this point later when we review progress in simulating solar convection.

2.3.3   Upper boundary layer

The physics of the uppermost part of the solar convection zone changes extremely rapidly with depth, and this physics must be represented fairly accurately in models in order to achieve believable simulations. In particular, the opacity of the solar gas increases by a factor of about 10^6 from just above the convection zone to 3×10^3 km into it, due to the production of the H^- ion. In the same distance, the specific heat at constant pressure first increases by about a factor of 10, and then decreases by a factor of 4, due to the thermodynamics of partial ionization of hydrogen (see Stix, 1989 Ch. 2 for a more detailed description). Density in this same interval increases by about three orders of magnitude which, when coupled with the order unity departures from the adiabatic gradient, implies a highly compressible medium, requiring fully compressible models to simulate. The so-called anelastic approximation, often used for atmospheric convection for which the density might vary by one order of magnitude and the departures from the adiabatic gradient might be no more than a few %, will not work well for this layer. Detailed treatment of radiative effects is important because of the rapid opacity changes, but also because the most detailed comparison with observations is with various radiative signatures, such as the shape of spectral lines formed in the inhomogeneous convecting layer.

2.3.4   Lower boundary layer

In contrast to the upper boundary layer, at the bottom boundary of the convection zone all the thermodynamic variables change with depth much more slowly and smoothly. There is a convection zone base because the opacity of the gas declines enough that all of the energy transport required by the luminosity can be achieved with a subadiabatic temperature gradient. The details of the transition are more complex because there must be some penetration of the convection below where the radiative gradient is exactly adiabatic. The exact nature of the transition is not known from observations, and varies with the penetration theory. A representative picture is given in Zahn (1992). There the penetration layer is described as nearly adiabatic (and could be slightly subadiabatic), in which the convective energy flux is actually inward, to compensate for the extra outward radiative flux implied by the difference between the adiabatic gradient and the smaller radiative gradient that would be present without penetration. This penetration layer is thought to be bounded by an extremely thin (perhaps 500 km) layer where the deceleration of the penetrating convective elements is completed and radiation takes over the heat transfer.

The above description takes no account of the rotation or differential rotation of the sun, or the possibility that strong magnetic fields may be produced and stored there by the solar dynamo. Even taking into account only rotation effects complicates the picture considerably. In particular there may be a sequence of layers spanning the base of the convection zone, in which the dominant dynamics evolves from layer to layer, which interact but are somewhat distinct. Table 2.1 depicts a schematic of these layers. Layer A, the uppermost, well within the convection zone, is a layer with modestly compressible convection influenced by rotation, and weakly by its latitude gradient. Next comes layer B in which the radial gradient of rotation becomes important. Within the radial shear layer occurs the penetration region and transition to an adiabatic or slightly subadiabatic gradient. Here then is a third layer, layer C, where baroclinic flow effects may become prevalent. Below the transition to a radiative gradient, but still within the radial shear layer, is a fourth layer, layer D, where barotropic dynamics, or flow that is nearly two dimensional on spherical shells, should be prominent. Here the subadiabatic gradient should be strong enough to suppress baroclinic waves. (The bulk Richardson number for this layer could be as high as 10^6.) Finally, layer E denotes the solidly rotating interior, in which barotropic and baroclinic waves and instabilities are absent, but gravity waves may still exist. Any or all of layers A-E could contain inertial oscillations.

A solar "tachocline" theory has been developed by Spiegel and Zahn (1992) to explain the existence and thickness of the radial shear layer, that invokes meridional circulation, baroclinic processes, and anisotropic eddy diffusion.

Table 2.1: Possible fluid dynamical regions near the base of the solar convection zone illustrating qualitatively how the dominant fluid dynamics may change with depth in this region. Inertial oscillations (frequency $\leq 2\Omega$) could be present in any or all layers.

Depth	Layer	Principal Fluid Dynamics
↓ ↓	A	convection influenced by rotation latitudinal shear only
— ↓ —		_____ *radial shear starts* ↓ _____
↓ ↓	B	convection influenced by rotation and radial shear baroclinic effects might be present
— ↓ —		_____ *subadiabatic temperature gradient starts* ↓ _____
↓ ↓	C	overshooting convection influenced by rotation and shear baroclinic flow and instability possible
— ↓ —		_____ *transition to radiative temperature gradient* ↓ _____
↓ ↓ ↓ ↓	D	barotropic flow and instabilities likely Kelvin-Helmholtz instability probable for unstable barotropic flow baroclinic effects supressed by subadiabatic gradient gravity waves highly likely
— ↓ —		_____ *radial and latitudinal differential rotation fade out* ↓ _____
↓	E	gravity waves if excitation mechanism exists

2.3.5 Waves and instabilities in and near the convection zone

Solar convection is vigorous enough that we should expect various wave motions to be excited by it. The most prominent oscillations observed in the sun by far are the acoustic modes now being exploited by helioseismology. These are excited by the near sonic convective flows–granulation–right near the surface. Within the convection zone, we should also expect to find inertial and Rossby waves, but these have not yet been found. They should be intimately connected to any giant cell convection that exists. At the base of the convection zone, penetrating elements are certain to excite gravity waves, but not enough amplitude from these may reach to the photosphere to be observed–though the very lowest longitude and latitude wave-number modes should not be attenuated much. There are occasional claims of observations of high wave-number gravity waves just above the photosphere, but overall these are insignificant compared to the acoustic waves–just the opposite of the situation in the earth's atmosphere and the oceans.

As for instabilities other than convection, if the description of the various layers near the base of the convection zone is even qualitatively correct, those depths could be fertile ground for Kelvin-Helmholtz, baroclinic and barotropic instabilities.

With respect to the last of these, Watson (1981) proposed that the observed differential rotation at the surface might be bounded by barotropic instability, even though the convection zone is decidedly not barotropic. He showed that profiles of solar type in a spherical shell are unstable for differential rotations above about 29%, close to the solar value. Applying this same result to the nearly barotropic layer below the transition to the radiative gradient underneath the convection zone, which we outlined above, implies that layer should be stable to 2D disturbances in spherical shells. But Gilman and Fox (in preparation) have shown that even a weak toroidal magnetic field destabilizes solar type profiles. Therefore it is likely that the barotropic layer contains 2D global disturbances of velocity and magnetic fields that are interacting with and constantly changing the differential rotation there. The strength of these global 2D disturbances could vary enough with depth to introduce small scale 3D turbulence in thin sheets, excited by local Kelvin-Helmholtz instability where the local Richardson number, due to differential horizontal motion of adjacent thin layers, becomes small enough.

2.4   Simulating solar convection

All geophysical fluid dynamicists are well (and sometimes painfully) aware that the relevant dimensionless numbers for convection in the atmospheres and oceans of planets are very large, in many cases too large to allow direct simulations, because the computational demands are simply too great. This is because the molecular diffusivities for momentum and temperature are small. Stars are particularly extreme in this respect. For the solar convection zone, the Rayleigh number is between about 10^{15} and 10^{28}, depending on the depth of the layer considered and its location in radius. A typical Reynolds number might be 10^{10}, a typical Taylor number 10^{20} to 10^{28}. In addition, since small scale heat transfer is governed by radiation rather than thermal conduction in stellar convection zones, the Prandtl number for the sun is in the range 10^{-5} to 10^{-9}.

Clearly numerical simulations of convection with these values is far beyond the reach of current supercomputers, and will be for several generations. And yet, there are obviously well ordered convective motions on the sun–the granulation and supergranulation–that intuition says should be modellable with available tools, as well as ordered global circulations–the differential rotation and meridional circulation–that must be driven by the convection. And clearly highly parametric theories such as mixing length models tell us almost nothing about the details of convective flows. Given this situation, solar convection modellers have resorted to the study of related or less extreme systems that are thought to have enough physics in common with the real sun to allow progress in physical understanding to be made. The best of these I would characterize as analogs for, rather than approximations to the solar convection zone. That is, they constitute in themselves completely well defined physical systems that could in principle be built and operate as simulated.

The feature common to all these analog systems is that the viscosity is taken to be much larger than in the real sun, so that the standard dimensionless parameters enumerated above take on computationally more manageable values. The assumption is that the high viscosity is related to the effects of small scale turbulence that the numerical models can not afford to resolve. As an approximation, we all know this is dubious, and needs to be replaced with more realistic representations. But as analogs, the systems that result are perfectly well defined physically. This basic approach is no different than that followed earlier in other geophysical fluid flow problems. In both cases it has been used successfully to develop physical understanding needed to interpret more advanced model results, which are more relevant to detailed prediction requirements.

For solar convection studies, assumption of a viscosity of, say 10^{12} cm^2/sec or so has been common, this value suggested again from mixing length estimates. Then typical Rayleigh numbers fall in the range 10^6 to 10^8, Reynolds numbers 10^2 to 10^5, Taylor numbers 10^4 to

10^8, and Prandtl number ~ 1. Within this framework attention is turned to the physics peculiar to stellar convection zones, namely compressibility and stratification, partial ionization and radiative transfer. For global scales of convection, spherical geometry and rotation are added. More recently linear viscosity terms have been replaced by nonlinear representations of various sorts, mostly appropriated from the geophysical literature, but even in these the de facto dimensionless numbers are still far below the true solar values.

In the more detailed review below of results from the past 25 years, it will be convenient to focus first on compressibility and related physics, which is more relevant to granulation and supergranulation scales, and then look at effects of rotation and spherical geometry together with compressibility, which combination is more relevant to understanding how deep convection in stars may sustain differential rotation.

2.5 Summary of methods and results for compressible convection

One advantage of the assumption of a linear viscosity is that it allows the studies of compressibility effects to be firmly tied to now classical results for Boussinesq convection. Historically the progression of compressible models has been as follows:

- convective stability criteria and diagrams, linear mode structure with depth for fluid layers containing one or more scale heights
- heuristic models to capture physics at expense of detailed structure
- truncated modal expansions for horizontal planforms (motivated by granulation and supergranulation pictures)
- 2 and 3D numerical simulations, some direct, some employing eddy diffusion; some fully compressible, some anelastic
- increasing complexity of physics, including particularly details of radiation and spectral line formation at the top, partial ionization, and formation of shocks.

The current result of this series of theoretical and model simulation efforts is that detailed comparisons with granulation-scale observations is quite good. But understanding the origin and physics of supergranules has lagged, despite a number of efforts. Some selected highlights along the way have been

- The most unstable linear modes in multi-scale height atmospheres extend over the whole layer. For constant kinematic viscosity and thermal diffusivity, relative amplitudes peak near the top of the layer; for constant dynamic values the amplitude peak is near the bottom. (Mixing length theory assumes an eddy extends over about one scale height before breaking up.)
- Early 2D nonlinear numerical simulations by Graham (1975) also showed cells extended the full depth in a multiscale height fluid.
- Heuristic models by Nelson and Musman (1978) explained the horizontal scale of granulation: in smaller scales lateral radiative transfer wipes out buoyancy; larger scales are relatively inefficient at extracting potential energy from the strongly superadiabatic gradient, and associated horizontal pressure variations alter the opacity in such a way as to reduce the contrast between bright and dark areas.
- 2D nonlinear simulations by Dupree (1976) clearly show that in a multiscale height atmosphere, the downdrafts are much more concentrated than the updrafts, presumably because the fluid in downdrafts is compressed by surrounding fluid, while updrafts expand against the surroundings.
- Limited modal expansion nonlinear models (Latour *et al.*, 1976; Toomre *et al.*, 1976)

capture the multiscale height vertical extent of convective cells predicted by linear theory, but do not clearly show asymmetry between downflows and upflows–because of the fixed planform with depth. This approach is superseded by full 3D simulations

• Nordlund and Stein (Nordlund, 1982, 1985; Stein and Nordlund, 1994) makes detailed simulations of granulation-scale convection including radiative transfer, that reproduces granule morphology and several radiative signatures.

• Simulations of deep, more turbulent 2D compressible convection show the coexistence of small scale velocity fields at the top with large scales near the bottom (Chan et al., 1982).

• In several 2 and 3D simulations, so called "buoyancy reversals" are seen in the broad updrafts, which act to weaken them relative to downdrafts (Massaguer and Zahn, 1980; Chan, 1983; Massaguer et al., 1984; Hurlburt et al., 1984, 1986). They arise because the fluid density in the updraft is controlled more by the pressure than the temperature. Since rising fluid in a medium in which the density is declining rapidly with height arrives at the higher elevation with higher pressure than its surroundings, it also has a higher density. Furthermore, when this fluid moves horizontally into the adjacent downdraft, its higher density enhances the downflow, contributing further to the asymmetry between downdrafts and updrafts.

• The organized concentrated downdrafts are shown to be primarily responsible for overshooting at the bottom of a multiscale height convecting layer (Hurlburt et al., 1986).

• Concentrated downdrafts are shown to transport enough kinetic energy downward substantially counteract or even cancel the enthalpy flux upward. This leads to the apparent paradox that the most organized flows–concentrated downdrafts–may not be responsible for the required net upward energy flux (Cattaneo et al., 1991).

• Some simulations (Cattaneo et al., 1991) show evidence of breakup into smaller scale structures near the bottom than the top of the convecting layer–counter to mixing length arguments and assumptions, and counterintuitive, at least to this author.

• Supersonic motions are reached in the upper thermal boundary layer, due to combined effects of stratification, pressure gradients, and nonadiabatic processes (Cattaneo et al., 1991).

• Strong downdrafts near the top are shown to be significantly enhanced by partial ionization, contributing to locally supersonic speeds (Rast et al., 1993; Rast and Toomre, 1993a, b). Is like a boundary layer instability, driven by the variation in specific heat arising from the ionization (discussed above).

Most compressible convection simulations have been done with the upper part of the solar convection zone in mind. Kim et al. (1995) have made detailed statistical comparisons between shallow and deep solar convection simulations, and with mixing length models. Recent attempts to include the lower part as well, in a single calculation (Nordlund et al., 1996) have resorted to the extreme measure of increasing the heat flux through the layer, or luminosity, by up to a factor of 10^6 beyond the solar value. This is done in order to overcome the severe computational demands arising from resolving acoustic frequencies in a fully compressible system, as well as the long initial transient that occurs as the bottom layer, which has high thermal capacity compared to the top, approaches thermal equilibrium. The hope with these calculations is that empirical extrapolations from the statistics of the convective flow can be made back down to solar conditions. This seems dubious to this author, given the large range of parameter space that must be crossed. Much of this difficulty (though not all) should be avoidable by using an anelastic rather than fully compressible model for the lower layer. But then a full zone calculation requires some mid-depth matching between anelastic and fully compressible models.

2.6 Convection as driver of differential rotation

The influence of rotation upon solar convection has never been measured directly from observations, because convection occurring with time and space scales for which that influence would be significant has been so difficult to see. But this effect must exist for differential rotation in the convection zone to be maintained. The influence should be felt in the formation of Reynolds stresses that redistribute angular momentum away from a profile of solid rotation. Two basic approaches have been used extensively to estimate the Reynolds stresses and the mean circulations that should result from them. One is a succession of 3D global convection models, that begin from the idealized case of Boussinesq convection in a rotating spherical shell. The other has been a succession of theoretical turbulence models, beginning with a formal development of the Reynolds stress tensor in spherical geometry coupled with heuristic arguments supporting the concept of an anisotropic eddy viscosity arising from the presence of buoyancy and coriolis forces. A variant on the second approach has assumed that convection generates eddy diffusivities that are functions of latitude, resulting in thermally driven meridional circulation that coupled with the eddy viscosity, drives differential rotation. All of these approaches have already been reviewed extensively in the literature, for example Gilman (1986) and Rudiger (1989), so we will confine ourselves here to a few points, and focus primarily on the 3D convection approach.

Global convection theory, if it applies at all to the sun, applies to the lower part of the zone where turnover times are long, assuming mixing length concepts are at least qualitatively correct. At these levels, the scale heights are of order 5×10^4 km, so the bottom half of the convection zone can be modelled with perhaps just 2 scale heights of density. Thus it is not implausible to start with the Boussinesq case. But the finite thickness of the shell as a fraction of the solar radius is very important, and the inclusion of the whole coriolis force, not just the horizontal components, as is done for global atmospheric dynamics, is crucial. Compressibility has been shown to modify the Boussinesq results quantitatively, but not fundamentally. As with the small scale convection models, all of the global convection calculations so far have also been in the realm of high viscosity analogs defined earlier.

Linear stability of Boussinesq convection in a rotating spherical shell predicts for a wide range of Taylor number or rotation that the first unstable mode is one for which the longitude and latitude spherical harmonic indices, m and l, are equal. This corresponds to a "cartridge belt" of rolls with north-south axes, centered on the equator. The Reynolds stress from correlation of longitude and latitude motions for this mode gives angular momentum transport toward the equator, which in principle could drive the observed equatorial acceleration. The same result obtains in the compressible case.

Nonlinear finite amplitude calculations have been necessary to determine how powerful this effect is in competition with others, namely radial angular momentum transport by the Reynolds stress component arising from correlation of east-west and radial motions, the induced axisymmetric meridional circulation, and viscosity itself. The basic result is that in order to achieve a finite amplitude equatorial acceleration, the influence of rotation must be strong enough that there is a strong preference for the $m = l$ mode. Otherwise equatorial deceleration results, as other modes, for which $m < l$, dominate, including the $m = 0$ mode. The difference is that in these modes, there is a strong tendency for fluid particles to conserve their angular momentum, thereby spinning up high latitudes compared to low. In the $m < l$ mode, longitudinal pressure torques, a consequence of the near geostrophic balance in the north-south rolls, break that conservation, allowing a preferential transport of angular momentum up the gradient toward the equator.

The rotational influence can not be so strong, on the other hand, that the dominant mode acquires a radial dimension substantially less than the shell thickness. In that case the equatorial acceleration becomes narrowly confined to equatorial latitudes, with higher latitudes showing distinct maxima and minima in rotation–more like Jupiter than the sun. Even with one roll filling the shell thickness, the thinner the shell, the more confined to equatorial latitudes is the equatorial acceleration. Therefore if this theory applies to stars, we should expect stars with deep convection zones to display broad equatorial acceleration zones if their rotation is at least as fast as the sun, while stars with shallow convection zones and the same rotation will have either narrow equatorial accelerations, or equatorial deceleration if the convective turnover time in the shallow layer is short compared to the rotation period of the star.

The extension of global convection simulations to the compressible case by Glatzmaier (1984, 1985a, b) and Gilman and Miller (1986), made using the anelastic equations, have provided the most quantitative comparisons with the solar case. In particular, there results have shown that using plausible eddy viscosity and thermal diffusivities, the solar rotation rate, and the solar luminosity to drive the convection, an equatorial acceleration of about the right surface magnitude and profile results. But these models also predict the radial gradient of angular velocity should be approximately constant on cylinders concentric with the axis of rotation (obviously the rotational constraint producing the Taylor-Proudman effect), while helioseismology results say the profile is more nearly constant with radius. At the very least this difference implies that the real sun is not as close to the asymptotic limit of strong rotation as are the convection models.

To make further progress, the next research efforts clearly should focus on understanding the origin of this difference. One obvious step to take is to try to move away from the high viscosity regime to allow more effects of small scale turbulence. Some recent simulations of compressible convection influenced by rotation in cartesian geometry suggest the formation of much smaller scale vortex structures in the flow, resulting in Reynolds stresses that are quite different from the global simulations. Such stresses still drive mean flows, but it is not clear that in spherical geometry they will produce a finite amplitude equatorial acceleration such as the global simulations have already achieved. Anisotropic eddy viscosity models can be made to give profiles close to the observed, but to a degree this approach assumes the answer rather than predicting it. Obviously detailed understanding of how the solar differential rotation in latitude and radius is produced and maintained has not yet been achieved.

2.7   Interaction of the convection zone with the solar surface and the shear layer at the base

There are obviously numerous effects that the convection has on the layers of the sun immediately above and below the convection zone. We comment here only on interactions involving angular momentum. At the photosphere, we see a poleward meridional flow of up to 20 m/sec coexisting at these levels with an equatorial acceleration. If no other angular momentum transport processes were at work, this meridional flow would quickly spin up the midlatitude photosphere relative to the equator, and the equatorial acceleration would be gone. That this is obviously not happening implies additional torques at work.

Torques associated with extraction of angular momentum by the solar wind from the whole sun are much too small to compete with the "coriolis torque" associated with the meridional flow. And because of the low density very little angular momentum is stored in the solar atmosphere above the photosphere. Therefore the torques must be coming from below. Magnetic fields may produce Maxwell stresses (analogous to Reynolds stresses) to supply such torques, but it also likely that there is a high degree of turbulent mixing of angular momentum by the convection itself, between the photosphere and the convection zone below.

The eddy viscosity needed to prevent midlatitudes from spinning up is actually larger than one would estimate from mixing length arguments, unless the required mixing occurs over only .01 or less of the solar radius. The thickness of the layer through which this strong mixing must occur is determined by the depth to which the meridional circulation observed at the surface penetrates, which is at present unknown.

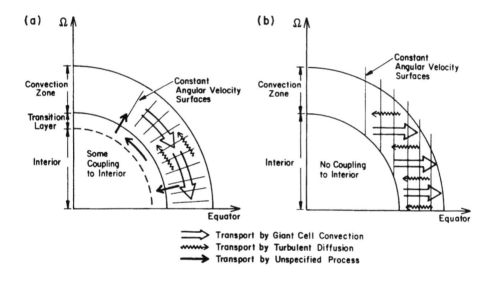

Figure 2.5: Schematic of two alternative descriptions of the angular momentum balance within the solar convection zone, and with the shear layer immediately below it. a) angular velocity constant along radial lines, and b) angular velocity constant on cylinders. Reprinted from Gilman *et al.* (1989), by permission of the author.

We also expect substantial angular momentum interactions at the base of the convection zone between the convecting layers and the shear layer below. The 3D global convection simulations summarized above generally were done assuming stress free boundaries at top and bottom. Under that assumption, the angular momentum balance is as depicted in Figure 2.5b (Figure 2.5 is reprinted from Gilman *et al.*, 1989). Angular momentum transport approximately perpendicular to the rotation axis (a combination of equatorward and radially outward momentum flux by Reynolds stresses, arising from the strong influence of rotation upon the global convection) maintains the cylindrical profile. A probably more accurate schematic of what is happening in the real sun is shown in Figure 2.5a. Here with angular velocity nearly constant along radial lines, the convection must still transport angular momentum from high latitudes to low. But since the more rapidly spinning low latitudes tend to drag the shear layer forward, and the high latitudes retard it, there is necessarily angular momentum leaking into the shear layer at low latitudes and leaking out at high. Thus in the steady state an angular momentum cycle must exist, with a poleward branch in the shear layer.

The time scale to complete the cycle is set by what fraction of the equatorward flux in the convection zone returns though the shear layer, and what fraction diffuses poleward in the zone. But for reasonable assumptions, we should expect the time scale to be years (as it is within the convection zone itself), not the millions of years associated with solar wind torques. Therefore we should expect the thickness of the shear layer to be governed by the efficiency of the poleward angular momentum transport process within the layer, rather than by the much weaker interactions with the interior below the shear layer.

Spiegel and Zahn (1992) invoked anisotropic eddy viscosity to accomplish this poleward transport, assuming 2D instabilities of the latitudinal differential rotation profile, but Watson's (1981) results imply this profile is stable to such disturbances. As remarked above, Gilman and Fox have shown that with a toroidal magnetic present, the differential rotation is again unstable, due the Maxwell stresses, so these stresses could then provide the poleward transport.

2.8   Concluding remarks

Obviously much remains to be done to gain a thorough understanding of solar convection and its role in maintaining the differential rotation. At the smallest scales, understanding the physics of granulation is relatively well in hand. However the origin of supergranulation as a distinct convective scale in the sun remains essentially unknown. Suggestions for example that it is due to the ionization zones of helium have not withstood quantitative test. On the global scale, the question remains as to whether global convection actually exists, at least in a form similar to that found in high viscosity analogs. But no alternative involving explicitly calculated 3D convection (as opposed to anisotropic viscosity or heavily parameterized turbulence models) has been demonstrated to generate the right equatorial acceleration, let alone the right radial gradient within the zone. Hopefully helioseismology can soon shed more light at least on the question of the existence of global scales of convection.

From the author's perspective, perhaps the most interesting question for the future is the exact nature of the global dynamics near the base of the convection zone. The outline of possible dynamical regimes in different overlapping layers in Table 2.1 suggests this is a problem particularly rich in complexity and subtly. Convection obviously plays a role, but the full range of dynamics characteristic of a global stably stratified layer is also present, and magnetic fields are bound to be important. Except for the magnetic effects this region of the sun has more in common with global atmospheric and ocean dynamics, than it does with stellar convection.

Bibliography

Cattaneo, F., Brummell, N.H., Toomre, J., Malagoli A., and Hurlburt, N.E., "Turbulent compressible convection," *Astrophys. J.* **370**, 282-294 (1991).

Chan, K.L., Sofia, S., and Wolff, C.L., "Turbulent compressible convection in a deep atmosphere, I, Preliminary two-dimensional results," *Astrophys. J.* **263**, 935-943 (1982).

Chan, K.L., "What causes the buoyancy reversal in compressible convection," *Astrophys. + Sp. Sci.* **95**, 453-457 (1983).

Christensen-Dalsgaard, J., Duvall, T.L., Gough, D.O., Harvey, J.W., Rhodes, E.J., "Speed of sound in the solar interior," *Nature* **315**, 378-382 (1985).

Christensen-Dalsgaard, J., Dappen, W., Ajukov, S.W., Anderson, E.R., Antia, H.M., Basu, S., Baturin, V.A., Berthomieu, G., Chaboyer, B., Chitre, S.M., Cox, A.N., Demarque, P., Donatowicz, J., Dziembowski, W.A., Gabriel, M., Gough, D.O., Guenther, D.B., Guzik, J.A., Harvey, J.W., Hill, F., Houdek, G., Iglesias, C.A., Kosovichev, A.G., Leibacher, J.W., Morel, P., Profitt, Provost, J., Reiter, J., Rhodes, Jr., E.J., Rogers, F.J., Roxburgh, I.W., Thompson, M.J., and Ulrich, R.J., "The current state of solar modelling," *Science* **272**, 1286-1292 (1996).

Dupree, R.G., "Nonlinear convective motion in shallow convective envelopes," *Astrophys. J.* **205**, 286-294 (1976).

Gilman, P.A., "The solar dynamo: Observations and theories of solar convection, global circulation, and magnetic fields," in *Physics of the Sun, Vol. I, The Solar Interior*, (P.A. Sturrock, T.E. Holzer, D.M. Mihalas, and R.K. Ulrich, eds.), pp. 95-160, D. Reidel Publishing Co., Dordrecht, (1986).

Gilman, P.A., and Miller, J., "Nonlinear convection of a compressible fluid in a rotating spherical shell," *Astrophys. J. Suppl. Ser.* **61**, 585-608 (1986).

Gilman, P.A., Morrow, C.A., and Deluca, E.E., "Angular momentum transport and dynamo action in the sun: Implications of recent oscillation measurements," *Astrophys. J.* **338**, 528-537 (1989).

Gough, D.O., and Weiss, N.O., "The calibration of stellar convection theories," *Mon. Not. Roy. Astron. Soc.* **176**, 589-607 (1976).

Glatzmaier, G.A., "Numerical simulation of stellar convective dynamos, I, The model and method," *J. Comput. Phys.* **55**, 461-483 (1984).

Glatzmaier, G.A., "Numerical simulations of stellar convective dynamos, II, Field propagation in the convection zone," *Astrophys. J.* **291**, 300-307 (1985a).

Glatzmaier, G.A., "Numerical simulations of stellar convective dynamos, III, At the base of the convection zone," *Geophys. Astrophys. Fluid Dyn.* **31**, 137-150 (1985b).

Graham, E., "Numerical simulation of two-dimensional compressible convection," *J. Fluid Mech.* **70**, 689-703 (1975).

Hathaway, D.H., Gilman, P.A., Harvey, J.W., Hill, F., Howard, R.F., Jones, H.P., Kasher, J.C., Leibacher, J.W., Pintar, J.A., and Simon, G.W., "GONG observations of solar surface flows," *Science* **272**, 1306-1309 (1996).

Howard, R., and La Bonte, B., "The sun is observed to be a torsional oscillator with a period of 11 years," *Astrophys. J. Lett.* **239**, L33-L36 (1980).

Hurlburt, N.E., Toomre, J., and Massaguer, J.M., "Two-dimensional compressible convection extending over multiple scale heights," *Astrophys. J.* **282**, 557-573 (1984).

Hurlburt, N.E., Toomre, J., and Massaguer, J.M., "Nonlinear compressible convection penetrating into stable layers and producing internal gravity waves," *Astrophys. J.* **311**, 563-577 (1986).

Kim, V-C., Fox, P.A., Sofia, S., and Demarque, P., "Modeling of shallow and inefficient convection in the outer layers of the sun using realistic physics," *Astrophys. J.* **442**, 422-433 (1995).

Kosovichev, A.G., "Helioseismic constraints on the gradient of angular velocity at the base of the solar convection zone," *Astrophys. J. Lett.* **469**, L61-L64, (1996).

Latour, J., Speigel, E.A., Toomre, J., and Zahn, J.-D., "Stellar convection, theory I, The anelastic model equations," *Astrophys. J.* **207**, 233-243 (1976).

Massaguer, J.M., and Zahn, J.-P., "Cellular convection in a stratified atmosphere," *Astron. Astrophys.* **87**, 315-327 (1980).

Massaguer, J.M., Latour, J., Toomre, J., and Zahn, J.-P., "Penetrative cellular convection in a stratified atmosphere," *Astron. Astrophys.* **140**, 1-16 (1984).

Nelson, G.D., and Musman, S., "The scale of solar granulation," *Astrophys. J. Lett.* **222**, L69-L72 (1978).

Nordlund, A., "Numerical simulations of the solar granulation, I, Basic equations and methods," *Astron. Astrophys.* **107**, 1-10 (1982).

Nordlund, A., "Solar convection," *Sol. Phys.* **100**, 209-235 (1985).

Nordlund, A., Stein, R.F., and Brandenberg, A., "Supercomputer windows into the solar convection zone," *Bull. Astron. Soc. India* **24**, 261- (1996).

November, L.J., Toomre, J., Gebbie, K.B., and Simon, G.W., "The detection of mesogranulation on the sun," *Astrophys. J.* **245**, L123-L126 (1981).

November, L.J., "The vertical component of the supergranular convection," *Astrophys. J.* **344**, 494-503 (1989).

November, L.J., "Inferring the depth extent of the horizontal supergranular flow," *Sol. Phys.* **154**, 1-17 (1994).

Rast, M.P., Nordlund, A., Stein, R.F., and Toomre, J., "Ionization effects in three-dimensional solar granulation simulations," *Astrophys. J.* **408**, L53-L56 (1993a).

Rast, M.P., and Toomre, J., "Compressible convection with ionization, I, Stability, flow asymmetries, and energy transport," *Astrophys. J.* **419**, 224-239 (1993b).

Rast, M.P., and Toomre, J., "Compressible convection with ionization, II, Thermal boundary layer instability," *Astrophys. J.* **419**, 240-254 (1993).

Rudiger, G., "Differential rotation and stellar convection," *Akademie-Verlag Berlin*, 328 pp. (1989).

Spiegel, E.A., and Zahn, J.-P., "The solar tachocline," *Astron. Astrophys.* **265**, 106-144 (1992).

Stein, R.F., and Nordlund, A. "Subphotospheric convection," in *Infrared Solar Physics*, (D.M. Rabin, J.T. Jefferies, and C. Lindsey, eds.), pp. 225-237, Proceedings IAU Symposium 154, Tucson, Arizona, March 2-6, 1992, Kluwer, Dordrecht (1994).

Stix, M., "The sun," *Springer Verlag*, 390 pp. (1989).

Toomre, J., Zahn, J.-P., Latour, J., and Spiegel, E.A., "Stellar convection theory, II, Single-mode study of the second convection zone in an A-type star," *Astrophys. J.* **207**, 545-563 (1976).

Watson, M. "Shear instability of differential rotation in stars," *Geophys. Astrophys. Fluid Dyn.* **16**, 285-298 (1981).

Zahn, J.-P., "Circulation and turbulence in rotating stars," *Astron. Astrophys.* **265**, 115-132 (1992).

Chapter 3

UNSTEADY NON-PENETRATIVE THERMAL CONVECTION FROM NON-UNIFORM SURFACES

RICHARD D. KEANE, NOBOYUKI FUJISAWA[*]

and

RONALD J. ADRIAN

Department of Theoretical and Applied Mechanics
University of Illinois at
Urbana-Champaign Urbana, IL 61801
[*] On leave from Gunma University, Gunma, Japan

The structure of steady, non-penetrative turbulent thermal convection over various two-dimensional, periodic roughness elements has been investigated experimentally by light-sheet visualization of the flow and temperature fields and by instantaneous measurements of the three-dimensional temperature fields to determine the interactions between the roughness elements and the turbulent convective sheets and plumes in the vicinity of the heated plate. Convective structures over smooth surfaces are similar to the 'spoke' pattern discovered by Clever and Busse (1989), which consists of sheets of buoyant plumes arranged in polygonal cell walls with thermals at the intersection of these polygonal cells. Roughness improves heat transfer, as shown by measurements of Nusselt number increasing as the wavelength and amplitude of the roughness elements increase. Increased surface roughness distorts the polygonal cells and affects the distribution of thermals in the vicinity of the bottom surface. It also produces thermals that penetrate significantly further into the core of the convective fluid layer. Over rough surfaces, the majority of the thermal plumes are generated in the depressions of the surface at the intersection of the polygonal cells. Three-dimensional quantitative visualization of the temperature, obtained by scanning thermochroic particle image thermometry, shows the structure and penetration into the core region of these thermals.

3.1 Introduction

Many important aspects of the inversion-capped, strongly unstable atmospheric boundary layer can be modeled in the laboratory by a fluid layer of wide horizontal extent between a non-uniform bottom surface and a stably stratified upper layer. If the temperature inversion above the atmospheric boundary layer is strong, it can be simulated approximately by a non-penetrative layer, i.e. a solid surface with no heat transfer. Although there have been many field observations of the structure of the atmospheric boundary layer, experimental studies of the effects of non-uniformity of the lower surface on the structure of the flow field or the temperature field are few in number. In modeling atmospheric turbulence, several researchers have analyzed and computationally simulated thermal convection in a fluid layer heated from below and compared results with those of field experiments. The simplest form of non-uniform surface that has been employed in investigations is a two-dimensional sinusoidal profile with varying amplitude and wavelength. It is the purpose of this study to provide some insights into the effects of surface roughness on thermal convection from the results of laboratory experiments on non-penetrative thermal convection over rough surfaces in a test section of sufficient size to limit artificial constraints imposed on the heat and mass transfer by the finite size of the enclosure. Unsteady, non-penetrative thermal convection above artificial roughness elements in the form of two-dimensional square waves is studied for various values of the amplitude and wavelength.

Adrian, Ferreira, and Boberg (1986) studied unsteady non-penetrative convection in water over a smooth surface with uniform heating to determine mean temperature profiles and joint statistics of turbulent temperature and velocity fluctuations for the core region of the convection layer. The results compared well with those of Willis and Deardorff (1974), and in the lower half of the convection layer, the statistics were similar to earlier results in Rayleigh convection. The highest Reynolds number experiments compared well with measurements of the planetary boundary layer by Telford and Warner (1964) and Lenschow (1970, 1974). These results demonstrated the similarity of the energy containing parts of the turbulent fields in relatively low Rayleigh number laboratory experiments and much higher Rayleigh number atmospheric flows.

Various investigations support a physical model of non-penetrative convection in which buoyantly active fluid in the form of warm thermals, plumes and thin sheets rise through relatively quiescent, slightly cooler-than-average fluid that slowly subsides. Laboratory experiments to study the structure and mechanisms of natural convective heat transfer in the vicinity of a smooth lower have also been reported by Tamai and Asaeda (1984). For a range of flux Rayleigh number (see Eq. 3.7), Ra_f, between 10^6 and 10^{12}, they described the polygonal structure of sheet-like plumes adjacent to the bottom plate and expressed the average length of the polygon's sides as a function of Ra_f. They confirmed that the sheet-like plumes consist of thermals that are discharged along the bursting lines of these plumes with a frequency proportional to Ra_f and a duration proportional to the square root of Ra_f. The behavior of such structures over non-uniform surfaces is less well understood.

Using direct numerical simulation, Krettenauer and Schumann (1989) investigated unsteady, non-penetrative, thermal convection over a lower surface whose elevation varied sinusoidally in one horizontal direction while remaining constant in the other. They concluded that in the turbulent regime, the statistics of the turbulent field exhibited little sensitivity to the waviness of the surface for a 10% amplitude, although roll cells did develop over the wavy surface.

In a later paper, Krettenauer and Schumann (1992) used large-eddy simulation to determine that the wavy lower surface did not significantly change the volume-averaged kinetic energy. The coherent structures of the turbulent field were comprised of large-scale primary rolls with axes parallel to the lower surface wave crests and secondary rolls with axes perpendicular to the wave crests. The latter became more pronounced for steep surface waves. They concluded that the wavy surface enforced three-dimensional motions such as the rolls perpendicular to the wave crests, but that it had rather small effect on the mean turbulence profiles and heat exchange at the surface. Using large eddy simulation, Dörnbrack and

Schumann (1993) studied unsteady, non-penetrative, thermal convection over a sinusoidally varying lower surface in the presence of a horizontal mean flow. They determined that the convective structure was similar to that over flat terrain investigated computationally by Schmidt and Schumann (1989) when the mean wind velocity was weak, i.e. of the order of the convection velocity scale, w_*, weak mean winds tended to destroy any coherent roll motions that were induced by the undulating surface, and hence they reduced the impact of terrain parameters on the turbulence statistics. The study failed to show any recirculating flow region in the mean fields, but it did reveal quasi-periodic formation of pools of stagnant cold fluid in the valleys which became heated and rose as bubbles. Their results agreed approximately with measurements of turbulence spectra in field observations over irregular terrain by Kaimal et al. (1982) These spectra were similar to observations over homogeneous surfaces, indicating that the terrain did not have sufficient variation to overcome wind effects. Further studies by Hechtel et al. (1990) using LES and based on measurements in the field concluded that realistic non- homogeneous surface fluxes have no effect on the simulation of the convective boundary layer (CBL) with light winds, but they also recommended improvements in computational resolution.

A simple model has been developed by Schumann (1988) for the surface layer of a convective boundary layer for zero mean wind velocity over homogeneous rough ground. The model predicts a decrease in the minimum friction velocity and an increase in the temperature difference between the mixed layer and the ground with increasing ratio of the boundary layer thickness to the roughness height. A new heat transfer relationship is proposed for rough surface thermal convection which is based on this ratio. Because the results show that the surface layer loses most of its circulation momentum to the mixed layer by asymmetrical large-scale convective motions, while the momentum loss at the rough surface is negligible, the turbulence in the mixed layer is only weakly dependent on the surface roughness. Experiments are needed to verify the results of the model. A more general model by Schumann (1991) concludes that the structure of the convective boundary layer is mainly controlled by its mean depth and the mean surface heat flux, while the surface roughness has little dynamical effect on the structure of the CBL, but strongly influences the relation between heat flux and temperature at the surface.

While the preponderance of available evidence indicates that surface roughness has relatively little effect on the mean statistical structure of the convective layer above the surface layer, it is also clear that it does affect the eddy structure. The first implication of this result is that similarity between the structures of low-order statistical moments of two flows does not guarantee that they have the same instantaneous eddy structure. Structural differences, observable by flow visualization, may manifest themselves as differences in the low order moments if the comparison is more stringent, or as differences in higher order moments and/or multi-point moments which are more sensitive to structural features such as spatio-temporal intermittency. The relative insensitivity of simple low order moments such as mean and root-mean square to perceptible changes in the eddy structure is an interesting feature that begs for explanation, and could suggest useful approximations once understood. Differences in the underlying eddy structure may also be important to the correct description and prediction of certain phenomena such as turbulent dispersion.

For these reasons the present work seeks to document some of the structural changes that occur in turbulent thermal convection over horizontal surfaces that contain a regular array of two-dimensional roughness elements.

3.2   Experimental apparatus and procedure

The convection test section consisted of a square container, 508 mm x 508 mm in interior horizontal dimension and 400 mm in depth with Plexiglas and glass side walls 10 mm thick. The upper boundary was a 13 mm thick transparent Plexiglas plate in contact with the upper water surface insulated on its top surface by a 100 mm layer of Styrofoam insulation. The lower surface was a 19 mm thick aluminum plate (Plate A) into which has been cut roughness elements in the form of two-dimensional square waves with dimensions, as tabulated in Table 3.1. The square waves had 50% duty cycle (i.e. 50% high and 50% low) and they were geometrically similar in the sense that the ratio of amplitude to wavelength was kept constant for the plates studied. As shown in Fig. 3.2, Plate A was attached to a second aluminum Plate B of thickness 19 mm, which in turn, was attached securely to an identical aluminum Plate C with a resistance wire thermometer on a layer of oil-soaked felt, between them. All plates were in close thermal contact with each other. An electrical resistance wire heating mat was bonded to the bottom of Plate C. It provided heat flux sufficient to produce a maximum flux Rayleigh number, Ra_f, of approximately 10^{10} for a fluid depth of 100 mm .

The mean temperature of the bottom of Plate B was measured by the resistance wire thermometer, T_{rw}. The mean temperature averaged over the surface of the roughness elements, $\overline{T_1}$ was estimated by subtracting the mean temperature drop that occurred due to heat conduction between the plane of the resistance wire thermometer, at elevation z_{rw}, and the upper surface of an equivalent smooth aluminum plate having a thickness equal to the mean thickness of the rough plate, $\overline{z_1} - z_{rw}$, where $\overline{z_1}$ is the mean elevation of the square wave. The estimated mean temperature of the lower surface depended on the heat flux H_0

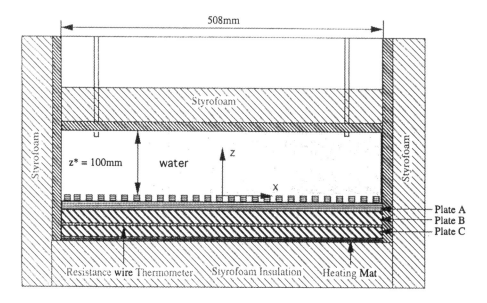

Figure 3.1: Test-section schematic.

(Wm^{-2}) according to

$$\overline{T_1} \cong T_{rw} - \frac{H_0(\overline{z_1} - z_{rw})}{k_{Al}}, \qquad (3.1)$$

where k_{Al} is the thermal conductivity of aluminum. The temperature drop correction ranged from 0.01C to 0.04C with a maximum uncertainty of less than 20%, estimated by using the maximum and minimum thicknesses of the rough plate as bounds for the thickness of the equivalent smooth plate.

The heat flux into the water was calculated from the rate of rise of the mean fluid temperature,

$$H_0 = \frac{d}{dt} \int_0^{z_*} \rho c T(z,t) dz \qquad (3.2)$$

where ρ is the density of the fluid and c is the thermal heat capacity. In statistically steady-state conditions the mean temperature $T(z,t)$ becomes linearly dependent on time (Adrian,

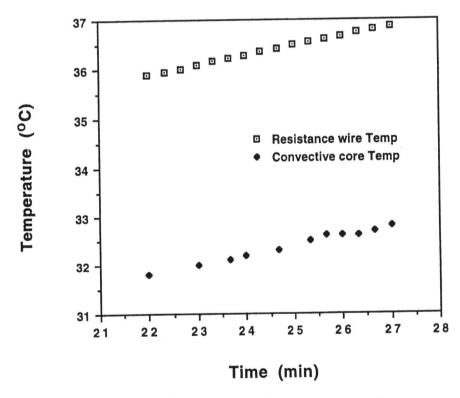

Figure 3.2: Temperature history for the resistance wire thermometer and the mean temperature of the fluid layer for turbulent thermal convection above plate A1 with $Ra_f = 3.6 \times 10^9$ as a function of time after the start of heating.

et al. 1986), so the rate of change in Eq. (3.2) becomes

$$H_0 = \rho c \frac{d\overline{T_1}}{dt} z_*$$ (3.3)

From Eq. (3.1) the rate of change of the mean lower surface temperature is, in turn, equal to the rate of change of the resistance wire thermometer temperature. This condition is demonstrated in Figure 3.2 which compares temperature in the bulk of the water to the resistance wire temperature. The heat flux H_0 through the lower surface was varied by changing the voltage applied to the resistance wire heating mat, while the water depth in the test section was maintained at 100 mm throughout the experiment. At this depth the flux Rayleigh numbers, Ra_f, ranged from 3.5×10^9 to 10.1×10^9 for the three heat fluxes that were used.

Flow visualization of the convection was accomplished by observing the temperature dependent color change of micro-encapsulated chiral nematic liquid crystals. The experiments used micro-encapsulated liquid crystals with diameters ranging from 50 to 100 μm, suspended in very low concentrations of 0.004% by volume in the water. To investigate the convective state when the mean temperature is increasing steadily i.e. when $\partial T/\partial t$ is constant, the temperature window over which the liquid crystals underwent their full range of color play was chosen to be 31.8°C to 33.8°C. As all flow visualization experiments began with the fluid temperature less than 25°C, this temperature window permitted a steady state to be reached before the crystals began to change color.

To illuminate the flow field, a light sheet of thickness ranging from 2 mm to 5 mm was oriented in the x-z plane, perpendicular to the roughness elements. The light sheet was inclined downward to enable visualization of the temperature field within the depressions of the rough surface where many thermal plumes originated. In addition, plan views (x-y plane) of the temperature field close to the lower plate were obtained by orienting the light sheet parallel to the lower plate.

Visualization of the flow field was performed by observing the light scattered at approximately 90° to the light sheet. Visualization was achieved using both photographic film and video recording. Because the color of the liquid crystals is dependent on the angle of view, the relatively wide-angle photographs of the light sheet exhibited a systematic color variation from left-to-right. The visualization was recorded on both photographic color film and slide film using an exposure time of 2 seconds and an f$^\#$4 exposure to allow adequate light on the 400 ASA film and to provide reasonable particle image streaks for determining the large scale motions within the flow field.

In all of the experiments, a quasi-steady state of constant temperature increase was achieved before the thermochroic crystals were visible. Flow visualization images were recorded when individual hot plumes became visible and continued until there was diffi-

Plate No.	Wavelength (mm)	Amplitude (mm)
A0	0	0
A1	6.35	1.27
A2	12.7	2.54
A3	25.4	5.08

Table 3.1: Dimensions of non-uniform surface plates in test section

culty distinguishing structure in the flow field due to uniform blue color of the crystals. Independent images of the flow field were obtained by choosing adequate time separation between successive photographs.

Quantitative temperature measurements on a plane were performed using a color CCD camera and a color frame grabber to record digital image output signals in RGB color space. Locating the camera 2m from the light sheet yielded a 1.4° angle of view which resulted in a negligibly small color change due to variation of the scattering angle (c.f. Gluckman et al. 1993, Ozawa, et al. 1992). Hence, the color of the liquid crystals depended solely on temperature. The RGB color image field was converted to a hue-saturation-intensity image field by Image Pro, an image processing package that estimated hue by using a linear approximation inside a 60° region of the color hexagon, and computed the intensity as (R+G+B)/3, and the saturation as [intensity-min(R,G,B)]/intensity (Media Cybernetics, 1996). Quantitative temperature fields were then obtained from a calibration of hue/intensity versus temperature.

Three-dimensional (i.e. volumetric) measurements of the 'instantaneous' turbulent temperature field were reconstructed from sequential sets of vertical cross-sections obtained by scanning the light sheet in the y-direction with a stepper motor at a speed of 18 mms^{-1}. Twenty-five vertical cross-sections of the flow field could be stored in the computer at a rate of 10 image fields per second, resulting in a physical distance between frames of 1.8 mm and a volume that was 45 mm thick in the y-direction. The total scanning time, 2.5 s, was short enough compared to the evolution time of the thermal field to permit interpreting the measured three-dimensional field as approximately instantaneous.

3.3 Results

3.3.1 Heat Transfer Characteristics

To determine the effect of surface roughness on the heat transfer, several experiments were conducted with plates A0,A1,A2 and A3 to determine the relationship between the Nusselt number and the Rayleigh number. The Nusselt number,

$$Nu = \frac{Q_0 z_*}{\kappa \Delta T} \tag{3.4}$$

is defined using the total layer depth z_*, the thermometric conductivity of the fluid κ, and the mean kinematic heat flux

$$Q_0 = H_0/\rho c. \tag{3.5}$$

For unsteady non-penetrative convection the temperature drop is defined to be

$$\Delta T = 2(\overline{T_1} - T_\infty) \tag{3.6}$$

where T_∞ is the mean temperature in the well-mixed region of the convection layer. When the upper surface is insulated, as in unsteady non-penetrative convection, the temperature difference between the upper and lower surface is the same as the difference between the lower surface and the well-mixed core, because most of the temperature drop occurs in the thin layer above the lower surface. On the other hand, in the classical case of Rayleigh convection between hot and cold plates, the temperature drop between the lower surface and the upper surface is essentially doubled because of the additional conduction layer that occurs just below the upper surface. Hence, in order to make fair comparison between non-penetrative

convection and Rayleigh convection, the temperature difference given in Eq. (3.6) is defined
to be twice the drop from the lower plate to the well-mixed core.

The flux Rayleigh number is given by

$$Ra_f = \frac{\beta g Q_0 z_*{}^4}{\kappa^2 \nu} \qquad (3.7)$$

where β is the coefficient of thermal expansion, g is the acceleration of gravity, ν is the
kinematic viscosity, and Q_0, κ and z_* are as above. The Rayleigh number, Ra is defined by

$$Ra = \frac{\beta g \Delta T z_*{}^3}{\kappa \nu}, \qquad (3.8)$$

where ΔT is given by Eq. (3.6). In Fig. 3.3.1 the Nusselt number is plotted as a function
of flux Rayleigh number, Ra_f, for the three rough surface geometries. The Nusselt number

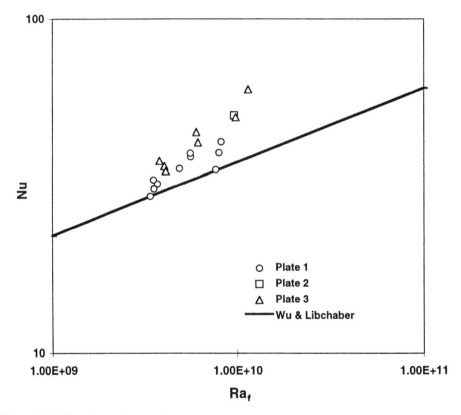

Figure 3.3: Nusselt number vs. flux Rayleigh number for three rough surface profiles ranging from plate A1 with the lowest amplitude and wavelength to plate A3 with the largest
wavelength and amplitude.The solid curve is the correlation found for Rayleigh convection
between smooth plates by Wu and Libchaber (1992).

increases approximately 25% as the wavelength and amplitude of the roughness increase. By analogy with forced convective turbulent transport in which the effect of wall roughness is to increase the skin friction and the heat transfer rate, the trend in Fig. 3.3.1 is not unexpected. For comparison, the heat transfer correlation of Wu and Libchaber (1992) for Rayleigh convection between smooth plates, $Nu = 0.146Ra^{2/7}$ has also been plotted against Ra_f in Fig. 3.3.1, and the Nusselt number is consistently below the Nusselt number for the rough surfaces.

3.3.2 Existence of Horizontal Mean Flow

In previous work Krishnamurti and Howard (1983) and Howard and Krishnamurti (1986) found by laboratory experiments and mathematical modeling that spontaneous generation of a large horizontal flow occurred in a horizontal layer of fluid heated below and cooled above. Yao (1983) showed that the presence of mean flows in an inclined test section did not extrapolate to zero mean flow when the test section was horizontal. In an attempt to determine if mean flows occurred in the present experiments, horizontal and vertical light sheet recordings were made with both long and short periods of time between successive images. The rapidly recorded images were made to discern possible preferred directions of mean flow by recognizing the same large-scale structures moving horizontally between successive exposures. The longest periods of recording images were limited by the temperature range for which the thermochroic crystals were visible. For the lowest kinematic heat flux, the maximum time was 12 minutes. Any large scale horizontal motions with a time constant larger than this value would be difficult to detect from the flow visualizations as they would be dominated by the eddies within the central region.

Although there were down flows in the test section along the sidewall of the test section through which the light sheet entered, there was no conclusive evidence for large scale horizontal flow in any group of visualizations of vertical light sheets for either the smooth or the rough lower surfaces. From plan views of the fluid flows just above the lower surfaces, polygonal cells produced by the interaction of large slow-moving down flows did not possess any regular lateral displacement to indicate the presence of a mean flow. The absence of mean horizontal flow may be due to the nature of the upper surface, which is thermally insulated and does not produce plumes to enhance the plume production of the lower surface.

3.3.3 Patterns of Convection

The experiments that used liquid crystals to visualize the flow were conducted at three heat fluxes with 24 to 36 nearly independent photographs taken at each heat flux. For purposes of exposition the results will be discussed with reference to the typical photographs presented in Figs. 3.3.3–3.3.3, but statements about general behavior are based on examination of all of the photographs. In Figs. 3.3.3–3.3.3 the sequence of colors observed at 90° to the light sheet over the 1.8°C range of color play is, from hottest to coldest: blue-green-yellow-orange-red, roughly 0.4° per color. The color does vary with view angle, so the color scale gradually shifts from left to right on each photograph. Photographs were taken with 2s exposure to create streak images of the particle trajectories for use in visualizing the velocity field. The flux Rayleigh numbers for the different roughness plates lie within $Ra_f = 3.4 \pm 0.4 \times 10^9$.

Figure 3.3.3 presents the temperature-velocity fields in a vertical (x-z) plane above the smooth surface, and each of the rough surfaces. Figures 3.3.3 and 3.3.3 present temperature-velocity fields in horizontal (x-y) planes located 5 mm above the lower surface. There are

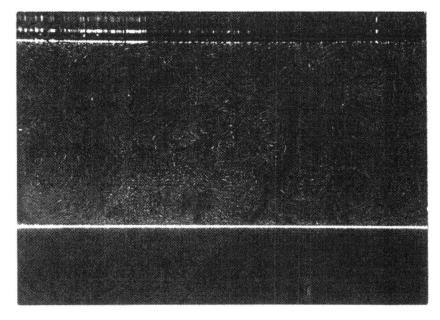

a

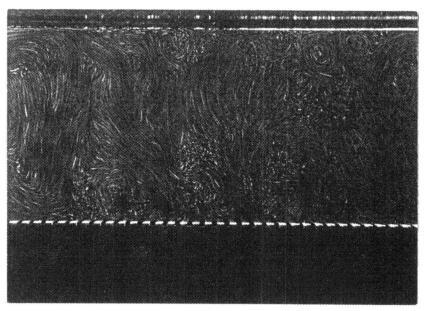

b

Figure 3.4: Flow visualization of non-penetrative thermal convection above rough surfaces with increasing roughness lengthscale. (a) smooth plate A0, $Ra_f = 2.9 \times 10^9$, (b) plate A1, $Ra_f = 3.6 \times 10^9$.

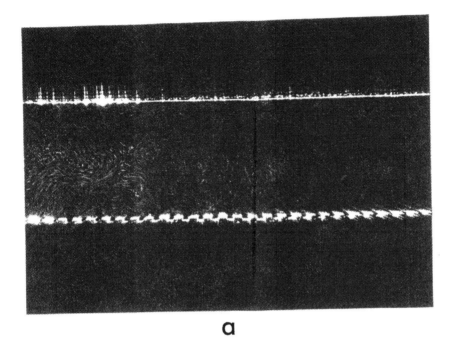

Figure 3.5: Flow visualization of non-penetrative thermal convection above rough surfaces with increasing roughness lengthscale. (a) plate A2, $Ra_f = 3.5 \times 10^9$, and (b) plate A3, $Ra_f = 3.8 \times 10^9$.

a

b

Figure 3.6: Planform flow visualization of convection 5 mm above the surface (a) smooth plate A0 ,$Ra_f = 3.0 \times 10^9$; (b) rough plate A1, $Ra_f = 3.5 \times 10^9$

Figure 3.7: Planform flow visualizations of convection 5 mm above the rough surface plate A3, illustrating the structure of the geometry of cells of downward flowing, colder fluid and their interaction with the roughness elements and the buoyant plumes. $Ra_f = 3.8 \times 10^9$. The time separation is 4.0 seconds.

many similarities between convection over the smooth surface and convection over the rough surfaces. Near the lower surface, blue, vertically oriented streaks corresponding to hot rising fluid are observed in all cases. Often, the small, hot plumes are part of mushroom-shaped thermals. These are particularly evident in Figs. 3.3.3(a) and 3.3.3(b). Other times, the small hot plumes seem to mix with the cooler fluid of the core. The plan view visualizations in Figs. 3.3.3 and 3.3.3 show a pattern much like the 'spoke' convection reported by Clever and Busse (1989). The blue lines represent long thin ridges of hot fluid, evidently rising upwards from the lower surface in thin sheets. The sheets form a random polygonal pattern, with various numbers of sides. The spokes are the vertices of the polygons, and being the intersection of several sheets, they are expected to produce somewhat stronger vertical motion.

The particle streak patterns in Fig 3.3.3 show clearly that the upward buoyant ridges are formed at the boundary between adjacent polygons of horizontally spreading fluid, caused, presumably, by cool, downward moving fluid hitting the lower surface. Upon reaching the lower surface, a down flow must spread horizontally, pushing fluid across the lower surface until it meets the horizontal flow created by an adjacent down flow, at which point the fluid motion stagnates horizontally and begins to rise vertically. Having been heated while crossing the plate, the fluid arriving at the cell boundary is buoyant and rises easily. From the plan views it is clear that the plumes visible in the vertical light sheets are frequently the cross-sections of the thin sheets of buoyant up flows at the edges of polygonal cells, rather than more circularly symmetric concentrated plumes.

Plate	Q_0 (°C cm/s)	z_* (mm)	Ra_f ($\times 10^9$)	θ_* (°C)	z_0 (mm)	θ_0 (0C)
A0	0.031	100.0	2.90	0.07	0.24	0.51
A1	0.033	101.3	3.56	0.07	0.24	0.54
A2	0.032	102.5	3.48	0.07	0.24	0.52
A3	0.032	105.1	3.84	0.07	0.24	0.52

Table 3.2: Thermal parameters

Over the smooth plate, buoyant plumes start anywhere, but over the rough plates they tend to start in the depressions. The horizontal width of a typical streak is about 2mm, and it is independent of the heat flux and the wavelength of the roughness, to within the accuracy of the flow visualization. This length scale should be associated with the molecular conductivity of the fluid since the hot fluid is derived from the thin thermal boundary layer on the lower surface. For reference, Table 3.2 gives Townsend's molecular scales of length, z_0 and temperature, θ_0, defined by

$$z_0 = \frac{\kappa}{[\beta g Q_0 \kappa]^{1/4}}, \qquad \theta_0 = \frac{Q_0}{[\beta g Q_0 \kappa]^{1/4}}, \tag{3.9}$$

and Deardorff's convection layer temperature scale defined by

$$\theta_* = \frac{Q_0}{[\beta g Q_0 z_*]^{1/3}}, \tag{3.10}$$

and the thickness of the thermal boundary implied by the two-sevenths power law for the Nusselt number, given by $\lambda = z_*/Nu$. In terms of these scales, the thermal sheets are about

Figure 3.8: Sections of thermal plumes above (a) smooth plate A0 and (b) rough plate A3 obtained by digital image recording using a color CCD camera.

$10z_0$ wide, or about one λ wide. (The ratio of λ/z_0 depends so weakly on Rayleigh number that it is effectively constant over the range of data presented here.)

The average linear transverse plume density from the lower plate has been estimated by counting the number of plumes in each realization of experimental parameters. It has been found that the density of plumes is independent of the roughness length scale and is only marginally affected by increasing Ra_f. For the lowest Ra_f considered here, 2.9×10^9, it has been estimated that plumes originate for plates 0, 1, 2 and 3 at intervals of 50, 47, 46 and 50 mm respectively. At the largest flux Rayleigh number in the experiments, $Ra_f = 10.1 \times 10^9$, the plumes originate at intervals of 55, 60, 44 and 45 mm respectively, not substantially lower than for the lower heat flux case.

The small hot plumes from different depressions often merge together, c.f. Fig 3.3.3(b) and 3.3.3(b) to form bigger plumes. The mergers may involve as many as four plumes, as in Fig. 3.3.3(b). Over the roughest surface, the plumes appear to be generated preferentially at the side walls of the depressions. Merger may occur between the plumes between one sidewall and the next. The merged plumes are, of course substantially wider, and most mergers seem to be complete by about 0.2-0.3 z_* above the lower surface. Above this height one can observe large plumes, having widths of order 0.2-0.3z_*, that extend almost across the entire layer. The geometry of the flow in this core region is not strongly dependent on the roughness of the plate.

Above the flat plate, the polygonal cells show no preferred orientation, but in the presence of roughness elements, the polygonal cell walls become more complex. When fluid flows horizontally across the roughness elements, it is influenced by thermal plumes which originate within the depressions of the surface, where they are unaffected by horizontal flows until they have sufficient buoyancy to leave the lower surface. At the edges of the convection cells, the horizontal flows interact with the upward buoyant flow from the lower surface to produce a strong three-dimensional disturbance, c.f. Fig. 3.3.3(b) and Fig. 3.3.3. These disturbances take the form of more concentrate plumes and they create a polygonal boundary that appears to be rather diffuse. Even so, the polygons still persist.

The three-dimensional structure of the buoyant flow in the neighborhood of the intersection of several cell boundaries, and in the depression of a rough surface, has been investigated by using scanning liquid crystal thermometry to make volumetric temperature measurements. Figure 3.3.3 shows two vertical sections of thermal plumes obtained by video recording and digitizing the liquid crystal color field on vertical light sheet sections that were perpendicular to the wave crests of the roughness elements and located near the center of the convection test section. The quantitative temperature fields of these two plumes obtained from a calibration of hue versus temperature are shown in Fig. 3.3.3. In each of the realizations, there is a hot plume rising from the lower surface on the right half of the plane. The arrows labeled **a–d** indicate the locations of planes that will be presented in ensuing figures.

Three additional sections of the plume over the smooth plate in Fig. 3.3.3 are shown in Fig. 3.3.3. The hottest part of the plume is only 10 mm wide in the y-direction, but there is a cooler portion that has significantly greater extent, more consistent with a hot ridge. Fig. 3.3.3 shows sections of the same plume in the y-z plane, verifying its narrow extent in the y-direction. Lastly, a plan view (x-y plane) in Fig 3.3.3(a) confirms that the thermal is at the intersection of at least two cell boundaries. The plan views at higher elevations in Fig. 3.3.3(c)-(d) show that the plume at the intersection grows while the cell walls tend to weaken and disperse.

The corresponding series of data for the rough plate realization is shown in Figs. 3.3.3, 3.3.3, 3.3.3 and 3.3.3. In this realization the thermal is relatively isolated and does not appear to be a part of a boundary, although a boundary does exist nearby, c.f. Fig. 3.3.3. the hot core of the thermal is approximately as wide as the depression (12 mm) and it contains considerably more temperature irregularity than the plume from the smooth plate, indicating small scale mixing.

3.4   Summary and conclusions

The temperature fields of unsteady, non-penetrative convection have been studied over a smooth surface and several rough surfaces containing two-dimensional square-wave profiles. Surface roughness reduces the temperature drop between the mean surface elevation and the center of the fluid, as manifested in a change in the Nusselt number-flux Rayleigh number relationship. The roughness also has several effects on the pattern of the coherent structures.

Over smooth surfaces, we observe thin sheets of hot fluid lifting from the surface that appear to form random polygonal patterns between the surface and a height of approximately three-tenths of the total layer depth. Above this depth, the temperature excess of the ascending fluid decreases to the point where the walls of the polygons are not readily visualized. In the surface region one can observe mushroom-shaped thermals that are hot and penetrate to approximately three-tenths of the layer depth. Above that, one observes strong, large thermals that can extend across the entire layer depth and carry significant heat, although they are relatively cool.

The random polygonal features above the smooth plate appear also in the case of surface roughness. The effects of the sharp-edged, two-dimensional roughness elements used in the present experiment are to disrupt the smoothness of the rising ridges of fluid and create substantial three-dimensional motion within them. However, the random polygonal shapes are roughly similar to those over smooth surfaces. Previous work by Schmidt and Schumann (1989) and others suggests that fluid is swept along the cell wall into the intersections where large thermals are formed, and these regions may be the origins of the large vertical motions that we see extending across the entire layer. The evidence that we have for this, obtained from 3-D reconstruction of the temperature field using a standing light sheet technique and thermochroic liquid crystals shows that indeed at the intersection of thin hot ridge-shaped regions it is possible to observe a strong, single thermal emerging. The cell walls are not oriented along a perpendicular roughness, and seem to depend weakly, if at all, on the direction of the two-dimensional roughness elements. We also do not observe large-scale horizontal motions as in the experiments of Wu and Libchaber (1992). The mushroom-shaped thermals appear to be generally ascending in a vertical direction although occasionally some large-scale randomly oriented motion may cause the thermals to tilt. However, there is no preferred tilting direction.

Acknowledgements

This work was supported by a grant from the National Science Foundation. NF received support from the Japan Ministry of Science. .

Bibliography

Adrian, R. J., R. T. D. S. Ferreira, and T. Boberg, 1986. Turbulent thermal convection in wide horizontal fluid layers, *Exp. Fluids*, **4**, 121-141.

Clever, R. M. and F. H. Busse, 1989. Three-dimensional knot convection in a layer heated from below, *J. Fluid Mech.* **198**, 345-363.

Dörnbrack, A. and U. Schumann, 1993. Numerical simulation of turbulent convective flow over wavy terrain, *Boundary Layer Meteor.* **65**, 323-355.

Gluckman, B. J., H. Willaime and J. P. Gollub, 1993. Geometry of isothermal and isoconcentration surfaces in thermal turbulence, *Phys. Fluids A*, **5**, 647-661.

Hallcrest Co. Inc, Illinois 1996. Private communication.

Hechtel, L. M., C.-H. Moeng and R. B. Stull, 1990. The effects of nonhomogeneous surface fluxes on the convective boundary layer: A case study using large-eddy simulation, *J. Atmos. Sci.*, **47**, 1721-1741.

Howard, L. N. and R. Krishnamurti, 1986. Large scale flow in turbulent convection: a mathematical model, *J. Fluid Mech.*, **170**, 385-410.

Kaimal, J. C., R. A. Eversole, D. H. Lenschow, B. B. Stankov, P. H. Kahn and J. A. Businger, 1982. Spectral characteristics of the convective boundary layer over uneven terrain, *J. Atmos. Sci.*, **39**, 1098-1114.

Krettenauer, K. and U. Schumann, 1989. Direct numerical simulation of thermal convection over a wavy surface, Meteorol. *Atmos. Phys.*, **41**, 165-179.

Krettenauer, K. and U. Schumann, 1992. Numerical simulation of turbulent convection over wavy terrain, *J. Fluid Mech.*, **237**, 261-299.

Krishnamurti, R. and L. N. Howard, 1983. Large scale flow in turbulent convection: Laboratory experiments and a mathematical model, *Papers in Meteorological Res.*, **6**, 143-159.

Lenschow, D. H., 1970. Airplane measurements of planetary boundary layer structure, *J. Applied Meteor.*, **9**, 874-884.

Lenschow, D. H., 1974. Model of the height variation of the turbulence kinetic energy budget in the unstable planetary boundary layer, *J. Atmos. Sci.*, **31**, 465-474.

Media Cybernetics, 1996. Private communication.

Ozawa, M., U. Muller, I. Kimura and T. Takamori, 1992. Flow and temperature measurement of natural convection in a Hele-Shaw cell using a thermo-sensitive liquid-crystal tracer, *Exp. Fluids* , **12**, 213-222.

Schmidt, H. and U. Schumann, 1989. Coherent structure of the convective boundary layer deduced from large-eddy simulation, *J. Fluid Mech.*, **200**, 511-562.

Schumann, U., 1988. Minimum friction velocity and heat transfer in the rough surface layer of a convective boundary layer, *Boundary Layer Meteor.*, **44**, 311-326.

Schumann, U., 1991. A simple model of the convective boundary layer over wavy terrain with variable heat flux, Beitr. *Phys. Atmosph.* (Contrib. to Atmos. Physics), **64**, 169-184.

Tamai, N. and T. Asaeda, 1984 Sheetlike plumes near a heated bottom plate at large Rayleigh number, *J. Geophys. Res.*, **89**, 727-734.

Telford, J. W. and J. Warner, 1964. Fluxes of heat and vapor in the lower atmosphere derived from aircraft observations, *J. Atmos. Sci.*, **21**, 539-548.

Willis, G. E. and J. W. Deardorff, 1974. A laboratory model of the unstable planetary boundary layer, *J. Atmos. Sci.*, **31**, 1297-1307.

Yao, C. S., 1983. High Reynolds Number, Unsteady Thermal Convection in a Shallow Layer, Ph. D. thesis, University of Illinois at Urbana-Champaign, Urbana, IL.

Wu, X. Z. and A. Libchaber, 1992. Scaling relations in thermal turbulence: The aspect-ratio dependence, *Phys. Rev. A*, **45**, 842-845.

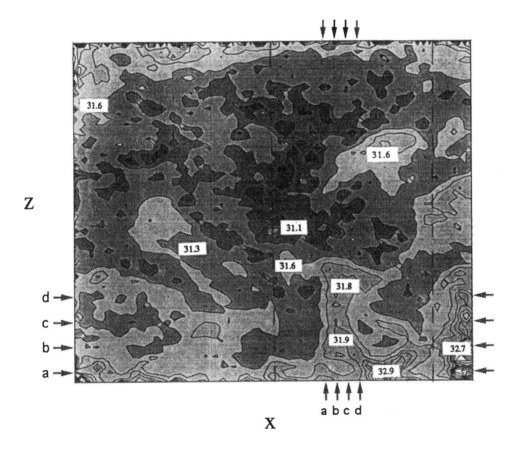

Figure 3.9: Temperature distribution in the x-z plane above the smooth plate A0. $y = 24$ mm for the thermal plume shown in Fig. 3.3.3(a).

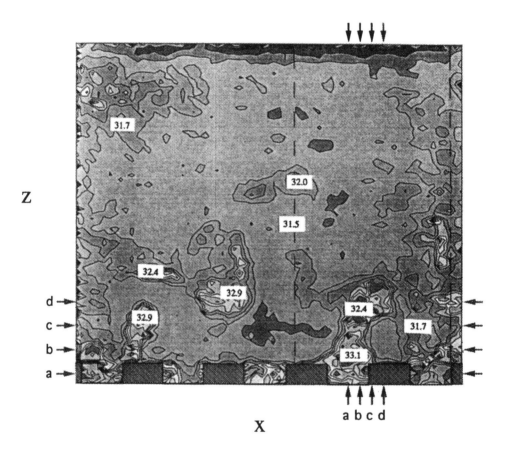

Figure 3.10: Temperature distribution in the x-z plane above the smooth plate A0. $y = 42$ mm for the thermal plume above the rough plate A3 shown in Fig. 3.3.3(b).

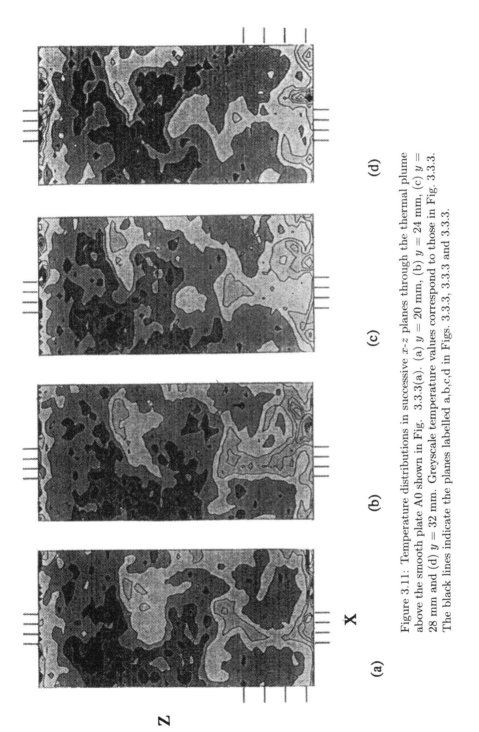

Figure 3.11: Temperature distributions in successive x-z planes through the thermal plume above the smooth plate A0 shown in Fig. 3.3.3(a). (a) $y = 20$ mm, (b) $y = 24$ mm, (c) $y = 28$ mm and (d) $y = 32$ mm. Greyscale temperature values correspond to those in Fig. 3.3.3. The black lines indicate the planes labelled a,b,c,d in Figs. 3.3.3, 3.3.3 and 3.3.3.

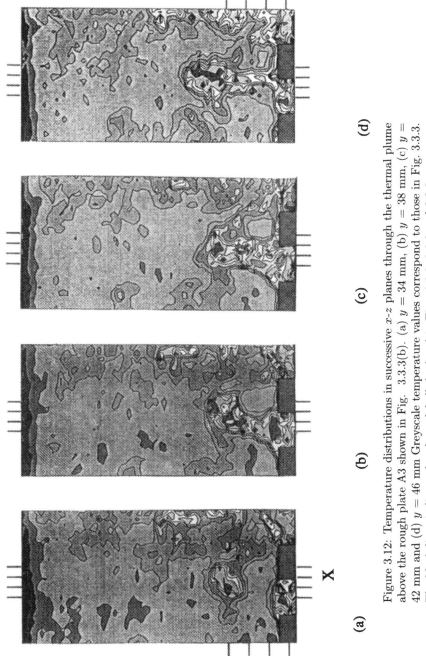

Figure 3.12: Temperature distributions in successive x-z planes through the thermal plume above the rough plate A3 shown in Fig. 3.3.3(b). (a) $y = 34$ mm, (b) $y = 38$ mm, (c) $y = 42$ mm and (d) $y = 46$ mm Greyscale temperature values correspond to those in Fig. 3.3.3. The black lines indicate the planes labelled a,b,c,d in Figs. 3.3.3, 3.3.3 and 3.3.3.

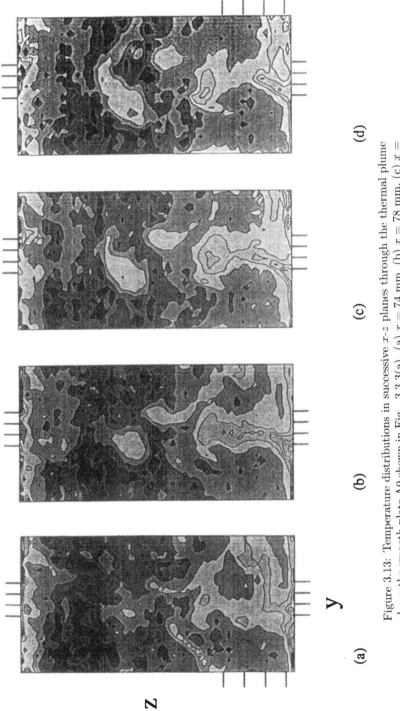

Figure 3.13: Temperature distributions in successive x-z planes through the thermal plume above the smooth plate A0 shown in Fig. 3.3.3(a). (a) $x = 74$ mm, (b) $x = 78$ mm, (c) $x = 82$ mm and (d) $x = 86$ mm. Greyscale temperature values correspond to those in Fig. 3.3.3. The black lines indicate the planes in Figs. 3.3.3, 3.3.3 and 3.3.3.

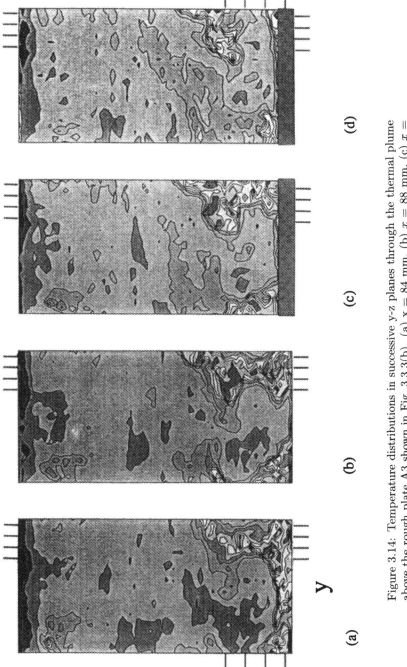

Figure 3.14: Temperature distributions in successive y-z planes through the thermal plume above the rough plate A3 shown in Fig. 3.3.3(b). (a) x = 84 mm, (b) x = 88 mm, (c) x = 92 mm and (d) x = 96 mm . Greyscale temperature values correspond to those in Fig. 3.3.3. The black lines indicate the planes in Figs. 3.3.3, 3.3.3 and 3.3.3.

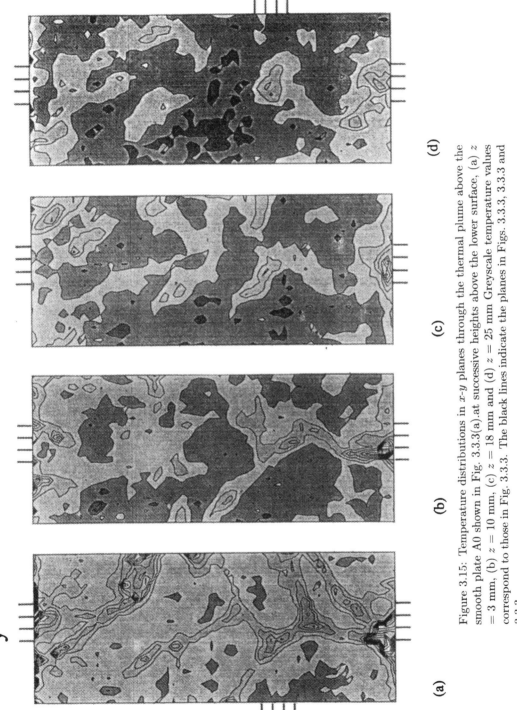

Figure 3.15: Temperature distributions in x-y planes through the thermal plume above the smooth plate A0 shown in Fig. 3.3.3(a). at successive heights above the lower surface, (a) $z = 3$ mm, (b) $z = 10$ mm, (c) $z = 18$ mm and (d) $z = 25$ mm Greyscale temperature values correspond to those in Fig. 3.3.3. The black lines indicate the planes in Figs. 3.3.3, 3.3.3 and 3.3.3.

KEANE, FUJISAWA, and ADRIAN

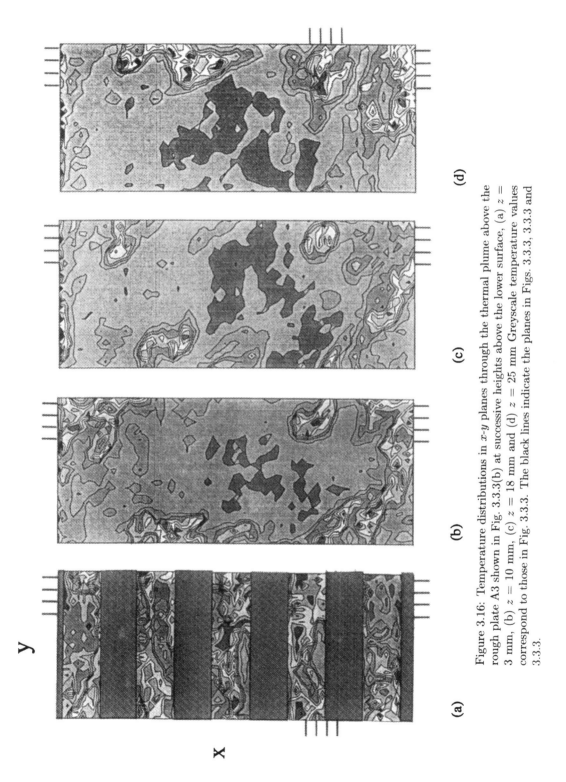

Figure 3.16: Temperature distributions in x-y planes through the thermal plume above the rough plate A3 shown in Fig. 3.3.3(b) at successive heights above the lower surface, (a) $z = 3$ mm, (b) $z = 10$ mm, (c) $z = 18$ mm and (d) $z = 25$ mm Greyscale temperature values correspond to those in Fig. 3.3.3. The black lines indicate the planes in Figs. 3.3.3, 3.3.3 and 3.3.3.

Chapter 4

ASTROPHYSICAL CONVECTION AND DYNAMOS

AXEL BRANDENBURG[1], ÅKE NORDLUND[2] and
ROBERT F. STEIN[3]

[1] *Department of Mathematics*
University of Newcastle upon Tyne NE1 7RU, UK
[2] *Copenhagen University Observatory*
Juliane Maries Vej 30, DK-2100 Copenhagen Ø, Denmark
[3] *Department of Physics and Astronomy*
Michigan State University, East Lansing, MI 48824

Convection can occur in various astrophysical settings. In this review some aspects of solar convection are highlighted. In deeper layers of the solar convection zones rotation becomes important and can lead to effects such as downward pumping of vorticity and magnetic fields. Rotation has the tendency to partially evacuate vortex tubes making them lighter. This effect can sometimes reverse the core of a downdraft and make it buoyant. The problem of different thermal and dynamical timescales is addressed and finally the formation of magnetic structures by convection is discussed.

4.1 Introduction

In astrophysics convection can occur in various settings. Convection in stars is ubiquitous. In stars of mass $M \lesssim 2M_\odot$ ($1\,M_\odot \approx 2 \times 10^{33}$ g is the mass of the sun) convection occurs in the outer layers of the star because the gas becomes relatively opaque. At the surface of the sun this convection is manifested by the granulation (Bray et al. 1984), which is a time-dependent cellular pattern similar to classical Rayleigh-Benard convection, such as may be seen by heating a shallow container of oil mixed with aluminum powder as tracer. A comparison with such a kitchen experiment is misleading, however, because it suggests that solar convection is a rather viscous phenomenon. While it is actually true that the viscosity of the gas in the sun (mostly hydrogen and helium) is comparable with that of honey (something like 10^2 cm^2/s), the Reynolds numbers Re $= UL/\nu$ are in fact rather large, because of the large length scales L. It is indeed quite typical for astrophysical bodies that such dimensionless numbers take astronomically large values. The Rayleigh number, for example, is around 10^{24} or larger, but we will come back to this below. Before that, let us discuss other examples of astrophysical convection.

Shallow surface convection zones also occur in white dwarfs, one of the end products of stellar evolution of stars with mass less than about eight solar masses. The end products of stars with larger mass are often neutron stars, the remnants of supernova explosions. Just after a neutron star has formed convection may develop for a very short period of a few seconds (see e.g. Thompson & Duncan 1993).

A typical (local) turnover time is

$$\tau_{\text{turnover}} = H_p/u_t. \tag{4.1}$$

Here, $H_p = |d \ln P/dr|^{-1}$ is the local pressure scale height, and u_t is the turbulent root-mean-square velocity. In solar like stars this ranges from minutes near the surface to months near the bottom. In neutron stars and white dwarfs the typical turnover times are more like $1-10$ ms. The integrated turnover time is

$$t_{\text{turnover}} = \int \frac{dr}{u_t(r)} \ . \tag{4.2}$$

For the sun, which has a rather deep convection zone (30% by radius, or 200 Mm), this is around a month.

The outer layers of stars of mass $\geq 2M_\odot$ are stably stratified, except for very shallow outer convection zones associated with the ionization zones of hydrogen and helium, but stars more massive than $1-2M_\odot$ have convective cores due to high temperature gradients arising from the high temperature sensitivity of nuclear reactions that become possible for stars hotter than the sun.

The energy source of convection is not always nuclear energy. In galaxies, a rather different type of convection occurs. Although galaxies are on average stably stratified, isolated events like supernova and superbubble explosions (collective supernova explosions triggering each other) lead to buoyant outflows. Those events heat the gas locally to 10^6 K, leading to buoyant bubbles that shoot through the gaseous galactic disc into the outer halo. Here the typical dimensions are kiloparsecs, where $1 \, \text{kpc} = 3 \times 10^{21}$ cm or 3000 light years.

Convection also occurs in another important class of objects – accretion discs – where the source of energy is gravitational potential energy. Matter gradually spirals onto a central object, which could be a protostar or stellar remnant (white dwarf, neutron star, black hole), or it could be a supermassive black hole of 10^8 solar masses or more. In the absence of friction matter would just stay in the same orbit, like the particles in Saturn's rings, and there would be no accretion onto the central object. However it is due only to turbulent friction and magnetic fields, that matter loses angular momentum. Associated with this is a conversion of potential energy into heat via viscous and Joule heating. This is a remarkable point. Although in astrophysics the microscopic viscosity ν and magnetic diffusivity η are generally very small (in the sense of huge Reynolds numbers), this does not imply that the total viscous and Joule dissipation are necessarily small.

One of the corner stones of Kolmogorov (1941) scaling in turbulence is that the average dissipation is of the order of $\rho U^3/L$, and does not depend on the value of the molecular viscosity. A similar result has recently been demonstrated for magnetic dissipation in driven, low-beta plasmas, where the dissipation scales as $B^2 U L/L_\text{B}^2$, L_B being the characteristic length of the field lines (Galsgaard & Nordlund 1996). In accretion discs, turbulent viscous and Joule dissipation can be extremely important. For instance, the most luminous objects in astrophysics are quasars, quasi-stellar objects, which are now believed to be gigantic accretion discs around supermassive black holes, where the source of radiation is viscous and Joule dissipation. So, although viscosity and magnetic diffusion may be "negligibly small", they cannot be neglected and may well be energetically very important. If convection occurs in accretion discs, it is probably not the main source of turbulence.

We conclude this introduction to the rather broad range of astrophysical convection by mentioning a few references. The extent and location of stellar convection zones is described in Kippenhahn & Weigert (1990). Gas flows and related magnetic fields in galaxies are described in Ruzmaikin et al. (1988), and the topic of accretion discs is introduced in the book by Frank et al. (1992). In the following we focus on solar convection.

4.2   Deep solar convection

In solar convection the Reynolds number, $R_e = UL/\nu$, is very large. The kinematic viscosity ν rises from about $1\,\mathrm{cm}^2/\mathrm{s}$ near the surface to about $10^4\,\mathrm{cm}^2/\mathrm{s}$ in deeper layers (due to radiative viscosity, i.e. momentum exchange by photons; e.g. Kippenhahn & Weigert 1990). The typical velocities U and vertical scale heights vary from $1\,\mathrm{km}/s$ and $1\,\mathrm{Mm}\ (= 1000\,\mathrm{km})$ near the surface to $0.1\,\mathrm{km}/s$ and $100\,\mathrm{Mm}$ deeper down, so the Reynolds number is then roughly

$$\mathrm{Re} = \frac{UL}{\nu} \approx \frac{10^5\,10^8}{1} \cdots \frac{10^4\,10^{10}}{10^4} = 10^{13}...10^{10}. \tag{4.3}$$

The Peclet number, $\mathrm{Pe} = UL/\chi$, that compares advective effects to thermal diffusion effects, is also very large. Thermal conduction is quite unimportant. Instead there is radiative diffusion of photons, whose mean free path is much larger than that of the electrons. Radiative diffusion can be described by Fick's law, so the radiative flux is

$$\boldsymbol{F}_{\mathrm{rad}} = -K(\rho, T)\boldsymbol{\nabla}T. \tag{4.4}$$

The radiative conductivity K depends strongly on density ρ and temperature T. In the deeper parts of the solar convection zone most of the opacity is due to close encounters of free electrons passing nearby an ion. This can absorb and emit radiation (free-free transition), and the opacity can be described by a power law (Kramers' opacity), so K is then

$$K = K_0 \left(\frac{\rho}{\rho_0}\right)^{-2} \left(\frac{T}{T_0}\right)^{6.5}. \tag{4.5}$$

The value of K_0 is well known ($K_0 = 5 \times 10^{-27}\,\mathrm{erg\,cm^{-1}\,s^{-1}\,K^{-1}}$, where $\rho_0 = 1\,\mathrm{g/\,cm^3}$ and $T_0 = 1\,\mathrm{K}$). In the following, however, we shall refer to Kramers' opacities even when the value of K_0 is modified (see below). K increases by about $\sim 10^2$ from the top to the bottom of the convection zone, so the radiative flux increases gradually towards the bottom of the convection zone (the temperature gradient is roughly constant), and takes over completely from convection in the radiative layers below the convection zone. Here, the further increase in conductivity is compensated for by a decrease in the temperature gradient (and a decrease in the surface area), to keep the total luminosity constant.

The radiative diffusivity coefficient, that characterizes the diffusion of temperature perturbations, is

$$\chi = \frac{K}{\rho c_p}, \tag{4.6}$$

where c_p is the specific heat at constant pressure. The factor ρc_p, the thermal energy per unit mass and Kelvin, increases by about five orders of magnitude through the convection zone (in the surface layers c_p is enhanced by about one order of magnitude relative to an ideal gas). Thus, the rate at which temperature fluctuations are smoothed out by diffusion *decreases* by about three orders of magnitude from the top to the bottom of the convection zone. Near the bottom of the solar convection zone χ is around $10^7\,\mathrm{cm}^2/\mathrm{s}$, so the Peclet number is

$$\mathrm{Pe} = \frac{UL}{\chi} \approx \frac{10^4\,10^{10}}{10^7} \sim 10^7, \tag{4.7}$$

whereas near the surface typical values based on Kramers' opacity are

$$\mathrm{Pe} \approx \frac{10^5\,10^8}{10^{10}} \sim 10^3. \tag{4.8}$$

Additional opacity in the surface layers, in particular from the negative hydrogen ion, increases the actual Peclet number there considerably.

It is worth noting that one may write the Peclet number as

$$\mathrm{Pe} = \frac{UL}{K/(\rho c_p)} = \frac{\rho c_p U \delta T}{KT/L} \frac{T}{\delta T} \approx \frac{F_{\mathrm{conv}}}{F_{\mathrm{rad}}} \frac{1}{\delta \ln T}; \tag{4.9}$$

i.e., as the ratio of the convective and radiative fluxes, divided by the relative temperature fluctuation. Near the surface, $\delta \ln T \sim 1$, but the radiative flux is only a tiny fraction of the total flux. Near the bottom, $F_{\mathrm{rad}} \sim F_{\mathrm{conv}}$, but $\delta \ln T \sim 10^{-6}$.

Thus, for both the momentum and the energy equation, it is clear that the real diffusive effects are much smaller than what can be handled by direct numerical simulations, even though the situation is less extreme in the energy equation. The ratio of the diffusion coefficients in the momentum or energy equations is the Prandtl number,

$$\mathrm{Pr} = \nu/\chi = \mathrm{Pe}/\mathrm{Re} \sim 10^3/10^{13}...10^7/10^{10} \sim 10^{-10}...10^{-3} \tag{4.10}$$

At this point there is a crucial decision to be made. One can either resort to some kind of subgrid scale modeling, or one can use scaled values for the viscosity and the radiative diffusivity, trying to preserve a reasonably small value for the Prandtl number, even though it is clear that one cannot come (even in a logarithmic sense) near solar values. Subgrid scale models, on the other hand, are normally insensitive to the value of Pr and resemble more the conditions for Prandtl number equal unity.

We have taken the attitude that on scales that can be resolved by the simulations, diffusive effects do not depend on the actual values of the microscopic diffusion coefficients, consistent with the Kolmogorov assumption mentioned earlier. Effects due to small Prandtl numbers are then expected only in the range of scales where diffusion of momentum is turbulent, but thermal diffusion is not. In the next section, we briefly mention some effects that may be expected there.

4.3   Low Prandtl number effects

The importance of the Prandtl number has been emphasized in the context of laboratory convection, but not so much in the context of astrophysical convection. There is one qualitative aspect that arises due to a small value of the Prandtl number. Small Prandtl number

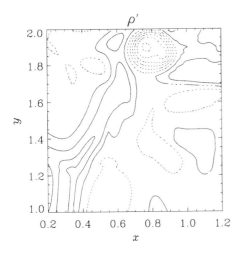
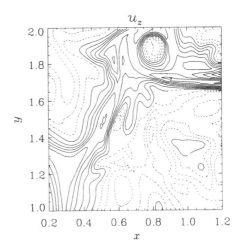

Figure 4.1: Contours of horizontal slices (just below the surface, $z = 0.1$) of the density fluctuation $\rho' = \rho - \langle \rho \rangle$ (dotted contours mean lighter material and solid contours heavier material) and the vertical velocity u_z (dotted contours mean upward, solid contours downward flow). The downdraft lanes are characterized by $\rho' > 0$ and $u_z > 0$, except in the vortex near the top of the figure, where $\rho' < 0$ (vortex buoyancy) and a small region with $u_z < 0$ has developed. Run D of Brandenburg et al. (1996).

means that radiative diffusion is large compared with viscosity. The viscosity determines the typical scale of vortex tubes (Constantin et al. 1995), so a small Prandtl number means that the temperature will rapidly equilibrate across such vorticity structures. The pressure is reduced inside vortex tubes, in proportion to the square of the speed of rotation. Using now the equation of state for a perfect gas,

$$p = \frac{\mathcal{R}T}{\mu}\rho, \qquad (4.11)$$

where $\mathcal{R}$ is the universal gas constant and μ the mean molecular weight, we see that for almost uniform temperature across the tube, the density in the tube must be lowered. This has consequences for convective downdrafts, that will be explained next.

First we need to emphasize that, because of vertical density stratification, the area occupied by the convective downdrafts is smaller than the area of the upwelling motions. This is simply because a fluid parcel contracts as it descends and it expands as it ascends. In order to conserve mass ($\rho v S = $ const), the downdrafts have to speed up (their cross-sectional area S decreases, so the velocity v increases); see Stein & Nordlund (1989). This, together with the Coriolis force, sets them into rapidly swirling motion. This is all not very surprising and quite similar to geophysical convection. Also, what does it have to do with small Prandtl numbers? In fact, in what we said above there seems to be a contradiction as to whether the density in the downdrafts is large or small. We started by saying that the density has to be small, because vortex tubes have low pressure. On the other hand, what generates those rapidly spinning downdrafts in the first place is of course the fact that they are cool and therefore dense, so the density cannot be low. This can be understood by considering the following sequence of events: material at the surface cools, becomes dense and begins to descend. As it descends, it contracts and is set into spinning motion. The spinning downdrafts then tend to eject material centrifugally, so the pressure is lowered, relative to what it would otherwise be. If the corresponding density reduction is larger than the density enhancement associated with the downdraft, the buoyancy may actually reverse, and eventually cause also the vertical velocity to reverse. This process has indeed been observed in numerical simulations of low Prandtl number convection in the presence of rotation (Brandenburg et al. 1996, Brummell et al. 1996; see also Fig. 4.1).

If the Prandtl number were of order unity this process would have been less pronounced, because the temperature would have been able to adjust to the smaller adiabatic value, so the density reduction would have been smaller by a factor γ.

4.4   The entropy gradient

In ordinary Boussinesq convection the degree of instability is characterized by the Rayleigh number

$$\text{Ra} = \frac{H^4}{\nu\chi}(\boldsymbol{g}\cdot\boldsymbol{\nabla}\ln T)_0, \qquad (4.12)$$

where the subscript 0 refers to the (unstable) nonconvective hydrostatic state, which is sometimes also the initial state used in simulations. Here, T is the temperature, H the height of the layer, and $\boldsymbol{g}$ is the gravitational acceleration. (Normally one would write $\alpha\boldsymbol{\nabla}T$ instead of $\boldsymbol{\nabla}\ln T$, where α is the expansion coefficient, but for a perfect gas $\alpha = 1/T$; see Chandrasekhar 1961.) This definition is readily extended to the regime of stratified convection by replacing the logarithmic temperature by the normalized specific entropy s/c_p, so

$$\text{Ra} = \frac{H^4}{\nu\chi c_p}(\boldsymbol{g}\cdot\boldsymbol{\nabla}s)_0, \qquad (4.13)$$

where $s/c_p = \ln T - (1 - 1/\gamma)\ln p$ is the normalized specific entropy, $(\boldsymbol{\nabla}s)_0$ is the entropy gradient of the unstably stratified nonconvective hydrostatic solution and c_p is the specific heat at constant pressure. For a perfect monoatomic gas, $c_p = \frac{5}{2}\mathcal{R}/\mu$.

In the inviscid limit, $\nu \to 0$, stability is simply governed by the sign of $\boldsymbol{g}\cdot\boldsymbol{\nabla}s$. In astrophysics this is known as Schwarzschild's criterion,

$$\boldsymbol{g}\cdot\boldsymbol{\nabla}s > 0 \quad \text{(instability)}. \qquad (4.14)$$

Let us now look at the profile of the horizontally averaged entropy, $\langle s \rangle$, as a function of z using data of a simulation. In Fig. 4.2 we show the entropy profile from a simulation where Kramers' opacity law is used. The vertical scale in the figure does not accurately represent the sun, especially near the surface which has artificially been pulled further down into the convection zone relative to the sun.

There are basically five regimes; see Fig. 4.2. The first regime (below 470 Mm) is the stably stratified radiative interior where the average entropy (here normalized by c_p/g) rises at a rate

$$-\boldsymbol{g}\cdot\boldsymbol{\nabla}\langle s/c_p \rangle \equiv N_{\text{BV}}^2 > 0, \qquad (4.15)$$

where N_{BV} is the Brunt-Väisälä frequency if the stratification were isothermal. (Because of temperature stratification the actual Brunt-Väisälä frequency is somewhat different.) Between 470 and 500 Mm (regime II) there is another layer where N_{BV}^2 is almost unchanged,

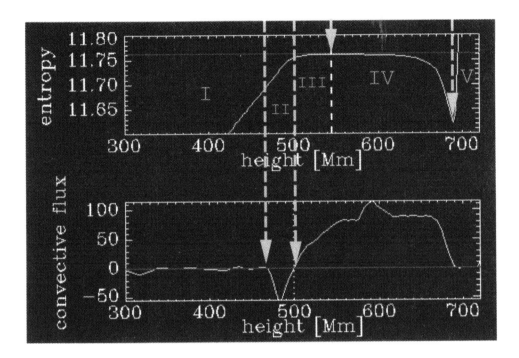

Figure 4.2: Profiles of entropy and convective flux in a single snapshot of the simulations with Kramers' opacities. Region I is the radiative interior, II the overshoot layer, III the radiative heating layer, IV the bulk of the convection zone, and V the surface layer.

but the convective flux is now negative. Here there are only sporadically entering plumes from the convection zone that "stir" the gas. In regime III (500–540 Mm) N_{BV}^2 is still positive, but much smaller than in the deep radiative interior. This regime should be considered as part of the convection zone, because there is a convective flux that is directed upwards. However, according to Schwarzschild this entire layer is still stable. At some point in the middle of the convection zone there is a point (near 540 Mm) where N_{BV}^2 finally turns negative and its magnitude increases until one reaches the surface layers (near 690 Mm in the plot), where there is a sharp increase of the entropy. This is then the fifth regime which is a stably stratified overshoot layer above the convection zone. Here, N_{BV}^2 is again positive and its value is larger than the value for the radiative interior.

To appreciate the properties of the slightly stable profile of $\langle s \rangle$ in the lower part of the convection zone (region III) we should first point out that in simulations with fixed vertical profiles of the conductivity, $K = K(z)$, having a narrow transition region between stable and unstable layers, this regime is very narrow. In Fig. 4.3 we show the profile of $\langle s \rangle$ from a snapshot of a simulation of Brandenburg et al. (1996), where there are three layers with different values of the conductivity $K = K(z)$. The unstable hydrostatic reference solutions have different polytropic indices in the three regions, with a continuous but rather abrupt change from region to region. As a consequence, the transition from transport by radiation

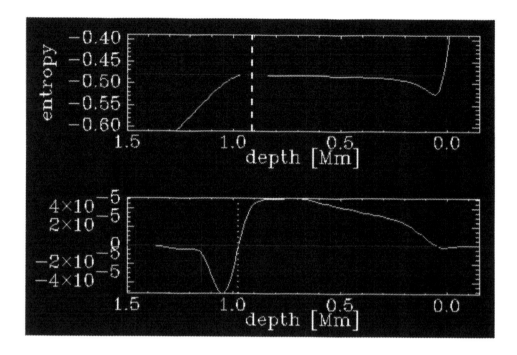

Figure 4.3: Profiles of entropy and convective flux in a single snapshot of the simulations with fixed piecewise constant conductivity $K(z)$.

to transport by convection is more abrupt than in the case with Kramers' opacity (Fig. 4.2).

The existence of a region with positive convective flux but stable average stratification is intimately related to the transition from radiative to convective energy transport (Nordlund et al. 1996). The decrease of the radiative energy flux with radius implies a negative divergence of the radiative energy flux, corresponding to heating of the fluid. Of course, the increase of the convective flux with radius corresponds to a cooling that, on the average, balances the heating from the decrease of the radiative flux. However, the heating and cooling have quite different horizontal distributions, and only cancel when averaged over horizontal area (and time). The radiative flux is approximately uniform, being proportional to the temperature gradient of a medium with very small relative temperature fluctuations. The convective flux is very non-uniform, being carried mostly by localized, cool downdrafts.

On the background of these considerations, it is easy to understand the reason why the average stratification must be weakly stable in this region (regime III in Fig. 4.2). The nearly uniform positive heating from the decreasing radiative flux gradually increases the entropy of the ascending material. Cool downdrafts that penetrate into this weakly stable region are decelerated and also gradually loose their entropy contrast relative to the background. In order to penetrate all the way through region III, they must arrive with a sufficient entropy deficiency and/or excess momentum. The gradually decreasing convective flux with depth is the horizontally averaged manifestation of both the decreasing entropy contrast of those downdrafts that manage to penetrate, and the decreasing number of downdrafts that

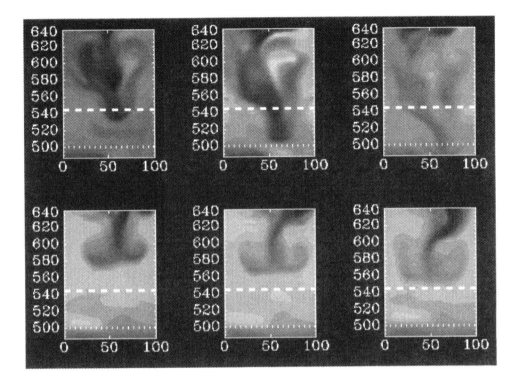

Figure 4.4: Evolution of two downdrafts shown as vertical slices of the entropy. Dark means low entropy.

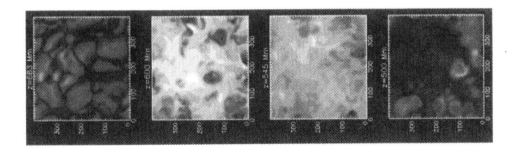

Figure 4.5: Horizontal slices of the temperature through a downdraft.

succeed.

In Fig. 4.4 we show the evolution of two plumes in two different vertical slices of the entropy for the run described in Fig. 4.2 (i.e. the run with Kramers' opacity). The boundaries between regions II, III, and IV are marked by dotted and dashed lines, respectively. Two different types of plumes are visible: one that plunges down through the entire convection zone into the stable layer below and another that develops a typical mushroom shaped head. The extent of plume penetration is an important problem that needs further investigation.

Figure 4.5 shows horizontal images of the temperature in four selected layers. Around the upper entropy minimum near the surface ($z = 683\,\mathrm{Mm}$) there is a clear granulation-like pattern with cool downdraft lanes between warm upwellings. This is qualitatively similar to the solar granulation, but here in the model the stratification is less than in reality, so this "surface layer" does not correspond to the real surface of the sun which is still further up in the atmosphere. In the bulk of the convection zone ($z = 600\,\mathrm{Mm}$ and $z = 545\,\mathrm{Mm}$) the temperature pattern is rather different with cool regions now being disconnected. In the lower overshoot layer the situation is different again. Even at the top of the overshoot layer ($z = 500\,\mathrm{Mm}$) there are now warm isolated spots. They are a consequence of cool downdrafts that shoot into the stable layer, where they appear hot in comparison to the surroundings, whose entropy decreases rapidly with depth (cf. Fig. 4.2).

4.5   The thermal time scale problem

In deeper layers of the sun the thermal time scale

$$\tau_{\mathrm{th}} = H_p^2/\chi = \mathrm{Pe}\,\tau_{\mathrm{turnover}} \tag{4.16}$$

becomes extremely long compared with the turnover time ($\mathrm{Pe} \approx 10^7$). Using Eq. (4.6), τ_{th} may also be written

$$\tau_{\mathrm{th}} = \frac{\rho c_p T H_p}{K T / H_p} \approx \frac{E_{\mathrm{thermal}}}{F_{\mathrm{rad}}} \tag{4.17}$$

where E_{thermal} is the thermal energy content and F_{rad} is the radiative flux. Near the bottom of the convection zone this is nearly the same as the local Kelvin-Helmholtz time scale

$$\tau_{\mathrm{KH}} = \frac{E_{\mathrm{thermal}}}{F_{\mathrm{tot}}}, \tag{4.18}$$

but in the bulk of the solar CZ and at the top, τ_{KH} is much smaller than τ_{th}, because F_{rad} is tiny in comparison to F_{tot}. When discussing thermal relaxation of simulations of deep solar convection, it is really τ_{KH} rather than τ_{th} that is relevant, but since the two are similar in the bottom layers, where thermal relaxation is the biggest problem, we continue to use τ_{th} in the discussion below.

Note that τ_{KH} varies by about eight orders of magnitude between the top and the bottom of the convection zone. On the other hand, the turnover time varies by only three orders of magnitude.

The increasing gap between thermal and turnover time scales in deeper layers has important consequences for the evolution of the flow. Any perturbation will take a time much longer than the turnover time to relax. The problem of the thermal relaxation of simulations covering deeper parts of the solar convection zone was first pointed out by Chan & Sofia (1986). One may think that by setting up an initial state sufficiently close to the final one

Table 4.1: Typical orders of magnitude of radiative diffusion χ, pressure scale height H_p, turbulent velocity u_t, as well as dynamical, thermal and Kelvin-Helmholtz time scales, respectively, at two different depths (cgs units).

location	r/R	χ	H_p	u_t	τ_{th}	τ_{KH}	τ_{turnover}
surface	1.0	10^{12}	10^8	10^5	10^8	10^5	10^3
bulk of CZ	0.8	10^7	10^{10}	10^4	10^{13}	10^{13}	10^6

the relaxation time may be reduced, but this does not seem particularly helpful in practice. There is always a little perturbation in the initial state, which still takes a long time to relax. Faster relaxation could be achieved using for example an implicit or semi-implicit scheme (Fox 1994) that allows a longer time step, or by just using lower resolution, but again the danger is that a new and time consuming relaxation process will be necessary when switching back to a shorter time step or to higher resolution. Furthermore, any changes in the physics of the simulation (e.g. adding magnetic fields or rotation) may result in a new long relaxation period.

One may then ask the question how the results change if one brings the thermal and turnover times closer together. The ratio between thermal and dynamical (or turnover) time scales is $\sim \text{Pe} \approx (F_{\text{conv}}/F_{\text{rad}})\,(T/\delta T)$; see Eq. (4.9). In order to simplify the discussion, we consider the scaling of the Peclet number near the bottom of the convection zone, where $F_{\text{conv}} \sim F_{\text{rad}} \sim F$. The scaling may be expressed either in terms of the total flux F, or in terms of the product of the Rayleigh and Prandtl numbers RaPr.

The scaling of the relative temperature fluctuation may be obtained from considering the convective flux F_{conv}, that is proportional to the product $u_t\delta T$, and the kinetic energy per unit mass u_t^2, that is proportional to δT (buoyancy), so

$$F_{\text{conv}} \sim u_t^3 \sim \delta T^{3/2} \tag{4.19}$$

The relation to the Rayleigh and Prandtl numbers may be obtained by observing that the radiative flux F_{rad} is proportional to χ, so the product RaPr ($\sim \chi^{-2}$) is proportional to F_{rad}^{-2}. Since $F_{\text{conv}} \sim F_{\text{rad}} \sim F$ in the layer we are considering, we obtain

$$\tau_{\text{th}}/\tau_{\text{turnover}} \sim \text{Pe} \sim (\text{RaPr})^{1/3} \sim F^{-2/3} \tag{4.20}$$

Thus, in order to bring the thermal and dynamical time scales closer together, one must increase the total flux or, equivalently, lower the Rayleigh number. The fact that the flux decreases with increasing Rayleigh number may appear counterintuitive, but is because a reduced radiative diffusivity corresponds to a reduced radiative energy flux.

Thus, the significance of increasing the Rayleigh number in numerical experiments is not just to lower the viscosity and diffusivity, but also to lower the total flux and thus bring the thermal time scale closer to that of the real system.

In Table 4.2 we have listed some relevant parameters for a few cases. From this we see that currently feasible models must have a flux that is of the order of a million times the solar flux, in order to avoid the problem with the thermal relaxation time scale. Such "toy models" may still be quite useful, because they display qualitative features that may also be expected to appear in models with lower fluxes (and ultimately in the real thing). To the

Table 4.2: Mach number, flux (relative to the solar value), and Peclet number for different values of the Rayleigh number (or rather the product RaPr). The degree of feasibility in subjective terms is also indicated.

RaPr	Ma	$F/F_\odot$	Pe	feasibility
10^6	10^{-1}	10^9	10	"trivial"
10^{12}	10^{-2}	10^6	10^3	feasible
10^{18}	10^{-3}	10^3	10^5	not feasible
10^{24}	10^{-4}	1	10^7	the sun

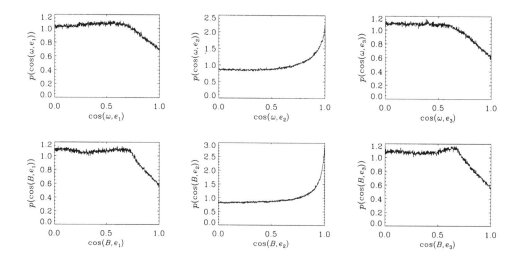

Figure 4.6: Upper row: histograms of the cosines of the angles between ω and the three eigenvectors e_1, e_2, e_3, of the rate of strain matrix. Lower row: same, as upper row, but for B. The direction of compression corresponds to e_1 and the direction of stretching to e_3. Note the enhanced probability of the intermediate eigenvector e_2 to be aligned with ω or B.

extent that one can understand and derive the scaling of these features with total flux, the toy models still serve a useful purpose.

It is expected that some features would be difficult to find scaling relations for. For example, the extent of the overshoot layer will be too large in the toy models and the braking of overshooting bubbles has a strongly nonlinear dependence on the overshooting distance. The extent of the overshoot layer could be reduced by making the subadiabatic gradient steeper, but this requires modifying the opacity law.

The main moral of all this is that it is currently impractical to model solar convection in deeper layers using realistic fluxes. Instead, we must rely on toy models with excess flux, i.e. with values of $F/F_\odot$ significantly larger than unity, and try to understand the model output as $F/F_\odot$ approaches the solar value. The scaling relations discussed here should be tested and deviations be understood in terms of the specific physical processes involved.

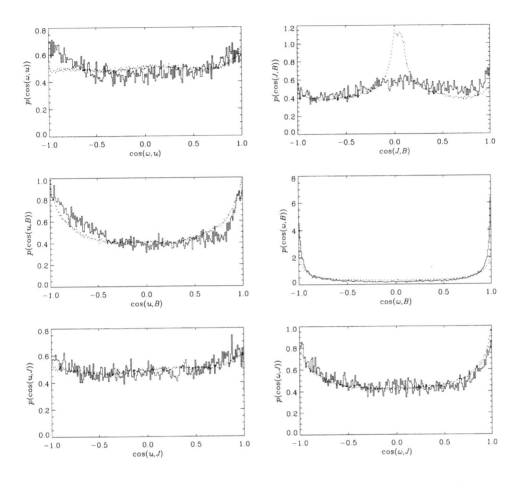

Figure 4.7: Cosines of the angles between different vector fields in the simulation. Note especially the strong tendency for alignment (or anti-alignment) between between the vorticity ω and the magnetic field B. The dotted lines refer to all points in a snapshot and the solid lines only to those where the magnetic field exceeds its root mean square value by a factor of three.

4.6   The formation of magnetic structures

In the sun the gas is hot enough to be ionized in the bulk of the convection zone, so it is electrically conducting. Consequently, the Maxwell equations are then coupled to the fluid equations via the Lorentz force and Ohm's law. It is generally found that turbulent convection can lead to dynamo action (e.g. Gilman 1983, Glatzmaier 1985, Meneguzzi & Pouquet 1989), which is the self-excited (spontaneous) conversion of kinetic energy into magnetic energy. Although this process is actually central to some of our work we are going to quote below, we only want to discuss those aspects that are of somewhat wider interest.

Important aspects concern the formation and advection of magnetic structures. This is generally relevant, because magnetic fields behave in many ways similar to vorticity.

In ordinary nonmagnetic turbulence the flow consists of a large number of vortex tubes (Siggia 1981, Kerr 1985, Vincent & Meneguzzi 1991). Something similar happens with the magnetic field, at least when the field is developing freely and not imposed externally. In the inviscid hydrodynamical case (from smooth initial conditions and withing finite time) the vorticity occurs rather in the form of sheets (Cao et al. 1996). In the magnetic case the situation is not quite clear yet. Visualisations of the strongest $\boldsymbol{B}$-vectors suggests mostly tube-like structures (Nordlund et al. 1992), although sometimes sheets can also be found. A multifractal analysis of the data also suggests that the strongest flux concentrations have the form of tubes (Brandenburg et al. 1992). However, the structure function exponents in MHD turbulence can best be understood using a model that assumes sheet-like structures (Politano & Pouquet 1995).

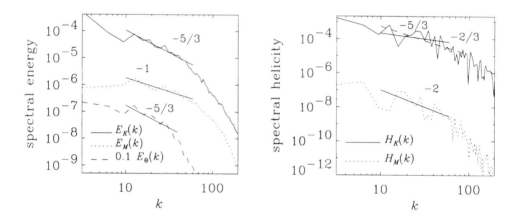

Figure 4.8: Left panel: power spectra of the kinetic energy (solid line), magnetic energy (dotted line) and temperature fluctuation (dashed line). Right panel: power spectra of the kinetic and magnetic helicities, $H_K(k)$ and $H_M(k)$, respectively.

In hydrodynamical turbulence the vortex tubes are aligned with the individual vorticity vectors $\boldsymbol{\omega}$. Likewise, in MHD turbulence the magnetic flux tubes are aligned with the individual magnetic field vectors $\boldsymbol{B}$. In both cases the tendency for alignment is thought to be a diffusive phenomenon (Constantin et al. 1995, Brandenburg et al. 1995a). In ordinary turbulence the vortex tubes are aligned with the intermediate eigenvector of the rate of strain matrix. The same is true of the magnetic field which is mostly parallel to the intermediate eigenvector (Fig. 4.6). It is then not surprising that the vorticity and magnetic field vectors are mostly parallel to each other (Fig. 4.7). There is also a tendency for $\boldsymbol{u}$ and $\boldsymbol{B}$ to be aligned, but this effect is quite weak in comparison with the enhanced alignment of $\boldsymbol{\omega}$ and $\boldsymbol{B}$. In the plot we have distinguished between all data points in the simulations (dotted lines) and those where the field is strong (solid line). The correlation plot of $\boldsymbol{J}$ and $\boldsymbol{B}$ indicated that the two vectors tend to be perpendicular on average (dotted line), but this is no longer true of those regions where the field is strong (solid line). Details of this analysis can be found in Brandenburg et al. (1996).

Nevertheless, despite their similarity, in other respects the magnetic field and the vorticity are actually quite different. The powerspectrum of the velocity has a short range roughly like a Kolmogorov $k^{-5/3}$ spectrum, so the vorticity has a $k^{+1/3}$ spectrum. However, a similar spectrum is not observed for the magnetic field, except very early in the evolution when nonlinear effects were still unimportant (Brandenburg et al. 1996).

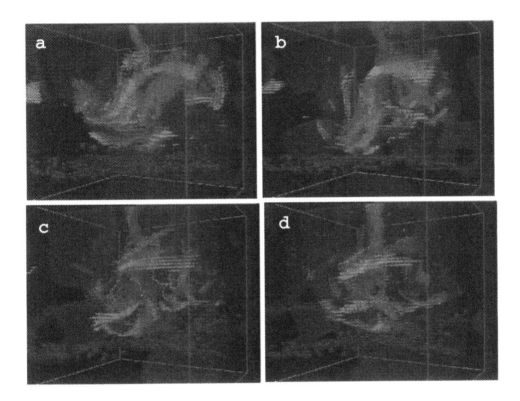

Figure 4.9: Snapshots of a video animation showing a strong magnetic flux tube (white) being wrapped around a spinning downdraft (dark grey, extending from the top of the box to the middle). As time goes on (frames a-d) the magnetic flux loop appears to be folded onto itself and is being pushed further down to the bottom the convection zone and into the lower overshoot layer (lower half of the box).

4.7   Magnetic dynamo action

Video animations of the strong magnetic field vectors show how magnetic flux tubes evolve. Some snapshots of such animations are reproduced in Fig. 4.9. One sees a tendency for magnetic loops to fold onto themselves. This appears to be similar to the stretch-twist-fold dynamo of Vainshtein & Zeldovich (1972), (see also a recent monograph by Childress & Gilbert 1995). In this type of dynamo constructive folding of tubes enhances the flux.

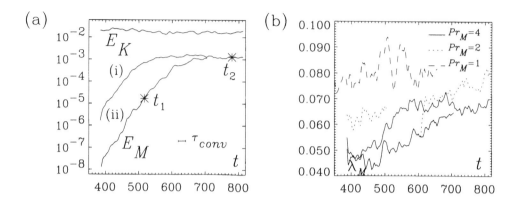

Figure 4.10: Left: evolution of the magnetic energy E_K and kinetic energy E_K for two different runs. Right: evolution of the magnetic Taylor microscale for different magnetic Prandtl numbers Pr_M.

Whether or not this is really the dominant process in the simulations remains to be seen. It is clear, however, that there is a dynamo acting in the simulations: magnetic energy increases exponentially in time over many orders of magnitude until saturation sets in (Fig. 4.10). For further details see Nordlund et al. (1992) and Brandenburg et al. (1996). After some period of exponential growth two solutions with different initial conditions settle at the same value. During the growth phase the size of magnetic structures, as measured by the magnetic Taylor microscale,

$$\lambda_M = \sqrt{5\langle \boldsymbol{B}^2 \rangle / \langle (\boldsymbol{\nabla} \times \boldsymbol{B})^2 \rangle}, \qquad (4.21)$$

increase with time. This suggests that the magnetic structures are intensifying not just by stretching, but in particular by growing in thickness, possibly via folding.

4.8 Downward pumping

In the presence of rotation and stratification there is a pronounced phenomenon of downward "pumping" of magnetic fields and vorticity. This is seen in images of the magnetic energy density $\boldsymbol{B}^2$ (Fig. 4.11) and the enstrophy density ω^2 (Fig. 4.12), showing that most of the field and vorticity accumulates near the lower overshoot layer. This is also clearly seen in video animations of the magnetic field, as described by Nordlund et al. (1992) and Brandenburg et al. (1996), see also Fig. 4.9. The amount of vorticity and magnetic field that has been pumped downwards is however much weaker if there is no rotation (Fig. 4.13). In that case most of the vorticity is seen near the top where fluid overturns into the downdraft lanes.

The nature of this pumping effect is not fully clarified. There are two forms of pumping that have been discussed in the context of magnetic fields. One is topological pumping of a horizontal magnetic field by convection in a connected network of downward or upward motion. The other is turbulent pumping. The two mechanism are quite distinct, but easily confused because both start with 't'. (See Brandenburg et al. 1995c for references and a recent discussion of the two effects.) In our simulations with rotation the pumping cannot be topological, because the downdrafts do not form a connected network. Turbulent pumping

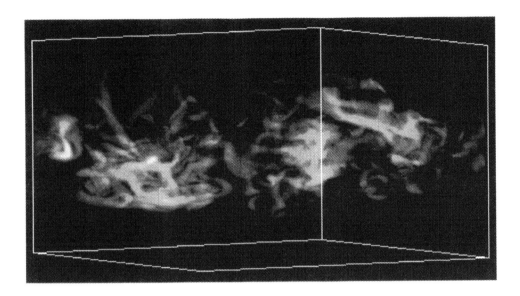

Figure 4.11: Three-dimensional image of the magnetic energy density $\boldsymbol{B}^2$. Note that most of the magnetic field is near the lower overshoot layer, which is located just a little below the middle of the box.

remains a possibility, but the details of the actual mechanism must be different from the original turbulent pumping, which corresponds to a net transport velocity $\boldsymbol{u}_{\text{pump}} = -\frac{1}{2}\boldsymbol{\nabla}\eta_t$, where $\eta_t \sim \frac{1}{3}u_t\ell$ is the standard expression for the turbulent magnetic diffusivity, where ℓ is the correlation length of the turbulence. This expression involves just ℓ and u_t, which is independent of rotation, so this expression is too simplistic in describing our results.

4.9 Outstanding problems

In the last three sections we have touched upon some aspects of magnetic field evolution in solar convection. This poses indeed a challenging problem. Among other reasons, we need to study solar magnetism in order to understand variations of the solar luminosity which, in turn, can influence the earth's climate.

A particularly important problem is to understand the origin of the large scale magnetic field and its 11 years cyclic variations (i.e. the sunspot cycle). Simulations are currently under way to investigate the effect of shear on a convective dynamo. Promising results have emerged in the context of accretion disc turbulence simulations where strong shear leads to a cyclic large scale magnetic field (Brandenburg et al. 1995b). In this case the magnetic field evolution is governed by flows that result from the magnetic field itself due to an instability. In the case of accretion discs relevant instabilities include the magnetic shear and buoyancy instabilities (for a review see Schramkowski & Torkelsson 1996). The same idea may also apply to stars where, in addition to magnetic buoyancy and shear instabilities, destabilized magnetostrophic waves have been discussed (Schmitt 1985). There is a related approach

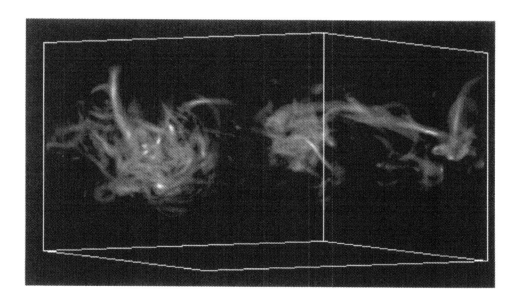

Figure 4.12: Three-dimensional image of the enstrophy density ω^2. Similar to Fig. 4.11 most of the vorticity is near the lower overshoot layer.

based on the Parker and other magnetic instabilities of flux tubes by Ferriz-Mas, Schmitt & Schüssler (1994). More recently, similar investigations have been carried out in connection with the dynamo effect in galaxies (Hanasz & Lesch 1997).

Future simulations will hopefully reveal whether the large scale field of the sun is really governed by magnetic instabilities, or whether cyclonic convection alone can do the job. An important prerequisite of realistic models of solar convection and magnetic field generation are high Rayleigh numbers and small fluxes. We hope that significant progress can be made in addressing these issues in the near future.

Bibliography

Brandenburg, A., Procaccia, I., Segel, D., Vincent, A., "Fractal level sets and multifractal fields in direct simulations of turbulence," *Phys. Rev.* **A 46**, 4819-4828 (1992).

Brandenburg, A., Procaccia, I., Segel, D., "The size and dynamics of magnetic flux structures in MHD turbulence," *Phys. Plasmas* **2**, 1148-1156 (1995a).

Brandenburg, A., Nordlund, Å., Stein, R. F., Torkelsson, U., "Dynamo generated turbulence and large scale magnetic fields in a Keplerian shear flow," *Astrophys. J.* **446**, 741-754 (1995b).

Brandenburg, A., Moss, D., Shukurov, A., "Galactic fountains as magnetic pumps," *Monthly Notices Roy. Astron. Soc.* **276**, 651-662 (1995c).

Brandenburg, A., Jennings, R. L., Nordlund, Å., Rieutord, M., Stein, R. F., Tuominen, I., "Magnetic structures in a dynamo simulation," *J. Fluid Mech.* **306**, 325-352 (1996).

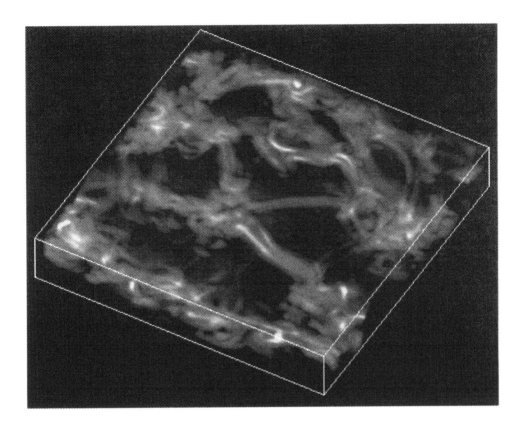

Figure 4.13: Three-dimensional visualisation of the vorticity in a simulation without rotation.

Bray, R.J., Loughhead, R.E., Durrant, C.J. *The solar granulation*, Cambridge University Press (1984).

Brummell, N. H., Hurlburt, N. E., & Toomre, J., "Turbulent compressible convection with rotation. I. Flow structure and evolution," *Astrophys. J.* **473**, 494-513 (1996).

Cao, N. Z., Chen, S. Y., & She, Z. S., "Scalings and relative scalings in the Navier-Stokes turbulence," *Phys. Rev. Letters* **76**, 3711-3714 (1996).

Chandrasekhar, S. *Hydrodynamic and Hydromagnetic Stability*. Oxford, Clarendon Press (1961). Dover Publications, Inc., New York

Chan, K. L., Sofia, S., "Turbulent compressible convection in a deep atmosphere. III. Tests on the validity and limitation of the numerical approach," *Astrophys. J.* **307**, 222-241 (1986).

Childress, S. & Gilbert, A. G. *Stretch, twist, fold: the fast dynamo*. Berlin: Springer (1995).

Constantin, P., Procaccia, I. & Segel, D., "Creation and dynamics of vortex tubes in 3-dimensional turbulence," *Phys. Rev.* **E**, 3207-3222 (1995).

Ferriz-Mas, A., Schmitt, D., Schüssler, M., "A dynamo effect due to instability of magnetic flux tubes," *Astron. Astrophys.* **289**, 949-949 (1994).

Fox, P. private communication (1994).

Frank, J., King, A. R., & Raine, D. J. *Accretion power in astrophysics.* Cambridge: Cambridge Univ. Press (1992).

Galsgaard, K. & Nordlund, Å., "Heating and activity of the solar corona: I. boundary shearing of an initially homogeneous magnetic-field," *J. Geophys. Res.* **101**, 13445-13460 (1996).

Gilman, P. A., "Dynamically consistent nonlinear dynamos driven by convection in a rotating spherical shell. II. Dynamos with cycles and strong feedbacks," *Astrophys. J. Suppl.* **53**, 243-268 (1983).

Glatzmaier, G. A., "Numerical simulations of stellar convective dynamos. II. Field propagation in the convection zone," *Astrophys. J.* **291**, 300-307 (1985).

Hanasz, M. & Lesch, H., "The galactic dynamo effect due to Parker-shearing instability of magnetic flux tubes. I. General formalism and the linear approximation," *Astron. Astrophys.*, in press (1997). astro-ph/9610167

Kerr, R. M., "Higher-order derivative correlations and the alignment of small-scale structures in isotropic numerical turbulence," *J. Fluid Mech.* **153**, 31-58 (1985).

Kippenhahn, R. & Weigert, A. *Stellar structure and evolution.* Springer: Berlin (1990).

Kolmogorov, A. N., "The local structure of turbulence in incompressible viscous fluid for very large Reynolds numbers," *CR Acad. Sci. USSR* **30**, 299-303 (1941).

Meneguzzi, M., Pouquet, A., "Turbulent dynamos driven by convection," *J. Fluid Mech.* **205**, 297-312 (1989).

Nordlund, Å., Brandenburg, A., Jennings, R. L., Rieutord, M., Ruokolainen, J., Stein, R. F., Tuominen, I., "Dynamo action in stratified convection with overshoot," *Astrophys. J.* **392**, 647-652 (1992).

Nordlund, Å., Stein, R. F., Brandenburg, A., "Supercomputer windows into the solar convection zone," *Bull. Astr. Soc. India* **24**, 261-279 (1996).

Politano, H. & Pouquet, A., "Model of intermittency in magnetohydrodynamic turbulence," *Phys. Rev.* **E 52**, 636-641 (1995).

Ruzmaikin, A. A., Sokoloff, D. D. & Shukurov, A. M. *Magnetic Fields of Galaxies.* Kluwer, Dordrecht (1988).

Schmitt, D. *Dynamowirkung magnetostrophischer Wellen.* PhD dissertation (1985).

Schramkowski, G. P. & Torkelsson, U., "Magnetohydrodynamic instabilities and turbulence in accretion disks," *Astron. Astrophys. Rev.* **7**, 55-96 (1996).

Siggia, E. D., "Numerical study of small-scale intermittency in three-dimensional turbulence," *J. Fluid Mech.* **107**, 375-406 (1981).

Stein, R.F., Nordlund, Å., "Topology of convection beneath the solar surface," *Astrophys. J. Letters* **342**, L95-L98 (1989).

Thompson, C. & Duncan, R. C., "Neutron star dynamos and the origins of pulsar magnetism," *Astrophys. J.* **408**, 194-217 (1993).

Vainshtein, S. I. & Zeldovich, Ya. B., "Origin of magnetic fields in astrophysics," *Sov. Phys. Usp.* **15**, 159-172 (1972).

Vincent, A. & Meneguzzi, M., "The spatial structure and statistical properties of homogeneous turbulence," *J. Fluid Mech.* **225**, 1-20 (1991).

Chapter 5

DYNAMICS OF CUMULUS ENTRAINMENT

WOJCIECH W. GRABOWSKI[1]

[1] *Mesoscale and Microscale Meteorology Division*
National Center for Atmospheric Research
P.O. Box 3000, Boulder, CO 80307-3000

Cumulus entrainment is discussed with the emphasis on dynamical aspects. It is argued that cumulus entrainment is similar to entrainment in other buoyancy–driven flows and it is associated with the presence of interfacial entraining eddies. It is suggested that the eddies are a direct consequence of an unstable nature of the interface which separates rising buoyant fluid from its environment. Development of interfacial instabilities is studied in detail with conclusions directly applicable to other buoyancy–driven flows.

An important feature of atmospheric moist convection is a possibility of buoyancy reversal. Buoyancy reversal results from the mixing of cloudy air and dry environmental air and subsequent evaporation of cloud droplets. Based on results from laboratory experiments, theoretical arguments and numerical simulations, it is argued that buoyancy reversal has a rather minor effect on the rate of entrainment despite the dramatic effects on the overall dynamics of buoyancy–reversing systems.

5.1 Introduction

Atmospheric moist convection is a natural example of a buoyancy–driven system. Atmospheric moist convection occurs in a stably stratified environment because the air density ρ decreases with height as a result of hydrostatic balance. Environmental stability is important not only for elementary stability analyses (e.g., linear analysis, parcel analysis), but it also allows propagation of buoyancy (or gravity) waves. The density perturbations associated with the moist convection are typically small, a few percent at the most, and fluid velocities are much smaller than the speed of sound (i.e., the Mach number is small). It follows that the anelastic approximation is sufficient, and for shallow convection an incompressible approximation is appropriate (shallow convection refers to convection with vertical extent much smaller than the density scale height $H \equiv [(1/\rho)(\partial\rho/\partial z)]^{-1}$; e.g., convection inside the planetary boundary layer, PBL). Density differences that drive moist convection are created mostly by the latent heating, although the weight of condensed water and its motion relative to the air are important as well. The Reynolds number based on the molecular viscosity is extremely high, in the range of 10^8–10^{10}, assuming velocity magnitude in the range of 1 to 10 m s^{-1} and horizontal extent of clouds in the range of 1 to 10 km.

The release of latent heat associated with phase changes of water substance allows creation of volumes (or parcels) of positively buoyant air, i.e., volumes having density smaller than the environmental air at a given level in an otherwise stably stratified unsaturated environment. In the meteorological community this is traditionally referred to as a conditional instability (e.g., Wallace and Hobbs 1977, section 2.7), where the word "conditional" emphasizes the role of the latent heating. Figure 5.1 illustrates parameters of the moist convection in a situation considered in Grabowski (1993). In this example, the atmosphere consists of three layers. The first one extends from the surface ($z = 0$, surface pressure and temperature of 1000 hPa and 293 K, respectively) up to 0.5 km and has the static stability of $\sigma = d(ln\theta)/dz = 0.6 \times 10^{-5}$ m^{-1} (where θ is the potential temperature) and a constant relative humidity of 80%. The second layer extends from 0.5 km up to 2.5 km with $\sigma = 1.35 \times 10^{-5}$ m^{-1}. The relative humidity decreases linearly from 80% at the bottom to 20% at the top of this layer. Above 2.5 km, $\sigma = 1.8 \times 10^{-5}$ m^{-1} and a constant relative humidity of 20% are assumed. This hypothetical sounding (shown in Fig. 5.1a,b) is supposed to imitate a typical situation in which the stratification increases and the relative humidity decreases as one moves from the boundary layer into the free atmosphere.

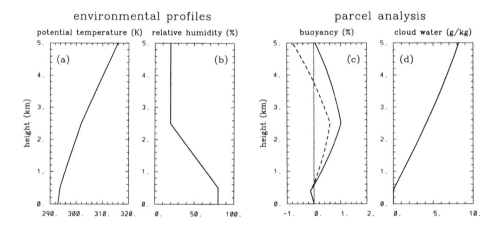

Figure 5.1: Example of the lower–tropospheric sounding and analysis of convection developing in this environment. Environmental profiles of the potential temperature and relative humidity are show in panels a and b. Profiles of the buoyancy of an adiabatic parcel (dashed line) and a pseudo–adiabatic parcel (solid line) rising from the surface are shown in panel c. Profile of the condensate mixing ratio for the adiabatic parcel is shown in panel d.

Figure 5.1c,d shows results of the adiabatic parcel analysis for this sounding. The buoyancy of the adiabatic parcel $B = (\rho_e - \rho_p)/\rho_e$ (where ρ_p and ρ_e is the air density of the parcel and the environmental density) rising from the surface is shown in Fig. 5.1c. The dashed and solid lines represent buoyancy with and without condensate loading, respectively, which correspond to situations referred to as the adiabatic (or reversible) and pseudo–adiabatic thermodynamical processes. The cloud water mixing ratio for the adiabatic parcel is shown in Fig. 5.1d. The lifting condensation level (LCL) for the parcel rising from $z = 0$ is at about 0.4 km, and the level of free convection (LFC; i.e., the level at which parcel buoyancy becomes positive) is about 0.6 km. The parcel reaches a maximum buoyancy (about 6×10^{-3} and 10^{-2} with and without condensate loading) at approximately 2.5 km, where the cloud water mixing ratio for the adiabatic parcel is about 4 g kg^{-1}. The parcel buoyancy decreases thereafter and becomes negative at about 3.8 km and 5 km for the parcel undergoing reversible and pseudo–adiabatic transformations, respectively. Because of the precipitation development which tends to remove condensate out from the rising parcel, the two cases represent limits for the parcel analysis.

The effects of mixing between the adiabatic parcel and the environmental air at a given level are shown in Fig. 5.2. Fig. 5.2a shows the profile of the mixing proportion for which minimum buoyancy of the mixture is obtained, whereas Fig. 5.2b shows the profile of the minimum buoyancy. The appropriate mixing of the cloudy air with dry environmental air and subsequent evaporation of cloud droplets leads to buoyancy reversal, i.e., the density of a resulting parcel is larger than both cloudy and cloud-free environmental air. The minimum buoyancy is of the same order of magnitude as the adiabatic parcel buoyancy, but of the opposite sign.

There is a consensus throughout the cloud physics community that buoyancy reversal illustrated above plays an important role in the cumulus entrainment (see reviews by Reuter 1986 and Blyth 1993). Concepts that are usually invoked are "cloud-top entrainment instability" and "penetrative downdrafts." The cloud-top entrainment instability is based on the hypothesis that a positive feedback exists between entrainment and buoyancy reversal due to evaporative cooling: entrainment results in formation of negatively buoyant parcels that sink into the cloud and bring more environmental air into the cloud; this leads to enhanced entrainment. Buoyancy reversal attracted considerable attention in the problem of stability of cloud-topped boundary layers; e.g., Lilly (1968), Randall (1980), Deardorff (1980), Mahrt and Paumier (1982), Albrecht et al. (1985), Kuo and Schubert (1988), Siems et al. (1990), Shy and Breidenthal (1990), MacVean and Mason (1990), Siems and Bretherton (1992), MacVean (1993), Krueger (1993), among others. The role of buoyancy reversal in cumulus dynamics has been considered in the context of conceptual models, laboratory experiments, and cloud measurements; e.g., Squires (1958b), Turner (1966), Telford (1975), Raymond (1979), Paluch (1979), Emanuel (1981), Haman and Malinowski (1984), Raymond and Blyth (1986), Johari (1992), Grabowski (1993, 1995).

This paper discusses cumulus entrainment from the perspective of the general problem of entrainment in buoyancy–driven flows. In particular, the role of latent heating for the entrainment process is clarified. It is argued that cumulus entrainment is an effect of interfacial instabilities, similar to entrainment in other buoyancy–driven flows, and that the latent heating is only required to maintain positive buoyancy of a rising cloudy parcel. It is also argued that evaporative cooling which results from the mixing of cloud and dry environmental air has only a minor effect on the entrainment rate. This is in dramatic contrast to the effect buoyancy reversal has on the overall cloud dynamics as documented in cloud observations, laboratory experiments and theoretical studies.

How to define entrainment in general, and entrainment rate in particular, is a critical issue. In high Reynolds number natural flows, turbulent entrainment is usually envisioned

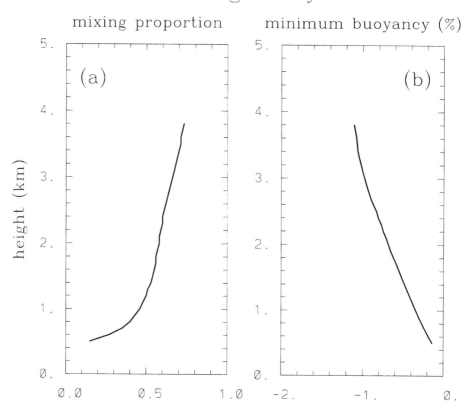

Figure 5.2: Results of the mixing analysis between the the cloudy air with parameters of the adiabatic parcel from Fig. 5.1c and 1d and environmental air shown in Fig. 5.1a and 5.1b. Panel *a* shows the profile of the mixing proportion which results in the minimum buoyancy at a given level. Panel *b* shows the profile of the minimum buoyancy.

as a sequence of events which begins when a large–scale structure engulfs the ambient fluid, and which proceeds with development of smaller and smaller structures down to the scale at which molecular homogenization occurs. In the classical fluid mechanics, the entrainment rate is defined based on the increase of mass or mass flux of a fluid element; see, for example, Turner (1986). Typically such a definition involves some elements of ensemble averaging in order to remove transient effects of entraining structures (eddies) whose dynamics is of limited interest as long as only average properties of the flow are considered. Changes of those average properties may be deduced, for example, from the increase of volume of a thermal, angle of spread of a plume or jet, or change of the average height of an interface. To deduce the average properties, similarity theory has proven very useful (Turner 1986).

Dramatically different definitions of the entrainment rate are often used in numerical and laboratory studies. For instance, in their study of the entrainment across a stratified interaface, Siems et al. (1990) used the rate of change of the average height of the selected mixing ratio contour as a proxy of the entrainment rate. The rate of molecular mixing between two fluids was used in Grabowski (1995) as a measure of the entrainmment rate in a numerical study of buoyancy–reversing convection. An increase of thermal volume (estimated from an increase of thermal spatial dimensions) is typically used to estimate the rate of entrainment in laboratory experiments with thermals (e.g., Sánchez et al. 1989). In this paper, the term "entrainment rate" will be used loosely to describe the rate of dillution of the buoyant fluid by nonturbulent environmental fluid. For instance, the statement "entrainment rate is only weakly affected by the presence of buoyancy reversal" implies that the volume of the cloud would not change dramatically if the buoyancy reversal was disallowed.

The next section briefly reviews theoretical and observational studies of entrainment in buoyancy–driven flow relevant to the problem of cumulus entrainment. Since the emphasis of this paper is on the dynamics of entrainment, understanding the physics of entraining eddies becomes a primary objective. This aspect will be discusssed in section 3. Section 4 will review the effects of evaporative cooling and buoyancy reversal on cumulus dynamics. Conclusions are presented in section 5.

5.2   Turbulent entrainment in cumulus clouds and in buoyancy–driven flows

It is beyond the scope of this paper to review the vast literature dealing with dynamics of buoyancy–driven flows accumulated over the past half century which is relevant to the problem of cumulus entrainment. Herein, only selected references will be used to show how development of ideas dealing with cumulus entrainment fits into development of the knowledge of the dynamics of buoyancy–driven flows. For a more extensive review of the cumulus entrainment problem, the reader is referred to Blyth (1993).

Turner (1986) discusses progress made in studies of a wide range of geophysical turbulent flows using entrainment hypothesis and the similarity theory. The entrainment hypothesis, which refers entrainment velocity to some measure of the mean velocity of a turbulent fluid, was originally proposed by G. I. Taylor (cf. Turner 1986); the usual starting reference is Morton et al. (1956). Morton (1957) extended this approach into a stratified environment and allowed for the effects of moisture. Laboratory experiments with thermals and plumes reveal enormous potential for this approach, not only for simple systems (like homogeneous thermals rising in unstratified environment), but for more complex systems in which effect of stratification and latent heating were considered (e.g., Scorer 1957, 1958; Woodward 1959;

Turner 1962, 1963, 1966, 1973; Johari 1992). These experiments also demonstrated that initial perturbation has to travel a considerable distance to enter the self–similar regime, e.g., for the rising thermal this distance is about 4 initial diameters. This aspect is relevant to the cumulus entrainment problem since the aspect ratio of small cumuli (i.e., cloud height divided by the cloud width) is usually not much greater than 1. As far as laboratory experiments with thermals are concerned, it might be mentioned that a very innovative release mechanism was used in Sánchez et al. (1989) to study thermals prior to the self-similar flow regime. Unfortunately, due to technical limitations, thermals were mostly laminar because of fairly small Reynolds numbers.

Early cloud observations (e.g., Stommel 1947; Malkus 1949, 1952, 1954; Warner 1955; Squires 1958a; Ackerman 1958, 1959) showed without doubt that small cumulus clouds are strongly diluted by environmental dry air. It was also apparent that cloud parameters (like temperature, moisture, and wind variables) vary strongly and erratically inside and around convective clouds and, *on the average*, the dilution increases as one moves upward away from the cloud base. These findings were later corroborated by more sophisticated aircraft observations which revealed subtle details of cumulus dynamics and microphysics. Similarity based concepts were used in early models of cumulus dynamics. Such models were able to explain some observed features of small convective clouds, but serious problems with these models were identified as well (see Blyth 1993, section 3f). It should also be emphasized that these models did not consider effects of evaporative cooling and buoyancy reversal on cumulus dynamics at all.

Squires (1958b) proposed that evaporative cooling associated with the mixing between cloudy air and dry environmental air may lead to formation of downdrafts which can penetrate deep into the cloud (i.e., penetrative downdrafts). This concept was later refined by Emanuel (1981) using similarity type arguments. Probably the most spectacular support for the idea of the cloud–top mixing and penetrative downdrafts came from the analysis presented in Paluch (1979). However, a recent study of the Paluch method by Taylor and Baker (1991) showed that similar results can be produced by continuous lateral mixing as well. Both cloud observations and more complete theoretical analyses revealed problems with the concept of penetrating downdrafts (e.g., Cooper and Rodi 1982, Haman and Malinowski 1984, Blyth et al. 1988, section 3e of Blyth 1993). There seems to be little doubt, however, that vertical transports of undiluted air upward and diluted air downward play an essential role in cumulus dynamics. This is supported by numerous cloud observations (cf. Reuter 1986, Blyth 1993), laboratory experiments (e.g., Turner 1966, Johari 1992), numerical experiments (e.g., Grabowski 1993, 1995; Carpenter and Droegemeier 1995), as well as simple conceptual models that are able to reproduce many observed features of cumulus convection (e.g., Raymond and Blyth 1986, Taylor and Baker 1991). It is not obvious, however, how evaporative cooling and buoyancy reversal influence entrainment rate. The significance of the distinction between effects of buoyancy reversal on the global dynamics and on the entrainment rate will be discussed in section 4.

Sophisticated techniques for flow visualization in laboratory experiments allow detailed analysis of the turbulent mixing over a wide range of scales, from the scale of entraining eddies down to Kolmogorov and Batchelor scales where molecular effects dominate (e.g., Papantoniou and List 1989, see also the discussion and references in Grabowski 1993). For convective clouds, the scale of entraining eddies is of the order of 100 meters, whereas the Kolmogorov and Batchelor microscales are of the order of a millimeter for typical turbulence levels (e.g., Baker et al. 1984, Grabowski 1993). Cloud in–situ observations using aircraft are not able to resolve such a span of scales because of the slow instrumental response when compared to the aircraft speed. It follows that thermodynamic fields on scales larger than, say, tens of meters can only be directly studied using aircraft. There is indirect evidence, however, that regions undergoing turbulent mixing are characterized by variations in thermodynamic fields down to scales of a few centimeters (e.g., Baker 1992, Brenguier 1993, Knight and Miller 1993). Recently some measurements with very fast response temperature sensors were applied to small cumulus studies (Haman et al. 1995). All of these observations seem consistent with the picture emerging from laboratory experiments with turbulent buoyancy–driven flows.

5.3 Entrainment as a result of interfacial instabilities

Klaassen and Clark (1985) suggested that cumulus entrainment is associated with dynamical instabilities that develop at the interface separating a convective cloud from its dry environment. They performed numerical experiments of small convective clouds with resolution high enough that the dynamics of the interface could be studied. The simulations were performed only in the two–dimensional framework and some concerns about numerical problems were raised. As a result, the suggestion that entrainment is an effect of essentially inviscid dynamics associated with interfacial instabilities was received by the cloud physics community with some skepticism (e.g., Reuter 1986). The experiments were later extended into the three–dimensional framework, Clark et al. (1988).

Grabowski and Clark (1991, 1993a, 1993b) performed a systematic study of the interfacial instabilities in both two– and three–dimensional frameworks using a simpler setup than the one used in Klaassen and Clark (1985). Moist thermals (i.e., thermals with effects of latent heating due to condensation and evaporation) rising from rest in a stably stratified environment were studied using an anelastic cloud model with grid nesting capabilities (Clark 1977, 1979; Clark and Farley 1984). A direct numerical simulation approach was adopted with Prandtl and Schmidt numbers equal to one. The global Reynolds number (i.e., the Reynolds number based on the thermal size, its rate of rise, and explicit viscosity used in the calculations) was in the range of $10^3 - 10^4$, i.e., similar to the Reynolds number of laboratory thermals.

To describe evolution of the thermal in nondimensional units, the length scale is taken as the initial thermal radius R, the velocity scale is taken as $U = (g \, R \, B)^{1/2}$, where g is the acceleration of gravity and $B = \Delta\rho/\rho$ is the thermal buoyancy, and the time scale is $T = R/U$. (In the setup considered, $R \approx 250$ m, $B \approx 3 \times 10^{-3}$, and $U \approx 3$ m s^{-1}. The time scale $T \approx 80$ sec, i.e., about 1 min). Using these scales, the nondimensional explicit viscosity ν applied in the calculations (equal the inverse of the global Reynolds number) varied from about 2×10^{-4} to 10^{-3}; the value used in most of the experiments was 6×10^{-4}.

A dual experiment approach was applied in which identical numerical experiments were performed with and without small–amplitude random perturbations passively applied to the buoyancy field at the time when the thermal achieved a near steady rate of rise. Development of unstable modes was sufficiently delayed in experiments without excitation that these experiments could be used to define the base state upon which instabilities develop. Direct comparison between experiments with and without excitation revealed the scale selection, growth rates and the evolution of the perturbation kinetic energy in the linear growth regime.

Figures 5.3 and 5.4 illustrate the development of unstable eddies on the cloud–environment interface for the two–dimensional and three–dimensional frameworks, respectively. Figure 5.3 shows evolution of the cloud water field and the perturbation streamfunction for the inner-

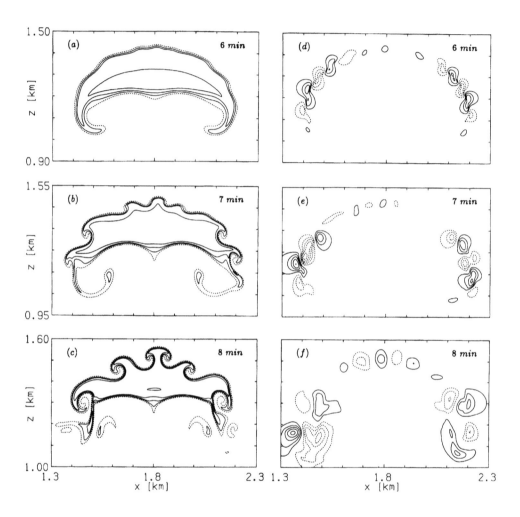

Figure 5.3: Fields of the cloud water (a, b, c) and the perturbation streamfunction (d, e, f) at $t = 6$ (a, d), 7 (b, e), and 8 min (c, f) for the two–dimensional simulation (Grabowski and Clark 1991). Contour interval is 0.3 g kg^{-1} for the cloud water and 2, 12, and 30 kg m^{-1} s^{-1} for $t = 6$, 7, and 8 min, respectively, for the streamfunction. **T** and **S** in (d) and (f) mark instabilities developing near thermal top and side, respectively.

most domain. The perturbation streamfunction is associated with the velocity field created by subtracting velocities from experiments with and without excitation. Figure 5.4 shows evolution of the cloud water field for the three–dimensional framework. Development of interfacial instabilities results in the transition from a smooth interface to an interface with complex structures in both two– and three–dimensional cases.

The paramount feature of the flow field near the interface is a shear layer of the velocity tangential to the interface which separates a cloudy thermal from its dry environment. The flow near the interface is a direct response to the buoyant cloudy air that pushes upward. Figure 5.5 shows profiles of the velocity tangential to the interface as a function of distance from the interface at several locations along the interface, from near the thermal top to near the thermal side, at two time levels. The velocity change across the shear layer (referred to as a shearing velocity) changes insignificantly with time at any given location, but the depth of the shear layer decreases. The collapse of the shear layer is a direct manifestation of the baroclinic vorticity production at the cloud–environment interface by the horizontal buoyancy gradients. Details of the evolution of the interfacial shear layer as the thermal continues to rise have been explored using a simple analytical model in Grabowski and Clark (1991, section 4b). The time scale of the shear layer collapse was approximately equal to the time scale associated with the thermal rise, T. The collapse of the shear layer plays a central role in the development of interfacial instabilities.

To understand the physics of the instability further, parameters like the Reynolds number, Richardson number and Rayleigh number (all formulated using parameters of the shear flow near the interface), together with the scale selection and growth rates have to be considered. It should be stressed that the evolution of the unperturbed flow near the interface (i.e., the base state upon which instabilities develop once excitation is provided) is very similar for both two–dimensional and three–dimensional experiments. As discussed above, the shear layer depth decreases with time and the magnitude of the shearing velocity depends on the position along the interface (cf. Fig. 5.5). Thus it is difficult to choose "typical" values of these parameters. The shear layer depth l and shearing velocity Δu are chosen as $l \approx 0.1R$ and $\Delta u \approx U$, respectively. These values correspond to the time when linear growth of unstable modes dominates and a position somewhere between thermal top and thermal side. The horizontal scale of unstable modes is estimated as $0.4R$, and the e-folding time scale (estimated from the rate of change of the perturbation kinetic energy; Fig. 9 in Grabowski and Clark 1993a) is approximately $0.3T$. However, it is more appropriate to scale growth rate and the horizontal scale (or wavenumber) using parameters of the shear layer. With this scaling, the scale of unstable modes is $\lambda \approx 4$ (the corresponding nondimensional wavenumber is $k = 2\pi/\lambda \approx 1.5$) and the e-folding time is $\tau \approx 3$ (the corresponding growth rate is $\sigma = 1/\tau \approx 0.3$).

The Reynolds number Re, the Richardson number Ri, and the Rayleigh number Ra may be estimated using depth (l), shearing velocity (Δu), buoyancy change (B) across the shear layer, viscosity (ν), and orientation of the interface with respect to the direction of gravity (represented by vertical component, $n_z \approx 0.5$, of the vector normal to the interface):

$$Re = \frac{l\,\Delta u}{\nu} \qquad Ri = -\frac{g\,B\,l\,n_z}{(\Delta u)^2} \qquad Ra = Re^2|Ri| \qquad . \qquad (5.1)$$

where Prandtl and Schmidt numbers of one have been assumed. Using the typical values as selected above, one gets:

$$Re \sim 10^2 \quad , \quad Ri \sim -0.1 \quad , \quad Ra \sim 10^3 \; . \qquad (5.2)$$

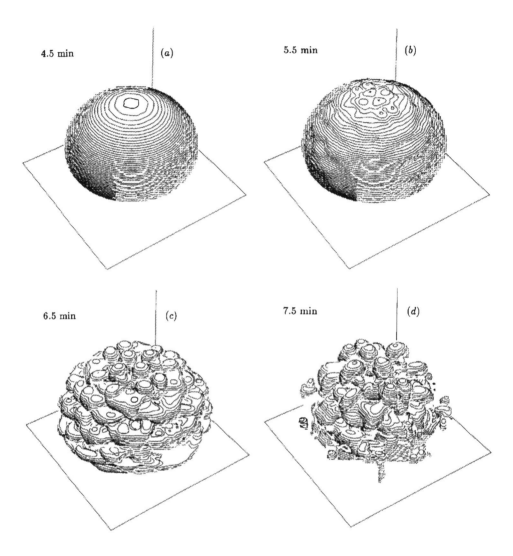

Figure 5.4: Three-dimensional perspective of the cloud water field greater than 0.01 g kg^{-1} for the three–dimensional experiment 3D3M (Grabowski and Clark 1993a) at $t = 4.5$ (a), 5.5 (b), 6.5 (c) and 7.5 min (d). The square at the bottom shows extent of the innermost computational domain which is 0.8 km × 0.8 km (i.e., about $3R$). Note that the experiment resolved only a quarter of the thermal and these data were used to plot the whole thermal with symmetries as assumed in the experimental setup.

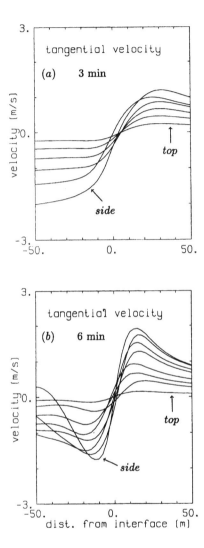

Figure 5.5: Profiles of the velocity tangential to the interface as a function of the distance from the interface at $t = 3$ min (a) and 6 min (b) for several positions along the interface. Profiles at locations close to thermal top and close to the side are marked accordingly. Data are from the three–dimensional experiment 3D6M (Grabowski and Clark 1993a).

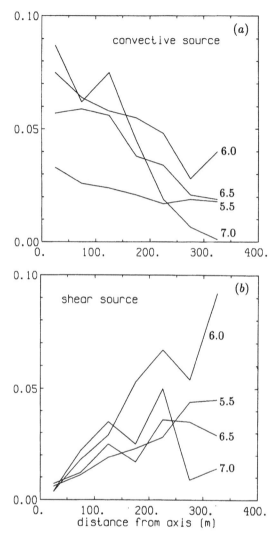

Figure 5.6: Average buoyancy production (a) and shear production (b) of the perturbation kinetic energy (PKE) (in kg m^{-1} s^{-3}) as a function of the distance from the central thermal axis (i.e., $x, y = 0$). Curves are marked according to time in minutes. Experiment 3D3M (Grabowski and Clark 1993a).

However, as mentioned above, these parameters change as the shear layer collapses, and they also vary between the top and the side of the rising thermal for a given time. For example, the Richardson number is as low as ~ -5 near the thermal top and rises to ~ 0 near the thermal side. Note that for a perfectly vertical interface $Ri = 0$, and for a point approaching the very top, $Ri \longrightarrow -\infty$. Thus, the above estimates for the Richardson and Rayleigh numbers should be treated with some caution.

Further insight into the physics of the instability is facilitated by the analysis of the perturbation kinetic energy. Perturbation kinetic energy is derived from momentum and continuity equations written for variables from experiments with and without excitation of the instability (see Grabowski and Clark 1991). The perturbation kinetic energy is defined as $PKE = \rho_o u_i' u_i'$ and its evolution equation is derived as:

$$\frac{\partial PKE}{\partial t} + \frac{\partial}{\partial x_j} \left[(\bar{u}_j + u_j') \, PKE + u_j' p' + u_i' \tau_{ij}' \right] =$$

$$= -\rho_o u_i' u_j' \frac{\partial \bar{u}_i}{\partial x_j} + g \rho_o u_3' B' - \tau_{ij}' \frac{\partial u_i'}{\partial x_j} \equiv S_S + S_B + S_D, \qquad (5.3)$$

where the overbar represents base–state variables (i.e., variables from the experiment without excitation), primes represent the difference between fields from experiments with and without excitation, and τ_{ij} is the stress tensor.

The first term on the left–hand side of (5.3) is the local rate of change of the perturbation kinetic energy. The second term represents divergence of the kinetic energy flux associated with advection, pressure work and diffusional transport. The three terms on the right–hand side are sources/sinks of the perturbation kinetic energy associated with the mean shear (S_S), buoyancy (S_B) and dissipation (S_D). Note that the shear term is the source of PKE in the case of the classical Kelvin–Helmholtz instability, whereas the buoyancy (or convective) term provides PKE in the classical Rayleigh–Taylor instability.

Figure 5.6 shows the shear and buoyancy (or convective) sources as a function of the distance from the thermal axis for several time levels for one of the three–dimensional experiments (Grabowski and Clark 1993a). The figure clearly shows that both the shear term and the buoyancy term contribute significantly to the production of PKE. The convective source dominates near the thermal top, whereas the shear source dominates near the side. This makes the instabilities developing near the top similar to the Rayleigh–Taylor instabilities, whereas the instabilities near the side are similar to the Kelvin–Helmholtz instabilities, at least from the PKE perspective.

Asai (1970) studied the linear stability of several profiles of plane shear flows with unstable stratification subject to three-dimensional perturbations. Two kinds of instability were found in flow possessing inflection points. The first was a thermal instability modified by shear. It occurred for Ri smaller than about -0.1 and had a maximum growth rate for perturbations elongated in the direction along the shear. However, for $Ri \sim -1$ perturbations elongated along the shear and transverse perturbations had similar growth rates. The maximum growth rate for perturbations with the same horizontal wave numbers and for $Ra = 10^4$ was observed for $k \sim 1.5$-2. The maximum growth rate was increasing from ≈ 0.05 for $Ri = -0.1$ to ≈ 0.5 for $Ri = -1$. Inertial (or "inflection-point", "shear") instability was found for Ri larger than about -0.1. The maximum growth rate was smaller (< 0.1), and corresponded to perturbations elongated in the transverse direction. Analysis of the perturbation kinetic energy associated with both types of the instability showed that the ratio of shear production and buoyancy production terms was ~ 0.5 for the thermal type and $\sim 10^4$ for the inertial type.

Extrapolation of these results to instabilities developing at the thermal interface is not straightforward. This is mainly due to variations of the governing parameters (both in time and along the interface at a given time) and is also due to differences in the base-state flow structure. For instance, the effect of contractional flow across the interface which tends to suppress development of unstable modes (Zel'dovich et al. 1980, Batchelor 1987, Dritschel et al. 1991) is not considered. Nevertheless, the scale selected and its growth rate fall within the limits set by the linear theory of Asai (1970). In general, three–dimensional linear theory developed for the case of planar geometry appears to provide reasonable estimates for both scales and growth rates of the cloud-environment interface instability.

Growth of small amplitude perturbations results in a transition from a smooth interface to one with complex structures. In the three–dimensional case, this transition may be associated with the transition to turbulence. Section 5 of Grabowski and Clark (1993a) presented analysis suggesting that the flow field near the end of the calculations indeed possessed features of turbulent flows. The transition from a laminar to a turbulent flow regime as a result of the instabilities was associated with a significant change in the vorticity dynamics, namely a replacement of the baroclinic production mechanism as a dominant one by the stretching mechanism. If this is generally true for buoyancy–driven flows, it follows that fully developed turbulence in such flows may be similar to isotropic homogeneous turbulence.

The picture of the entrainment process that emerges from the above discussion is the following. As the buoyant thermal rises, the shear layer of the velocity tangential to the interface develops. The shear layer separates the thermal from its environment. The depth of the shear layer collapses as the thermal continues to rise. The collapse is directly related to the vorticity production by horizontal buoyancy gradients. The time scale of this collapse is approximately equal to the time scale of the thermal rise, R/U, where R is the thermal radius and U is the thermal rise rate. The instabilities that develop at the interface are similar to unstable modes of the three–dimensional linear stability theory for the planar shear layer with unstable stratification (Asai 1970). As the discussion above illustrates, the scale of most unstable modes is estimated as $0.4R$, and the e-folding time is about $0.3R/U$. As the amplitude of unstable modes grows with time, large–scale entraining eddies develop at the interface. These eddies provide kinetic energy for smaller–scale motions that ultimately develop in the three–dimensional case. The smaller–scale motions transfer the kinetic energy toward the microscale (e.g., Kolmogorov and Batchelor scales) where it is dissipated by molecular processes. Effects of buoyancy reversal associated with evaporation of cloud droplets as a result of entrainment of dry environmental air into the cloud become significant only when scales not much larger than the microscale are reached. This is directly related to the fact that buoyancy reversal is realized by molecular processes, namely the diffusion of heat and water vapor and the sedimentation of cloud droplets (Grabowski 1993).

The above discussion describes entrainment dynamics which is relevant to a single laboratory thermal or a single turret of a convective cloud. The discussion in section 3c of Brenguier and Grabowski (1993) illustrates (albeit using a two–dimensional framework and with the emphasis on the microphysical aspects) how entrainment associated with a single thermal or turret relates to the overall pictures of the interaction between small cumulus and its environment. A similar picture of the entrainment process seems to emerge from the fully three–dimensional simulations (Carpenter and Droegemeier 1995).

Finally, it should be stressed that the general picture presented above does not depend upon the presence of the latent heating. Just the buoyancy difference, which is needed to set the flow pattern near the interface, is important. This is supported by limited tests using thermals without effects of latent heating, i.e., dry thermals. In these tests interfacial instabilities developed in a manner similar to the moist thermal experiments discussed above.

5.4   Entrainment and buoyancy reversal

As stated in the introduction, buoyancy reversal associated with evaporative cooling has long been considered as an important factor in cumulus entrainment. Although the discussion of the previous section did not consider buoyancy reversal, the long history of cloud physics research tends to emphasize the role of concepts like the cloud top entrainment instability and penetrative downdrafts (see the introduction). This section attempts to clarify this apparent controversy.

The picture of the entrainment process discussed at the end of the previous section considers formation of reversed buoyancy as a final product of a chain of events associated with entrainment. The fundamental question is whether reversed buoyancy can feed back into cloud–scale flow to the extent that enhancement of entrainment is possible. For instance, the concept of the cloud top entrainment instability assumes a strong feedback between buoyancy reversal and entrainment. The key aspect is a distinction between effects of buoyancy reversal on the overall flow evolution (e.g., formation of negatively buoyant downdrafts) and effects on the entrainment rate.

Before proceeding, it is worthwhile to mention basic parameters considered in buoyancy–reversing flows. The two basic parameters are: (i) the ratio D between the positive buoyancy of the undiluted buoyant fluid and the minimum negative buoyancy that can be generated in the mixing process, and (ii) the mixing proportion χ for which the minimum negative buoyancy occurs (c.f. Siems et al. 1990). As far as cumulus convection is concerned, the mixing proportion χ is shown in Fig. 5.1a, and the ratio D can be deduced by combining information in Figs. 5.1c and 5.2b. Laboratory experiments with buoyancy–reversing flows (e.g., Siems et al. 1990, Shy and Breidenthal 1990, Johari 1992) suggest that overall effects of buoyancy reversal depend strongly on the magnitude of the ratio D. For the stratocumulus case, the typical value is $D \sim 0.1$ due to low cloud water content in these clouds and strong capping inversion (e.g., Siems et al. 1990). For cumulus clouds, $D \sim 1$ or larger (Siems et al. 1990, Grabowski 1993). There are usually a few other nondimensional parameters needed to fully describe flow regime in the cumulus case (cf. Grabowski 1995).

In general, some enhancement of the entrainment in buoyancy–reversing convection is expected after negative buoyancy reversal is released. This is based on a heuristic argument which considers kinetic energy of the flow and its relation to the entrainment. In classical fluid mechanics, the entrainment velocity (on which entrainment rate directly depends) is assumed proportional to the velocity of a convective element (cf. Turner 1986). Presence of buoyancy reversal is very likely to result in the enhancement of the total kinetic energy since additional buoyancy source is present (cf. section 4b of Grabowski 1993). This in turn may lead to enhanced entrainment, but the magnitude of this enhancement is not clear. Below, three aspects of this problem are discussed which together suggest that the enhancement of the entrainment is actually fairly limited despite the dramatic impact of buoyancy reversal upon the overall flow evolution.

1. Strong support for the distinction between effects of buoyancy reversal on the overall flow evolution and on the entrainment rate comes from laboratory experiments with classical and buoyancy–reversing thermals discussed in Johari (1992). Johari considered the rate of dilution of thermals as measured by the distance the thermal has to rise before all released

material is diluted to a prescribed degree. Despite dramatically different behavior of classical and buoyancy–reversing thermals, the distance was very similar for both types of thermals (Johari 1992, Fig. 9 and 14). This implies that the rate of dilution of unmixed cores is very similar in classical and buoyancy–reversing thermals.

2. The laboratory results discussed above seem consistent with the notion that the influence of buoyancy reversal on a single entraining eddy in high Reynolds number flow should be fairly small. This is a direct effect of the extreme separation between scales of entraining eddies and scales at which homogenization occurs. Because the homogenization at the microscale occurs at the time when the initial eddy is already disintegrated into small-scale motions and a substantial part of the eddy's kinetic energy has already cascaded toward smaller scales, it is too late for the strong direct influence on eddy kinetic energy. This seems to be supported by numerical results of Siems and Bretherton (1992).

3. Grabowski (1993, 1995) considered effects of buoyancy reversal on entrainment in a simple two-fluid model of cumulus convection. The effects of buoyancy reversal on the overall flow evolution and on the rate of mixing between the two fluids were considered. The simplified treatment of cumulus dynamics allowed the consideration of two cases in an otherwise identical dynamical setup. Buoyancy reversal was not allowed in the first case: the buoyancy of a mixture between the two fluids (i.e., the pure environment and the pure cloud) was given by a linear combination of cloud buoyancy (at a given level) and environmental buoyancy (which was assumed zero). The entrainment in this case (referred to as a no-buoyancy–reversal case, or NBR case) was only reducing the cloud buoyancy. Buoyancy reversal was allowed in the second case and the buoyancy of a mixture was given by the bilinear function with buoyancy reversal parameters D and χ deduced from analysis of the mixing between a cloudy parcel rising adiabatically from the cloud base and environmental air at this level (i.e., similar to Fig. 5.2). In this case (referred to as a buoyancy–reversal, or BR case), the entrainment and mixing between a cloud and its environment produced negative (reversed) buoyancy. Grabowski (1993) discussed effects of buoyancy reversal on the overall flow evolution (e.g., formation of negatively buoyant downdrafts, intermittent cloud evolution) which were consistent with observations of buoyancy–reversing laboratory jets and plumes (Turner 1966). Grabowski (1995), on the other hand, presented a detailed analysis of the effects of buoyancy reversal on the rate of mixing between the two fluids. The main conclusion of Grabowski (1995) was that despite the fact that buoyancy reversal had a dramatic impact on the overall flow evolution, its effect on the rate of mixing between the two fluids was minor, i.e., similar amounts of mass were mixed in BR and NBR cases.

The above discussion suggests that a clear distinction has to be made between effects of buoyancy reversal on the global flow evolution and on the rate of entrainment. Buoyancy–reversing thermals were not diluted earlier during their rise than classical thermals in laboratory experiments (Johari 1992), although once diluted their behavior was dramatically different. The amount of mixing produced in a simple two-fluid analog of moist convection (Grabowski 1993, 1995) was similar in experiments which allowed and disallowed buoyancy reversal, although dramatic differences in the overall flow pattern were apparent. In summary, buoyancy reversal modifies the global flow (for instance, by generating large–scale downdrafts), but its effect on entrainment dynamics or entrainment rate seems rather minor.

5.5   Conclusions

This paper provides a systematic overview of the cumulus entrainment with the emphasis on dynamical aspects. It is argued that cumulus entrainment is an example of entrainment in a buoyancy–driven flow and that phase changes of water substance are important only as a source of either positive buoyancy in undiluted (or slightly diluted) cloudy volumes or negative buoyancy in strongly diluted volumes. It follows that the results discussed in section 5.3 of this paper are applicable to entrainment in other buoyancy–driven flows. This is supported by striking similarities in the visual appearance of convective clouds and high-Reynolds number (turbulent) thermals and plumes.

Section 5.3 discusses instabilities that develop at the interface separating rising buoyant fluid from its environment. It is argued that these instabilities play a fundamental role in the entrainment. The essential aspect of the instability is the shear layer that develops near the interface as the buoyant thermal pushes upward. Because of the deformational flow near the interface, the shear layer collapses (with the time scale similar to that of the thermal rise) and becomes unstable. Unstable modes in the linear growth regime are characterized by spatial scales and growth rates that are similar to those predicted by linear stability analysis of a planar shear layer with unstable stratification (Asai 1970). Both shear and buoyancy production terms are important for the kinetic energy of the instability, with the buoyancy source dominating near thermal top and shear source dominating near the side (i.e., away from the top). The interfacial instabilities can be thought of as a combination of Rayleigh–Taylor and Kelvin–Helmholtz instabilities occurring in a complex geometrical setup.

Evolution of instabilities results in the development of the turbulent transition layer that separates undiluted fluid from the environment. This seems to agree with some interpretations of early observations of laboratory thermals (e.g., Scorer 1958, p. 156). Analysis presented in section 5 of Grabowski and Clark (1993a) suggests that at this stage the flow associated with rising thermal has strong similarities with homogeneous and isotropic numerical turbulence. Thus, according to numerical results, these are the interfacial instabilities that are directly responsible for the generation of turbulence observed in both cloud measurements and laboratory experiments with thermals and plumes. As the turbulent mixing between a cloud and its environment progresses, negative (or reversed) buoyancy due to evaporation of cloud droplets is generated. Formation of negative buoyancy leads to generation of large–scale downdrafts, as observed in laboratory experiments (e.g., Turner 1966, Johari 1992) and simulated in the simple two–fluid model of Grabowski (1993, 1995).

Although not considered in the context of this paper, weight and evaporation of precipitation play a similar, if not a more important, role in formation and maintenance of buoyancy–driven downdrafts. Whatever the actual mechanism, deep ("penetrative") downdrafts are a unique feature of the atmospheric moist convection. It follows that a convective cloud is likely to contain at any level both the undiluted air rising from the cloud base and strongly diluted (possibly precipitation–loaded), negatively–buoyant air descending from higher levels. As this effect cannot be represented in simple one–dimensional models, it is not surprising that such models cannot reproduce many observed features of cumulus clouds (e.g., Warner 1970). This is clearly in contrast with the success such models had when applied to classical (i.e., without buoyancy reversal) buoyancy–driven flows (e.g., Turner 1986).

Although buoyancy reversal has a strong influence on the overall cumulus dynamics, it is argued that the actual entrainment rate is not influenced significantly. This is clearly at odds with the long tradition of cloud physics which seems to emphasize quite the opposite.

Both laboratory experiments with buoyancy–reversing thermals (Johari 1992) and numerical experiments with buoyancy–reversing convection (Grabowski 1993, 1995) support the distinction between effects of buoyancy reversal on the global dynamics and on the rate of entrainment and mixing.

As far as cumulus entrainment is concerned, effects of buoyancy reversal on convection are similar to effects of environmental vertical shear of the horizontal wind. The strong influence of environmental shear on the dynamics of small cumuli is well documented, e.g., small cumuli developing in shear have a downshear tilt, the diluted cloudy material is usually found on the downshear part of the cloud (e.g., Malkus 1952, Ackerman 1958, Rodgers et al. 1985). However, environmental shear has a rather small direct effect on the interfacial instabilities that drive the entrainment (Grabowski and Clark 1993b). This can be explained as a result of the large difference between typical magnitudes of environmental shear (usually smaller than 10^{-2} s^{-1}) and the magnitude of baroclinically generated interfacial shear (typically around 10^{-1} s^{-1} for small cumuli). Thus, dramatic effects of environmental shear on the overall dynamics do not modify dynamics of entrainment in a significant way.

Acknowledgements

Comments on the manuscript by Terry Clark are gratefully acknowledged, so is the editorial help of O. Bortfeld and L. Miranda. The National Center for Atmospheric Research is sponsored by the National Science Foundation.

Bibliography

Ackerman, B., "Turbulence around tropical cumuli.", *J. Meteor.*, **15**, 69–74 (1958).

Ackerman, B., "The variability of the water content of tropical cumuli." *J. Meteor.*, **16**, 191–198 (1959).

Albrecht, B. A., R. S. Penc, and W. H. Schubert, "An observational study of cloud-topped mixed layers." *J. Atmos. Sci.*, **42**, 800–822, (1985).

Asai, T., "Stability of a plane parallel flow with variable vertical shear and unstable stratification." *J. Meteor. Soc. Japan*, **48**, 129–138 (1970).

Baker, B. A., "Turbulent entrainment and mixing in clouds: A new observational approach." *J. Atmos. Sci.*, **49**, 387–404 (1992).

Baker, M. B., R. E. Breidenthal, T. W Choularton, and J. Latham, "The effects of turbulent mixing in clouds." *J. Atmos. Sci.*, **41**, 299–304 (1984).

Batchelor, G. K., "The stability of a large gas bubble rising through liquid." *J. Fluid Mech.*, **184**, 399-422 (1987).

Blyth, A. M., "Entrainment in cumulus clouds." *J. Appl. Meteor.*, **32**, 626–640 (1993).

Blyth, A. M., W. A. Cooper, and J. B. Jensen, "A study of the source of entrained air in Montana cumuli." *J. Atmos. Sci.*, **45**, 3944–3964 (1988).

Brenguier, J.-L., "Observations of cloud microstructure at the centimeter scale." *J. Appl. Meteor.*, **32**, 783–793 (1993).

Brenguier, J-L. and W. W. Grabowski, "Cumulus entrainment and cloud droplet spectra: A numerical model within a two-dimensional dynamical framework." *J. Atmos. Sci.*, **50**, 120–136 (1993).

Carpenter, R. L., and K. K. Droegemeier, "Entrainment and detrainment in numerically simulated cumulus congestus clouds." *Preprints, Conf. on Cloud Physics*, Dallas, American Meteorological Society, 569–574 (1995).

Clark T. L., "A small-scale dynamic model using a terrain following coordinate transformation." *J. Comput. Phys.*, **24**, 186–215 (1977).

Clark T. L., "Numerical simulations with a three-dimensional cloud model: Lateral boundary condition experiments and multicellular severe storm simulations." *J. Atmos. Sci.*, **36**, 2191–2215 (1979).

Clark T. L. and R. D. Farley, "Severe downslope windstorm calculations in two and three spatial dimensions using anelastic interactive grid nesting: A possible mechanism for gustiness." *J. Atmos. Sci.*, **41**, 329–350 (1984).

Clark T. L., P. K. Smolarkiewicz and W. D. Hall, "Three-dimensional cumulus entrainment studies." *Preprints, 10th International Cloud Physics Conf.*, Bad Homburg (FRG), 88–90 (1988).

Cooper W. A. and A. R. Rodi, "Cloud droplet spectra in summertime cumulus clouds." *Preprints, Conf. on Cloud Physics*, Chicago, American Meteorological Society, 147–150, (1982).

Deardorff, J. W., "Cloud top entrainment instability." *J. Atmos. Sci.*, **37**, 131–147 (1980).

Dritschel, D. G., P. H. Haynes, M. N. Juckes, and T. G. Shepherd, "The stability of a two-dimensional vorticity filament under uniform strain." *J. Fluid Mech.*, **230**, 647-665 (1991).

Emanuel, K. A., "A similarity theory for unsaturated downdrafts within clouds." *J. Atmos. Sci.*, **38**, 1541–1557 (1981).

Grabowski, W. W., "Cumulus entrainment, fine-scale mixing and buoyancy reversal." *Quart. J. Roy. Meteor. Soc.*, **119**, 935–956 (1993).

Grabowski, W. W., "Entrainment and mixing in buoyancy reversing convection with application to cloud–top entrainment instability." *Quart. J. Roy. Meteor. Soc.*, **121**, 231–253 (1995).

Grabowski, W. W., and T. L. Clark, "Cloud-environment interface instability: Rising thermal calculations in two spatial dimensions." *J. Atmos. Sci.*, **48**, 527–546 (1991).

Grabowski, W. W., and T. L. Clark, "Cloud-environment interface instability. Part II: Extension to three spatial dimensions." *J. Atmos. Sci.*, **50**, 555–573 (1993a).

Grabowski, W. W., and T. L. Clark, "Cloud-environment interface instability. Part III: Direct influence of environmental shear." *J. Atmos. Sci.*, **50**, 3821–3828 (1993b).

Haman, K. E., and S. P. Malinowski, "Formation of downdrafts in cumulus clouds." *Preprints, 9th International Cloud Physics Conf.*, Tallinn (USSR), 527–530 (1984).

Haman, K. E., A. Makulski, S. P. Malinowski and R. Busen, "Temperature measurements in clouds on a centimeter scale." *Preprints, Conf. on Cloud Physics*, Dallas, American Meteorological Society, 98–101 (1995).

Johari., H., "Mixing in thermals with and without buoyancy reversal." *J. Atmos. Sci.*, **49**, 1412–1426 (1992).

Klaassen, G. P., and T. L. Clark, "Dynamics of the cloud-environment interface and entrainment in small cumuli: Two-dimensional simulations in the absence of ambient shear." *J. Atmos. Sci.*, **42**, 2621–2642 (1985).

Krueger, S. K., "Linear eddy modeling of entrainment and mixing in stratus clouds." *J. Atmos. Sci.*, **50**, 3078–3090 (1993).

Knight, C. A., and L. J. Miller, "First radar echoes from cumulus clouds." *Bull. Amer. Meteo. Soc.*, **74**, 179–188 (1993).

Kuo, H.-C., and W. H. Schubert, "Stability of cloud-topped boundary layers." *Quart. J. Roy. Meteor. Soc.*, **114**, 887–916 (1988).

Lilly, D. K., "Models of cloud-topped mixed layers under a strong inversion." *Quart. J. Roy. Meteor. Soc.*, **94**, 292–309 (1968).

MacVean, M. K., "A numerical investigation of the criterion for cloud-top entrainment instability." *J. Atmos. Sci.*, **50**, 2481–2495 (1993).

MacVean, M. K., and P. J. Mason, "Cloud-top entrainment instability through small-scale mixing and its parameterization in numerical models." *J. Atmos. Sci.*, **47**, 1012–1030 (1990).

Malkus, J. S., "Effects of wind shear on some aspects of convection." *Trans. Amer. Geophys. Union*, **30**, 19–25 (1949).

Malkus, J. S., "The slopes of cumulus clouds in relation to external wind shear." *Quart. J. Roy. Meteor. Soc.*, **78**, 530–542 (1952).

Malkus, J. S., "Some results of trade–cumulus cloud investigation." *J. Meteor.*, **11**, 220–237 (1954).

Mahrt, L., and J. Paumier, "Cloud-top entrainment instability observed during AMTEX." *J. Atmos. Sci.*, **39**, 622–634 (1982).

Morton, B. R., "Buoyant plumes in a moist atmosphere." *J. Fluid Mech.*, **2**, 127–144 (1957).

Morton, B. R., G. I. Taylor, and J. S. Turner, "Turbulent gravitational convection from maintained and instantaneous sources." *Proc. R. Soc. Lond.*, **A 234**, 1-23 (1956).

Paluch, I. R., "The entrainment mechanism in Colorado cumuli." *J. Atmos. Sci.*, **36**, 2467–2478 (1979).

Papantoniou, D., and E. J. List, "Large–scale structure in the far field of buoyant jets." *J. Fluid Mech.*, **209**, 151–190 (1989).

Randall, D. A., "Conditional instability of the first kind upside-down." *J. Atmos. Sci.*, **37**, 125–130 (1980).

Raymond, D. J., "A two-scale model of moist, non-precipitating convection." *J. Atmos. Sci.*, **36**, 816–831 (1979).

Raymond, D. J., and A. M. Blyth, "A stochastic mixing model for nonprecipitating cumulus clouds." *J. Atmos. Sci.*, **43**, 2708–2718 (1986).

Reuter, G. W., "A historical review of cumulus entrainment studies." *Bull. Amer. Meteor. Soc.*, **67**, 151–154 (1986).

Rogers, D. P., J. W. Telford, and S. K. Chai, "Entrainment and the temporal development of the microphysics of convective cloud." *J. Atmos. Sci.*, **42**, 1846–1858 (1985).

Sánchez O., D. J. Raymond, L. Libersky and A. G. Petschek, "The development of thermals from rest." *J. Atmos. Sci.*, **46**, 2280–2292 (1989).

Scorer, R. S., "Experiments on convection of isolated masses of buoyant fluid." *J. Fluid Mech.*, **2**, 583–594 (1957).

Scorer, R. S., *Natural aerodynamics*. Pergamon Press, 312 pp, (1958).

Shy, S. S. and R. E. Breidenthal, "Laboratory experiments on the cloud-top entrainment instability." *J. Fluid Mech.*, **214**, 1–15 (1990).

Siems, S. T., C. S. Bretherton, M. B. Baker, S. Shy, and R. B. Breidenthal, "Buoyancy reversal and cloud-top entrainment instability." *Quart. J. Roy. Meteor. Soc.*, **116**, 705–739 (1990).

Siems, S. T. and C. S. Bretherton, "A numerical investigation of cloud-top entrainment instability and related experiments." *Quart. J. Roy. Meteor. Soc.*, **118**, 787–818 (1992).

Squires, P., "The spatial variation of liquid water and droplet concentration in cumuli." *Tellus*, **10**, 372–380 (1958a).

Squires, P., "Penetrative downdraughts in cumuli." *Tellus*, **10**, 381–389 (1958b).

Stommel, H., "Entrainment of air into a cumulus cloud." *J. Meteor.*, **4**, 91–94 (1947).

Taylor, G. R., and M. B. Baker, "Entrainment and detrainment in cumulus clouds." *J. Atmos. Sci.*, **48**, 112–121 (1991).

Telford, J. W., "Turbulence, entrainment, and mixing in cloud dynamics." *Pure Appl. Geophys.*, **113**, 1067–1084 (1975).

Turner, J. S., "The 'starting plume' in neutral surroundings." *J. Fluid Mech.*, **13**, 356–369 (1962).

Turner, J. S., "Model experiments relating to thermals with increasing buoyancy." *Quart. J. Roy. Met. Soc.* 89, 62–74 (1963).

Turner, J. S., "Jets and plumes with negative or reversing buoyancy." *J. Fluid. Mech.*, **26**, 779–792 (1966).

Turner J. S., *Buoyancy Effects in Fluids*. Cambridge University Press, 368 pp, (1973).

Turner, J. S., "Turbulent entrainment: the development of the entrainment assumption, and its application to geophysical flows." *J. Fluid. Mech.*, **173**, 431–471 (1986).

Wallace, J. M., and P. V. Hobbs, *Atmospheric science: An introductory survey*. Academic Press, 467 pp, (1977).

Warner, J., "The water content of cumuliform cloud." *Tellus*, **7**, 449–457 (1955).

Warner, J., "On steady–state one–dimensional models of cumulus convection." *J. Atmos. Sci.*, **27**, 1035–1040 (1970).

Woodward, B., "The motion in and around isolated thermals." *Quart. J. Roy. Met. Soc.* 85, 144–151 (1959).

Zel'dovich, Ya. B., A. G. Istratov, N. I. Kidin, and V. B. Librovich, "Flame propagation in tubes: Hydrodynamics and Stability." *Combustion Sci. Technol.*, **24**, 1-13 (1980).

Chapter 6

THE 2/7 LAW IN TURBULENT THERMAL CONVECTION

STÉPHANE ZALESKI

Modélisation en Mécanique, CNRS URA 229,
Université Pierre et Marie Curie, 4 place Jussieu
75252 Paris Cedex 05, France. zaleski@lmm.jussieu.fr

This communication reviews the theoretical understanding of $Nu - \mathrm{Ra}$ heat flux scaling laws in an enclosure heated from below. Various types of theories are presented. Simple scaling theories are mostly based on dimensional analysis. More refined mechanistic theories involve physical mechanisms such as plume production in the boundary layer. Recent theories yield $Nu \sim \mathrm{Ra}^{2/7}$, close to the experimental results.

6.1 Introduction

The study of Rayleigh-Bénard convection has a long history. Rather recently, attention has focused on scaling results for various quantities as a function of the Rayleigh number. It was confirmed that the simplest scaling laws could not be obtained repeatably from experiments. These simple or "classical" theories propose a scaling $Nu \sim \mathrm{Ra}^{1/3}$ or *1/3 law* and are well verified only in some of the experiments. Starting with Threlfall (1975) it became increasingly clear that measured heat flux laws may in some cases have exponents definitely smaller than 1/3, especially at $Pr = 0.7$. New theories and experimental scalings have emerged, of which the most definite aspect is the law $Nu \simeq \mathrm{Ra}^{2/7}$, or *2/7 law*. These experiments and the corresponding theories have been reviewed by Siggia (1994) and there is no attempt in this communication to supersede his thorough account of the subject, except insofar as to summarize a few experiments that have been reported since 1994. As a result this communication is not a comprehensive review. Rather, as one of the original proponents of the 2/7 law, I would like to give an exposition of the derivation of this theory and of some of the other theories of the $Nu - \mathrm{Ra}$ relation. In the discussion I confine myself to a few issues still in debate such as: what is the range of validity of the law? What is the status of the various theories that yield it?

6.2   Problem definition

Heslot, Castaing and Libchaber (1987) made experimental observations of a scaling regime and of characteristic fluctuations which they called "hard turbulence". In that work, hard turbulence refers to both the 2/7 scaling law for Nu and to exponential temperature distributions (as observed in time series of the temperature T measured at a single probe). Here we will focus on the Nu scaling alone. Heslot et al. (1987) used a cryogenic apparatus with thick copper plates on top and bottom, while heat flux was minimized on the lateral sides. The fluid, helium at 5K, is considered to be well described in the Boussinesq approximation (Tritton, 1988). This experiment is typical of a large number of similar ones that preceded or followed it, and which together motivated the mathematical problem. To define the mathematical problem, we shall use slightly unusual notations that emphasize some important features of the dimensional analysis of the equations.

We define the buoyancy rate as

$$\theta = g\alpha \left[T - (T_1 + T_2)/2\right], \tag{6.1}$$

where g is the gravity acceleration, α the thermal expansion coefficient and T_1 and T_2 the temperatures of the top and bottom plates. Although θ has the dimensions of an acceleration ($[\theta] = [L][T]^{-2}$), it is proportional to temperature in the Boussinesq approximation and hence we shall indifferently call θ the temperature or the buoyancy in what follows.

The Boussinesq equations read with the usual notations:

$$\partial_t \mathbf{u} + \mathbf{u} \cdot \nabla \mathbf{u} = -\frac{1}{\rho}\nabla p + \theta \hat{\mathbf{z}} + \nu \nabla^2 \mathbf{u}, \tag{6.2}$$

$$\partial_t \theta + \mathbf{u} \cdot \nabla \theta = \kappa \nabla^2 \theta \tag{6.3}$$

and

$$\nabla \cdot \mathbf{u} = 0. \tag{6.4}$$

Notice that g and α have disappeared from the equations: this expresses the fact that the temperature scale $1/\alpha$ plays no role in the problem. Boundary conditions, in agreement with experimentally imposed conditions, are defined on both horizontal and vertical walls. The vertical walls are separated by a distance of order $L_h = AL$ where A is the *aspect ratio*. The geometry of the box is usually not relevant to high Rayleigh number scaling theories. (Experiments have been performed in both cylindrical boxes (Heslot et al. 1987) and in rectangular ones.) Then $\mathbf{u} = \mathbf{0}$ on all walls and $\theta = \pm \Delta/2$ for $z = 0, L$ where $\Delta = |T_1 - T_2|$ is the temperature difference across the box, and $\mathbf{n} \cdot \nabla \theta = 0$ on horizontal walls where $\mathbf{n}$ is the normal vector to the wall.

The control parameters are the Rayleigh and Prandtl numbers

$$\mathrm{Ra} = \frac{\Delta L^3}{\kappa \nu}, \qquad Pr = \frac{\nu}{\kappa}. \tag{6.5}$$

The aspect ratio may A also be a relevant parameter. In the Boussinesq approximation and with our definition of θ, buoyancy flux, or equivalently, heat flux is proportional to

$$H = \langle \theta w \rangle + \kappa \partial_z \langle \theta \rangle \tag{6.6}$$

(Brackets indicate averages, which here are taken over time and over a horizontal section of the box). The dimensionless measure of heat flux is the Nusselt number

$$Nu = \frac{LH}{\kappa \Delta}. \tag{6.7}$$

The equations have a number of useful exact properties. The heat flux is related to the dissipation of the temperature squared

$$Nu = \langle (\nabla \theta)^2 \rangle \frac{L^2}{\Delta^2},$$ (6.8)

and to the dissipation of kinetic energy

$$(Nu - 1)\mathrm{Ra} = \epsilon L^4/(\kappa^2 \nu) \quad \text{where} \quad \epsilon = \nu \langle (\nabla \mathbf{u})^2 \rangle.$$ (6.9)

The above equation relates heat flux to energy dissipation and hence to momentum transport. It is thus a form of the Reynolds analogy between momentum and heat transport. See Siggia (1994) for a derivation of the previous two equations. Another useful exact result is the upper bound on heat flux (Howard, 1972; Busse 1978)

$$Nu \le \left(\frac{\mathrm{Ra}}{1035} \right)^{1/2}.$$ (6.10)

6.3   Simple approaches to scaling

Experiments indicate that scaling relations hold, of the form

$$Nu \simeq \mathrm{Ra}^{\beta_N},$$ (6.11)

where β_N is usually obtained from experiments. In the elementary approach to scaling, we derive laws of the form (6.11) by assuming that the scaling relation is independent of a given dimensionless number in the limit of that number going to zero or infinity. Another simple approach is to posit the marginal stability of the boundary layers. We describe each in turn.

6.3.1   Similarity arguments based on dimensional analysis

The simplest approach to the derivation of the heat flux law is to consider that the convection problem of section 6.2 has solutions that may be approached by matching two flows in half-infinite space. Thus we keep the boundary conditions on the horizontal plane $z = 0$ but extend the domain to infinity laterally and vertically. Usually such a problem has at least one physically meaningful solution. For instance the heat *conduction* problem, i.e. the heat equation Eq. (6.3) with $\mathbf{u} = \mathbf{0}$ always and initially $T = 0$ has a well-known solution, with a thermal boundary layer developing near the wall. In a turbulent boundary layer, the well-known *law of the wall* involves only one boundary and is well verified in ducts.

In the half-infinite space geometry, the length scale L disappears from the problem. Dimensional analysis then shows that (Priestley, 1954; Spiegel, 1971)

$$H = \Delta^{4/3} \nu^{1/3} f(Pr)$$ (6.12)

where f is an unknown dimensionless function. In dimensionless form

$$Nu = f(Pr)\mathrm{Ra}^{1/3}.$$ (6.13)

Most publications report exponents $\beta_N \simeq 0.29 \pm 0.02$ not too far from this 1/3 exponent. In view of the simplicity of the theory this is a remarkable success. We shall return to this *1/3 scaling* in what follows. A similar approach has been suggested by Spiegel (1971). If we now consider that heat flux should be independent of the molecular diffusivity coefficients ν and κ, we obtain

$$H = \Delta^{3/2} L^{3/2} \tag{6.14}$$

or

$$Nu = (Pr\,\mathrm{Ra})^{1/2}. \tag{6.15}$$

Dimensional analysis also predicts the variation of average temperature with height. In the same spirit as the reasoning leading to (6.15) the flow is assumed to "remember" only the flux H and to be independent of molecular coefficients. Then the only choice is (Prandtl, 1934)

$$\langle \theta \rangle \sim z^{-1/3} H^{2/3}. \tag{6.16}$$

The experimental validity of this law is debatable. Alternative laws have been proposed for the temperature dependence (Long, 1975), most of them in the form

$$\langle \theta \rangle \sim z^{-s}. \tag{6.17}$$

The determination of s is partially empirical and partially based on other theories of turbulence such as the optimum theory (Malkus 1954b, Spiegel 1962) .

6.3.2 Marginal stability and boundary layer similarity

Most of the theories of thermal convection involve a *thermal boundary layer* of thickness δ, over which most of the temperature drop occurs. Moreover the temperature in the boundary layer is considered linear so that

$$H \sim \frac{\kappa \Delta}{2\delta} \tag{6.18}$$

or

$$Nu \sim L/2\delta. \tag{6.19}$$

From another point of view, (6.19) states that near the wall, the average temperature follows an analogue of the universal "law of the wall" in turbulent boundary layers. To state this law, we *define* δ by (6.18). Then (6.19) is an identity. The universal law for the boundary layer could then be

$$\langle \theta \rangle = f_T(z/\delta)\Delta/2, \tag{6.20}$$

where f_T is a universal function independent of the Rayleigh and Prandtl numbers. This law should be valid not only in the boundary layer but far from it (for $z \gg \delta$) if a universal scaling exists. In that case, the universal law (6.20) should match the algebraic law (6.17) for large z. Unfortunately the experimental study of relation (6.20) is more difficult than the study of the Nu/Ra relation and has not been pursued as relentlessly.

The marginal stability theory states that the mean turbulent flow is marginally stable (Malkus 1954b, Spiegel 1962). The detailed argument is complex but I offer here a simple derivation (see also Howard, 1964). The boundary layer may be viewed as a layer of thickness δ subject to the temperature difference $\Delta/2$. The corresponding Rayleigh number is

$$\mathrm{Ra}_{(\mathrm{bl})} = \frac{\Delta \delta^3}{2\kappa\nu} = \mathrm{Ra}/(16 Nu^3) \tag{6.21}$$

Such a layer will be marginally stable if

$$\mathrm{Ra}_{(\mathrm{bl})} = \mathrm{Ra}_c \tag{6.22}$$

where Ra_c is some Rayleigh number appropriate for the peculiar boundary condition on the top of the boundary layer. One may argue that since the boundary layer has a rigid boundary condition at its bottom and a constraint more like a stress-free boundary condition at its top, the estimate $Ra_c \simeq 1000$ should be in the right ball park. This yields

$$Nu \sim 0.040 Ra^{1/3}, \tag{6.23}$$

which is not far from the experimental result of Goldstein et al. (1990)

$$Nu \sim 0.0659 Ra^{1/3}. \tag{6.24}$$

6.4   Mechanistic approaches to scaling

In more mechanistic approaches, we go beyond reliance on just dimensional analysis and similarity hypotheses, or on a priori principles such as marginal stability. Interestingly most similarity theories share a common view of the interior of the convecting flow, which we describe first. The difference is in their views of the boundary layers.

6.4.1   Inviscid interior scaling

The physics of the interior are common to the classical and to the 2/7 theories. Moreover they are in part exact, as we shall see. We consider that there is a typical interior velocity u_c defined as, say, the root-mean-square velocity, and a typical interior temperature θ_c.

1. A simple estimate of the heat flux is that there are correlated fluctuations of temperature and velocity, i.e. regions of space where the typical temperature is θ_c and the typical velocity is u_c. If these regions fill a fraction f of space then

$$H \sim \theta_c u_c f \tag{6.25}$$

   but in most developments one simplifies to assume $f \sim 1$ so

$$H \sim \theta_c u_c. \tag{6.26}$$

2. Another simple scaling hypothesis is that the interior velocity is related to the interior temperature by a relation that does not involve the molecular transport coefficients ν and κ. It yields

$$u_c \sim (\theta_c L)^{1/2} \tag{6.27}$$

   This assumption is similar to the assumption that yields (6.15) or the turbulent drag law.

Equations (6.26) and (6.27) together yield

$$H \sim u_c^3 / L \tag{6.28}$$

Since the inviscid scaling for dissipation

$$\epsilon \sim u_c^3 / L \tag{6.29}$$

is generally accepted in turbulence, relation (6.28) becomes

$$Nu \, Ra \sim \epsilon L^4 / (\kappa^2 \nu) \tag{6.30}$$

which is equivalent to (6.9) at large Nu.

6.4.2 Plume theories with a single length scale

Mechanistic approaches to scaling simply assume that the boundary layer emits plumes when it becomes marginally stable (an idea nicely expressed in Howard, 1964). A plume is a piece of boundary layer, of temperature T_1 rising in the interior fluid of temperature near $(T_2 + T_1)/2$. The plume has a typical size of the order of the boundary layer thickness δ. At Prandtl number of order 1 or larger, the Reynolds number of such an object is relatively low and it is thus submitted to viscous drag only. Kraichnan (1962) obtained several scalings laws based on this plume structure. The terminal velocity of the rising plume is

$$w_p \sim \Delta \delta^2 / \nu \tag{6.31}$$

This velocity is reached quickly, within a distance of the order of the boundary layer thickness δ. At this height the plumes fill a fraction $f \sim \mathcal{O}(1)$ of space and still have temperature Δ. Then from (6.25)

$$H \sim w_p \Delta \sim \Delta^2 \delta^2 / \nu \tag{6.32}$$

Using the heat flux estimate in (6.32) together with the conductive heat flux in the boundary layer given by (6.18) one finds $\Delta \delta^3 \sim \kappa \nu$, which in terms of dimensionless numbers is again the $1/3$ law in the form with no Pr dependence:

$$Nu \sim \mathrm{Ra}^{1/3}. \tag{6.33}$$

Things change at low Pr, because the Reynolds number of the rising plume becomes large. The plume then accelerates but does not reach its terminal velocity on boundary layer scales δ. One nevertheless makes the assumption of a single length scale δ and estimates the flux at height δ. There the accelerating plume velocity is

$$w_p \sim (\Delta \delta)^{1/2} \tag{6.34}$$

(compare with the interior scaling for velocity (6.27)) and the heat flux is

$$H \sim w_p \Delta \sim \Delta^{3/2} \delta^{1/2} \tag{6.35}$$

Using this new heat flux estimate together with the conductive heat flux in the boundary layer given by (6.18) we get

$$Nu \sim (\mathrm{Ra}Pr)^{1/3} \tag{6.36}$$

Notice that this contradicts marginal stability, since now

$$\mathrm{Ra}_{(\mathrm{bl})} \sim Pr^{-1} \tag{6.37}$$

and at some low Prandtl number the boundary layer Rayleigh number $\mathrm{Ra}_{(\mathrm{bl})}$ is bound to become large enough for instability.

The heat flux scalings obtained by the plume theory yield also predictions for the interior temperature scale and the interior Reynolds number $Re_c \sim u_c L / \nu$. Substituting the heat flux scalings into the interior scaling laws (6.26) and (6.27) yields at large Pr (estimated by Kraichnan to be $Pr > 0.1$)

$$Re_c \sim u_c L / \nu \sim \mathrm{Ra}^{4/9} Pr^{-2/3} \qquad \theta_c / \Delta \sim \mathrm{Ra}^{-1/9} Pr^{-1/3} \tag{6.38}$$

and at low Prandtl number ($Pr \leq 0.1$)

$$Re_c \sim u_c L / \nu \sim \mathrm{Ra}^{4/9} Pr^{-5/9} \qquad \theta_c / \Delta \sim (\mathrm{Ra}Pr)^{-1/9} \tag{6.39}$$

6.4.3   Plume theory with several length scales

Motivated by discrepancies between the observations and the above predictions, Castaing et al. (1989) and several others proposed new mechanistic scaling theories. In a plume picture at large Pr, one again assumes rising plumes of size δ, obtaining law (6.31). However one assumes that the fraction of space filled by the plumes near the horizontal plates is no longer $f \sim \mathcal{O}(1)$, but decreases with Rayleigh number. In other words, we picture a two-phase flow, with a dispersed, hot phase of plumes of size δ and temperature $\theta \sim \Delta$ floating in a background phase of temperature $\theta \sim 0$. To estimate the fraction f we assume that the plumes are vertically stacked (obviously a very simplified picture) and separated by a distance δ_2. We thus introduce a second length scale, δ_2 (called by Castaing at al. the *mixing zone thickness*) and obtain $f = \delta/\delta_2$. How do we estimate δ_2? A simple way is to consider that the boundary layer emits plumes with a frequency $\omega_{\rm bl} \sim \kappa/\delta^2$. Then two plumes emitted at times $1/\omega_{\rm bl}$ apart are separated by a distance

$$\delta_2 \sim w_p/\omega_{\rm bl} = \frac{\Delta \delta^4}{\kappa \nu} \tag{6.40}$$

The two phases will mix producing a fluid of average temperature $\theta_c = f\Delta = \Delta\delta_2/\delta$ or

$$\theta_c \sim \frac{\kappa \nu}{\delta^3} \tag{6.41}$$

Then from the conductive heat flux in the boundary layer (6.19), the interior scaling relations (6.26) and our refined estimate for the mixing zone average temperature (6.27) and (6.41)

$$Nu \sim \mathrm{Ra}^{2/7} Pr^{-1/7} \tag{6.42}$$

although the Prandtl dependence was not given by Castaing et al. As in the single length scale theory, there is a low Prandtl number version (proposed by Cioni et al., 1997) obtained by assuming that the plume accelerates in the boundary layer. The acceleration is simply Δ over a time $1/\omega_{\rm bl}$ thus

$$\delta_2 \sim \frac{\Delta \delta^4}{\kappa^2} \tag{6.43}$$

They same reasoning as before yields

$$Nu \sim (\mathrm{Ra}Pr)^{2/7}. \tag{6.44}$$

The reader will have by now noticed the recurrence of relations of the form $(\mathrm{Ra}Pr)^n$. This originates from the fact that viscosity disappears in the combination $\mathrm{Ra}Pr$, and the theories that yield it are obtained by letting viscosity disappear from the equations.

The 2/7 theories, in the present form, also predict velocities and temperature scales in the bulk. At large Pr:

$$Re_c \sim u_c L/\nu \sim \mathrm{Ra}^{3/7} Pr^{-5/7} \qquad \theta_c/\Delta \sim \mathrm{Ra}^{-1/7} Pr^{-3/7}, \tag{6.45}$$

while at low Prandtl number

$$Re_c \sim u_c L/\nu \sim \mathrm{Ra}^{3/7} Pr^{-4/7} \qquad \theta_c/\Delta \sim (\mathrm{Ra}Pr)^{-1/7}. \tag{6.46}$$

6.4.4 2/7 scaling: Shraiman-Siggia theory

Shraiman and Siggia (1990) proposed another derivation of the 2/7 law, based on a picture of the boundary layer that involves, among other ingredients, a similarity solution due to Lévêque (1928) (see also Schlichting, 1968, chapter 12). We will label in short 'SSL' this theory. It considers how the thermal boundary layer is affected by a shearing wind over a significant portion of the length of the bottom plate L_h. Whenever convection is in the form of cells that do not stretch over the whole length of the bottom plate, as in the numerical experiments of Kerr, for instance, L_h should be considered to be the half horizontal extent of the large scale convection cell. The velocity field is simply modeled by a linear shear flow parallel to the horizontal $u(z) = \omega z$. The buoyancy forces and velocity fluctuations are neglected inside the boundary layer, and vertical derivatives are much larger than horizontal derivatives so that (6.3) reduces to

$$\omega z \partial_x \theta = \kappa \partial_{zz}^2 \theta. \tag{6.47}$$

This equation can be solved (Lévêque, 1928). Here we obtain the sought result by dimensional analysis. We simply remark that setting $z' = z/\delta$, $x = x/L_h$, all parameters disappear from (6.47) if

$$\delta \sim (\kappa L_h / \omega)^{1/3} \tag{6.48}$$

From classical turbulent momentum boundary layer theory, ω is related to the *friction velocity* $u_* = (\nu\omega)^{1/2}$, which in turn is related to ϵ by $\epsilon = u_*^3/L$. There should be a logarithmic difference between u_* and u_c (6.27), but this author doubts that the effect is observable. Next one either identifies u_c with u_* and uses the inviscid interior scaling relation (6.26) and (6.27) which yields again (6.41) when $L_h = L$, or one uses the Reynolds analogy relation (6.9) and the inviscid scaling for dissipation (6.29). In either case

$$Nu \sim \mathrm{Ra}^{2/7} Pr^{-1/7} A^{-3/7} \tag{6.49}$$

where we introduce the aspect ratio $A = L_h/L$. Here and it what follows, the aspect ratio dependence is introduced on the assumption that there is a single large scale circulation. If there are several large scale cells, L_h should be understood as the wavelength of that cell. Note that the plume theory of section 6.4.3 and the SSL theory differ by their aspect-ratio dependence.

6.4.5 Range of validity

Few of the above theories have a clearly defined range of validity in the parameter space defined by Ra, Pr and aspect ratio A. We first review upper limits of validity. They are of two kinds. A large Ra bound that has often been suggested is related to the crossing of two length scales of the flow: several authors have argued that as the length scale $\lambda^* = \nu/ = u_c$ which varies with Re as $\mathrm{Ra}^{-3/7}$, decreased and became smaller that the thermal boundary layer width $\delta \sim \mathrm{Ra}^{-2/7}$ the scaling would be modified. This was predicted both by Kraichnan (1962) for the theory of section 6.4.2 and by Siggia (1994) for the SSL theory of section 6.4.4. Siggia estimates that the transition occurs when $\delta \simeq 7 - 12\lambda^*$ and obtains

$$\mathrm{Ra} < (3 \times 10^{12} - 2 \times 10^{14})Pr^4. \tag{6.50}$$

Above these values one expected to see the $\mathrm{Ra}^{1/2}$ scaling (6.15). However the length λ^* is considerably less than the boundary layer size δ even at moderate Ra in the numerical simulations of Kerr (1996 and this volume). While this is only moderate Ra, if confirmed it would be surprising if scaling where

$$\lambda^* \sim Ra^{-3/7} < \delta \sim Ra^{-2/7} \qquad (6.51)$$

reversed itself at higher Ra. This would have two consequences (i) the estimate of λ^* by Siggia (1994) must be incorrect (either because of a fundamental difficulty with the SSL theory and its scalings or through incorrect estimates of prefactors) and (ii) a high Ra crossover is less likely.

Another kind of upper limit would occur if the boundary layer Rayliegh number $Ra_{(bl)}$, which grows like $Ra^{1/7}$ in the 2/7 theory reaches a stability limit. Indeed from (6.21) and (6.49)

$$Ra_{(bl)} \sim 16\,Ra^{1/7} Pr^{3/7} A^{9/7}. \qquad (6.52)$$

If $Ra_{(bl)}$ has an upper limit, this yields

$$Ra < C Pr^{-3} A^{-9}, \qquad (6.53)$$

where the A dependence occurs in the SSL theory only. As in the case above, the prefactor C is difficult to estimate so I prefer to omit it. At large Pr the 1/3 should be much easier to observe.

The predictions for a lower Ra limit for the 2/7 regime is less clear. Siggia (1994) has argued that the Reynolds number of the interior should be large enough so that a fully developed turbulent boundary layer is established, but if one disregards logarithmic terms, the scaling theory of section 6.4.4 only requires a large enough Reynolds number in the interior for the inviscid scaling of section 6.4.1 to hold. Whatever the precise physical argument, requiring $Re_c > 3 \times 10^3$ yields using the 2/7 theory

$$Ra > 10^8\,Pr^{5/3}, \qquad (6.54)$$

where the prefactor is from Siggia (1994).

In Zaleski (1991) I argued that the Rayleigh number should be large enough so that the shear stabilizes the boundary layer, thus allowing the boundary layer Rayleigh number $Ra_{(bl)}$ to grow beyond the limits imposed by the stability considerations of section 6.3.2 flow. This yields (Zaleski 1991)

$$Ra > 5 \times 10^7\,Pr^4 A^n. \qquad (6.55)$$

where the exponent is $n = 1/3$ for the SSL theory and $n = 5$ for the plume theory. The strong dependence on Pr may explain why the 2/7 regime is not observed in some higher Pr experiments.

There is a serious difficulty with this argument since the linear stability calculations in Zaleski (1991) are based on a purely two-dimensional flow. In a three-dimensional flow, the base conducting solution remains unstable with respect to rolls parallel to the shear flow (Clever and Busse 1991, 1992). However, buoyancy and shear may still interact non-linearly, or in the presence of fluctuating shear. The assumption of the stability of the boundary layer is not well verified in experiments. For instance (6.37) shows an unbounded rise of $Ra_{(bl)}$, and the corresponding rise in Nu is observed experimentally (Cioni et al. 1997).

To me the alternative is as follows (i) either at higher Ra and lower Pr than the experimentally observed values $RaNu^3$ is bounded after all, so the boundary layer is stable (but is not marginally stable for all cases). (ii) or the boundary layer is unstable, which more precisely stated means that the analogy between the mean temperature profile and a base flow in linear stability theory cannot be made.

It may be interesting to add that some researchers have studied the effect of an imposed shear on turbulent convection. While a reduction in the Nusselt number was observed in experiments by Ingersoll (1966) an increase was observed by Solomon and Gollub (1990, 1991). I shall return to this issue later on in this paper.

6.5 Comparison with experiments

The most impressive experimental fact in the field is the finding of Libchaber and his coworkers (Castaing et al., 1989; Sano et al., 1989; Wu and Libchaber, 1992) in gaseous helium ($Pr \simeq 0.7$). Data gathered over 4 to 5 decades of Ra, appeared to fit with a $Nu \sim \text{Ra}^{0.283}$ close to the 2/7 exponent. However, a thorough discussion must take into account a broader range of experimental findings where the situation is less clear.

Additional experiments yield results in four broad classes (see table in Siggia 1994): (i) For an experiment believed to be equivalent to very large Pr (Goldstein et al. 1990) $\beta_N = 1/3$ is observed (ii) For water ($Pr = 4$ to 7) a large number of other experiments give exponents smaller than 1/3 with various aspects ratios. However Goldstein and Tokuda (1980) observe an increase in the exponent towards 1/3 near the high end of the Rayleigh number range. (iii) Experiments in helium gas yield $\beta_N \simeq 2/7$ except for the recent experiment by Chavanne et al. (1996) which shows a much faster increase of Nu for $\text{Ra} > 10^{11}$. The results directly contradict those obtained by Wu (1991) (for the same fluids, liquid and gaseous helium, same Prandtl and same geometry, an $A = 1/2$ cylinder !) The Chavanne experiment was very carefully constructed, difficulties related to the vicinity of the critical point in helium were addressed and the new results should be thus be taken seriously. (iv) Finally experiments in mercury ($Pr \simeq 10^{-2}$) show a complex sequence of power laws (Cioni et al. 1997) but are too limited in Reynolds number to test for a crossover to the $\text{Ra}^{1/2}$ scaling (6.15).

On the electronic side of things, numerical experiments do not yet cover a range of Ra large enough to offer unassailable falsification of 1/3, although they inch towards $\beta_N < 1/3$ (see the contributions of Kerr and Werne in this volume). The stronger evidence in the simulations comes from velocity, temperature and length scales that more closely follow the expectations from theories for 2/7 than their 1/3 counterparts.

However the comparison of exponents for Nu involves more subtle issues. Exponents measured on a log-log graph may converge slowly as Ra is increased. Moreover one order of magnitude of Re corresponds to more than two orders of magnitude of Ra (compare for instance Ra and Re in (6.38)). For example, a range of 3 decades in Ra, which may seem very large, is not that impressive any mire when it is found to correspond to a little over 1 decade in Reynolds number. Thus one is tempted to allow for so-called *corrections to scaling*. For instance, matching the universal law (6.17) with the interior temperature, Long (1975) argued that the corrections had the form

$$Nu \sim C_1 \text{Ra}^{1/3}(1 + C_2(\text{Ra}Nu)^{-s/3} + \cdots) \qquad (6.56)$$

where the dots indicate asymptotically negligible corrections for large Ra. Long replotted the experimental data of Garon and Goldstein (1973), which were fit by these latter authors to

$$Nu \simeq 0.130\text{Ra}^{0.293} \qquad (6.57)$$

over the range $10^7 \leq \mathrm{Ra} \leq 3\,10^9$ for water. Long obtained a good agreement with (6.56) with the Prandtl temperature profile exponent $s = 1/3$. It is interesting to note that the same Garon and Goldstein data were replotted in a different form by Siggia (1994) together with data from the experiments of Goldstein and Tokuda (1980). This second replotting shows a lot of variability at high Ra.

The difficulty of discriminating between the theories on the basis of the $\mathrm{Ra} - Nu$ scalings motivates the investigation of other types of scalings, for instance using measurements of θ_c or u_c. The velocity is often not measured in experiments but related to a frequency ω_p, clearly visible on the time series, and thought to be related to the circulation of plumes around the box (Villermaux 1995) and thus to the core velocity u_c. Unfortunately, measurements of either u_c or ω_p only create new questions. Most measurements report high exponents β_u. For instance Ciliberto et al. (1996) and Sano et al. (1989) obtain $\beta_u \simeq 0.49$ in water. However from (6.9), (6.10) and (6.29) one gets $\beta_u \leq 0.5$. Thus experiments show u_c close to the upper bound while Nu is far from it, which seems paradoxical. On the other hand Cioni et al (1997) measure $\beta_u \simeq 0.424$ in mercury, not far from the estimate (6.45). Takeshita et al. (1996) obtain $\beta_u \simeq 0.46 \pm 0.002$ and similar results were obtained by Tanaka and Miyata (1980). In two- and three-dimensional simulations (Werne et al., 1991; Kerr, 1996), $\beta_u > 3/7$ is also observed.

To explain the large values of β_u that are observed, Cioni et al (1987) suggested that the inviscid dissipation scaling (6.29) does not hold at moderate Pr because of the effect of buoyancy, but does hold at low Pr as in mercury. This introduces fresh complications into the problem and would require independent testing.

Thus, measurements of β_u indicate inadequacies of all the existing theories for the 2/7 regime rather than supporting one over the other. The situation for Pr scaling is more promising. Data for mercury $Pr = .025 << 1$ (Cioni et al., 1997) indicates consistency with (6.44), while for $Pr > 4$ helium experiments (Liu and Ecke, 1997) support a weaker decrease with Pr, even weaker than the predictions (6.42) and (6.49). Whether these experiments are at a sufficiently high Ra to be complete tests of theories that might only be asymptotically correct can be questioned, but the experimental Ra for $Pr=.025$ and 7 are sufficiently high that it is clear that SSL, if valid, is only true for the the high Pr regime.

Finally, measurements of θ_c are generally in agreement with (6.45). However the difference with the classical scaling (6.38) is too small for the small experimental uncertainties to discriminate between the two theories.

6.6 Critique

The first question to be addressed by a critique of the above theories is whether, as provocatively asserted in the title, the 2/7 scaling has the status of a scientific law. Although it is close to the exponents β_N measured in most experiments, claiming that experiments confirm a particular fractional number would stretch this point. In this vein it is perhaps useful to recall the remark of Massaguer (1990) who noticed that 2/7 is the simplest fraction between 1/3 and 1/4.

On the purely theoretical side, it is apparent to practioners of similarity theory that when several length scales are present in a problem it is possible to devise almost any theory to fit the experiments. To avoid this pitfall, it is in my opinion best to try to construct theories with the most detailed possible picture of the fluid mechanics involved.

In this spirit the SSL theory seems to offers a more detailed view of the mechanisms in the boundary layer: for instance a difficulty with the plume theory is that nothing indicates why plumes should detach once the boundary layer reaches height δ. The SSL theory on the other hand seems incomplete because when the box aspect ratio is not close to or smaller than one, nothing fixes the cell aspect ratio A. (Recall that there is an important distinction between the cell and the box geometry. A box may contain several cells of variable size, which allows for an additional degree of freedom of the system.) A wavelength selection mechanism is needed to fix the cell size as a function of Ra, for instance $A \sim \mathrm{Ra}^{\beta_A}$. Such a dependence would be hard to see in a numerical experiment where the cell size is usually fixed by the horizontal box size. But if it exists, it would affect all the exponents in the theory.

A interesting point in the SSL theory is the strong dependence of the boundary layer stability on the aspect ratio A, as seen in equation (6.52). Indeed as the thermal boundary layer develops, it thickens and may become unstable. This instability will be maximal closest to the upwellings. This is precisely where one would expect to see instability. Because shear stabilizes rolls perpendicular to the flow direction (Castaing et al. 1989, Zaleski 1991, Zaleski 1992) these structures should be parallel to the flow direction. Such rolls have been observed by Weiss and McKenzie (Dan McKenzie, personal communication) at $Pr = \infty$, in an aspect-ratio 8×8 box in three dimensions, with stress-free boundaries and fixed-flux boundary conditions on both boundaries. The simulations showed a rising plume at one corner and a sinking one at the opposite diagonal, at Ra = 50,000. Both the upper and lower boundary layers showed weak rolls developed in the direction of the flow, which is parallel to the diagonal. The structures observed in the thermal boundary layer in the simulation of Kerr (1996) may be also be these streamwise unstable rolls. These structures are perhaps also responsible for the velocity fluctuations observed near the wall in the same numerical simulations. These fluctuations cannot be accounted for in the SSL theory which assumes no vertical velocity in eq. (6.47).

Some authors have argued that the SSL theory and the plume theory are distinguished by the existence of the large scale circulation as a critical assumption of the theory. I attempt to qualify this statement through three remarks. First, the interior scaling derivation may be obtained indifferently for a steady large scale circulation or for a fluctuating flow. Second, the 2/7 plume theory of section 6.4.3 is compatible with both types of flow, just as the Kraichnan plume theory is. Third in the SSL picture of the flow near the boundary, a coherent large scale flow is also not necessary. The shearing flow needed in the SSL theory may be the large scale flow or smaller-scale flows coming from intermittently established large cells. However these cells should not be much smaller than the box size. In that respect, there are recent experiments by Ciliberto et al. (1996) in which the flow is blocked by screens near the boundary. In that experiment the blocking does not change the scaling. This does introduce an argument against the SSL picture.

6.7   Conclusion

We have reviewed some of the theories and some of the experimental results relevant to the 2/7 scaling for heat transport in Rayleigh-Bénard convection. It is not yet fully determined, in this reviewers opinion, what the accurate extent of the 2/7 scaling is in terms of Pr and A. However the theories give an interesting series of predictions such as Prandtl number or aspect ratio dependence, velocity scaling, or limits of validity. The description of the large scale flow and its effects, the study of low Pr remain important open problems. As the theories point to interesting connections between various experimental facts, they indicate directions in which real or numerical experiments should be performed.

Bibliography

Busse F. H. "The optimum theory of turbulence", *Adv. Appl.Mech.* **18**, 77-121 (1978).

Castaing, B., Gunaratne G., Heslot, F., Kadanoff, L., Libchaber A., Thomae, S., Wu X.-Z., Zaleski S. and Zanetti, G. "Scaling theory of hard thermal turbulence in Rayleigh-Bénard convection", *J. Fluid Mech.* **204**, 1–30 (1989).

Chavanne, X., Chilla, F., Chabaud, B., Castaing, B., Chaussy, J. and Hébral B. "High Rayleigh number convection with gaseous helium at low temperature", *J. Low Temp. Phys.*, **104** 109 (1996).

Ciliberto, S., Cioni, S. and C. Laroche "Large scale flow properties of turbulent thermal convection", *Phys. Rev. E*, **54**, 5901-5904 (1996).

Cioni, S. Ciliberto, S. and J. Sommeria "Strongly turbulent Rayleigh-Bénard convection in mercury: comparison with results at moderate Prandtl number", *J. Fluid Mech.* **335**, 111-140 (1997).

Clever, R. M. and F. H. Busse "Instabilities of longitudinal rolls in the presence of Poiseuille flow", *J. Fluid Mech.*, **229**, 517–529 (1991).

Clever, R. M. and F. H. Busse "Three-dimensional convection in a horizontal fluid layer subjected to constant shear", *J. Fluid Mech.*, **234**, 511-527 (1992).

Garon A. M. and Goldstein R. J., "Velocity and heat transfer measurements in thermal convection", *Phys. Fluids*, **16**, 1818-1825.

Goldstein R. J., Chiang, H. D., See D. L. "High-Rayleigh-number convection in a horizontal enclosure, *J. Fluid Mech.*, **213**, 111-126 (1990).

Goldstein R. J. and Tokuda, S. "Heat transfer by thermal convection at high Rayleigh numbers", *Int. J. Heat Mass Transf.* **23**, 738-740 (1980).

Heslot, F., Castaing, B. Libchaber, A. "Transition to turbulence in helium gas", *Phys. Rev.* **A 36**, 5870–5873 (1987).

Howard, L. N. "Convection at high Rayleigh number" *Applied Mechanics, Proceedings of the 11th ICAM, Munich (Germany)*, Henry Görtler, Ed., 1109-1115 (1964).

Howard, L. N. "Bounds on flow quantities", *Annu. Rev. Fluid Mech.* **4**, 473-494 (1972).

Ingersoll, A. P. "Thermal convection with shear at high Rayleigh number", *J. Fluid Mech.* **17**, 405–432. (1966).

Kerr, R. M, "Rayleigh number scaling in numerical convection", *J. Fluid Mech.* **310**, 139–179 (1996).

Kraichnan, R. H., "Turbulent thermal convection at arbitrary Prandtl number", *Phys. Fluids*, **5**, 1374–1389 (1962).

Lathrop, D. P., Fineberg J. and Swinney, H. L. "Transition to shear driven turbulence in Couette-Taylor flow" Phys. Rev. A. **46**, 6390-6405 (1992).

Lévêque, M. A. "Les lois de la transmission de la chaleur par convection", Thèse de la Faculté des Sciences de Paris soutenue le 30 mars 1928, published as an article in *Annls. Mines, Paris* **13**, 201–409 (1928) (see page 285).

Liu Y., and Ecke, R. E., Heat transport scaling in turbulent Rayleigh-Bénard convection: effects of rotation and Prandtl number. *Phys. Rev. Lett*, **79**, 2257-2260 (1997).

Long, R. R. "Relation between Nusselt number and Rayleigh number in turbulent thermal convection", *J. Fluid Mech.* **73**, 445–451 (1975).

Malkus, W. V. R. "Discrete transitions in turbulent convection", *Proc. Roy. Soc.*, **A 225**, 185-195, (1954a).

Malkus, W. V. R. "The heat transport and spectrum of thermal turbulence", *Proc. Roy. Soc.*, **A 225**, 196-212 (1954b).

Massaguer, J. M. "Stellar Convection as a Low Prandtl number flow", proceedings of the conference *The Sun and Cold Stars*, Helsinki (1990).

Prandtl, L. "Meteorologische Anwendungen der Strömungslehre", *Beitr. Phys. fr. Atmos.*, **19**, 188–202 (1932).

Priestley, C. H. B. *Turbulent heat transfer in the lower atmosphere*, U. of Chicago Press, (1959).

Sano, M., Wu, X. and Libchaber, A. "Turbulence in helium-gas free convection", *Phys. Rev. A* **40**, 6421–6430 (1989).

Schlichting, H. *Boundary Layer Theory*, Mc Graw Hill (1968).

Shraiman, B. I. and Siggia, E. D. "Heat transport in high Rayleigh number convection", *Phys. Rev. A* **42**, 3650–3653 (1990).

Siggia, E. D. "High Rayleigh number convection", *Annu. Rev. Fluid Mech.* **26**, 137–168 (1994).

Solomon, T. H. and Gollub, J. P. "Sheared boundary layers in turbulent Rayleigh-Bénard convection", *Phys. Rev. Lett.* **64**, 2382–2385 (1990).

Solomon, T. H. and Gollub, J. P. "Thermal boundary layers and heat flux in turbulent convection: the role of recirculating flows", *Phys. Rev. A* **45**, 1283 (1991).

Spiegel, E. A. "On the Malkus theory of turbulence", in *Mécanique de la Turbulence. Colloque International du CNRS à Marseille*, Éditions du CNRS, Paris, p. 181–201, (1962).

Spiegel, E. A. "Convection in stars", *Annu. Rev. Astron. Astrophys.* **9**, 263–301 (1971).

Tanaka, H. and Miyata H. Turbulent natural Convection in a horizontal water layer heated from below, *Int. J. Heat and mass Transfer*, **23** 1273–1281 (1980).

Takeshita, T., Segawa, T., Glazier, J. A. and Sano, M. "Thermal turbulence in mercury", *Phys. Rev. Lett.*, **76**, 1465–1468 (1996).

Threlfall, D. C. . Free convection in low-temperature gaseous helium. *J. Fluid Mech.* **67**, 17-28 (1975).

Tritton, D. J. (1988). *Physical Fluid Dynamics, 2nd edition*. Clarendon Press, Oxford.

Villermaux, E. "Memory-induced low frequency oscillations in closed convection boxes", *Phys. Rev. Lett.* **75**, 4618-4621 (1995).

Werne, J., DeLuca, E.E., Rosner, R. and Cattaneo, F. "Development of hard–turbulent convection in two dimensions: Numerical Evidence." *Phys. Rev. Lett.* **67**, 3519–3522 (1991).

Wu, X. Z. "Along a road to developed turbulence: free thermal convection in low temperature helium", Ph. D. thesis, The University of Chicago (1991).

Wu, X.-Z. & Libchaber, A. "Scaling relations in thermal turbulence: The aspect-ratio dependence." *Phys. Rev. A* **45**, 842–845 (1992).

Zaleski S. "La transition vers le régime de turbulence thermique dure en convection de
Rayleigh-Bénard", *C. R. Acad. Sci. Paris*, **313, Sér. II**, 1099-1103 (1991) (*in French,
extended abstract in English*).

Zaleski S. "Thermal convection at high Rayleigh number in two-dimensional sheared layers",
proceedings of the Workshop *The global geometry of turbulence*, Rota, Spain, July 1990,
p 167-179. J. Jimenez Ed., Plenum Press (1992).

Chapter 7

ORGANIZATION OF ATMOSPHERIC CONVECTION OVER THE TROPICAL OCEANS: THE ROLE OF VERTICAL SHEAR AND BUOYANCY

MARGARET A. LEMONE

Microscale and Mesoscale Meteorology
National Center for Atmospheric Research[1]
P.O. Box 3000, Boulder, CO 80307-3000

Observational, theoretical, and numerical-modeling work on atmospheric dry and precipitating convection provide several examples of the roles of shear and buoyancy in creating organized convective flows. In general, convective structures are more random when there is less vertical shear of the horizontal wind. Increased shear generally leads to more linear organization, although for precipiting convection, the most extreme shears lead to continuously-propagating, long-lived thunderstorms known as a 'supercells.' Atmospheric scientists have developed several forms of the Richardson number to describe regimes for dry and precipitating convection; two are introduced herein.

7.1   Introduction

Organized flows in convective boundary layers have long been documented in the laboratory (e.g., Chandra 1938), and theoretical work of Kuo (1963), Asai (1970), and others indicate that the change in mean flow across the convecting layer has a strong effect on convective organization. Although astute observers of clouds (Malkus and Riehl 1964) or soaring habits of gulls (Woodcock 1941) saw evidence of organization, it took the first pictures from space and displays from radars (e.g., Anderson et al. 1966, Konrad 1968) to convince us just how often atmospheric convection is organized.

Here, I will show several examples of organized convection in Earth's atmosphere, starting with convection in the lowest km of the Earth's atmosphere in fair weather, and conclude with some thoughts on the organization of deep, precipitating convection. I will relate this

---

[1]NCAR is sponsored by the National Science Foundation.

to various forms of the Richardson number Ri, a measure of the ratio of the energy input into a flow from buoyancy to the energy input from shear.

$$Ri = \frac{\frac{g}{\Theta_v}\frac{\partial \Theta}{\partial z}}{(\frac{\partial U}{\partial z})^2 + (\frac{\partial V}{\partial z})^2} \qquad (7.1)$$

where $\Theta \approx T(\frac{1000}{p})^{\frac{R}{c_p}} = T + \frac{g}{c_p}z$ is proportional to specific entropy, z is the height above the pressure surface $p = 1000$ mb, U and V are orthogonal components of the horizontal wind, the ratio of the ideal gas constant for air to the specific heat for air $R/c_p \approx 0.287$ and the vertical gradients are typically taken over the depth of the convecting fluid.

In the dry atmosphere, $\frac{\partial \Theta}{\partial z} = 0$ implies neutral stratification. However, since the fraction of water vapor in air decreases with height, and water vapor has a molecular weight (18 gm mole^{-1}) much lower than the effective molecular weight of dry air ($\sim$28 gm mole^{-1}), water-vapor stratification decreases the static stability. I thus replace Θ with Θ_v in (7.1) to obtain:

$$Ri = \frac{g}{\Theta_v}\frac{\frac{\partial \Theta_v}{\partial z}}{(\frac{\partial U}{\partial z})^2 + (\frac{\partial V}{\partial z})^2}, \qquad (7.2)$$

where $\Theta_v = T_v(1000/p)^{\frac{R}{c_p}}$, the virtual temperature $T_v = T(1 + 0.61q)$, the water vapor mixing ratio q is the mass ratio of dry to moist air (here in g g^{-1}), and p is the air pressure (mb or hPa). As we shall see, the depth over which the gradients are calculated departs significantly from the depth over which the convection occurs.

7.2   Convection in the fair weather mixed layer

Over the tropical oceans, the lowest few hundreds of meters of the atmosphere in undisturbed weather is typically convective, well-mixed due to turbulent motions, and about 1K cooler than the underlying surface. Figure 7.1 shows profiles of horizontally-averaged virtual potential temperature Θ_v and water-vapor mixing ratio $\bar{q}$ as a function of height in a typical fair-weather mixed layer over the tropical ocean. When Θ_v doesn't vary with height, the layer is considered well-mixed on average. Hence we call this layer the *mixed layer*, whose depth we denote as h. The mixed layer is separated from the statically stable atmosphere above by a *transition layer*, whose top lies roughly at cloud base (Fitzjarrald and Garstang 1981). Typical values of h are 450-600 m; the transition layer is typically around 150-200 m thick (Nicholls and LeMone 1980, Fitzjarrald and Garstang 1981). Since the ocean is $\sim$1 K warmer than the mixed layer, and continuously evaporating, strong heat and moisture fluxes occur, realized in the form of buoyant plumes that merge as they rise through the mixed layer (Fig. 7.1). The spacing between the groups of plumes increases with height, reaching a maximum value of between 1.5 and 3 times the mixed layer depth h. This is the typical horizontal scale of large eddies in the mixed layer; I will refer to the ratio of the horizontal to vertical dimension as the *aspect ratio*. It is interesting to note that thermals over the tropical oceans are warmer than their surroundings only near the surface. By about halfway up the mixed layer, the thermals are actually *cooler* than the surrounding air, their buoyancy supplied by a relatively high concentration of water vapor.

The larger scale structure of the transporting currents changes as a function of the wind speed. Figure 7.2 shows radar-reflectivity patterns that reveal the structure of the fair

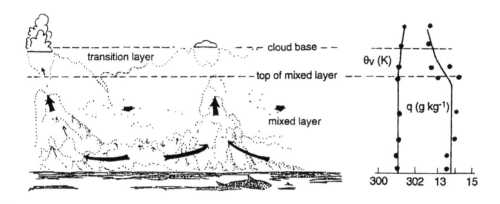

Figure 7.1: Structure of the fair weather tropical atmospheric boundary layer over the ocean. Left: schematic of structure and circulation. Stippling outlines eddies, thin arrows indicate small-eddy motion, wide arrows large-eddy motion. Right: Vertical profiles of virtual potential temperature θ_v and water-vapor mixing ratio q, based on aircraft data averaged over 30-km flight legs (data). Based on Pennell and LeMone (1974). Terms from Malkus (1958). Not shown is the surface layer, through which wind, temperature, and moisture have large vertical gradients which are well-described by semi-empirical theory (see e.g., Fairall et al. 1996).

weather mixed layer over Florida. The high reflectivities (dark areas) reveal high concentrations of bugs in upward-moving air currents. Figure 7.2a shows horizontal roll vortices, which are associated with stronger winds. Rolls are parallel to the mean mixed-layer wind and typically surmounted by long lines of cumulus clouds. Air trajectories travel in opposing helices whose axes are alongwind, with the upwelling portion of the circulation transporting plumes of warm, moist air upward to form the cloud lines. Typically, the spacing between updrafts and cloud lines is 2-3h. Figure 7.2b shows cellular convection, which is associated with lighter winds. In this case, the vertical mass flux results from convergence in both the along- and cross-wind directions.

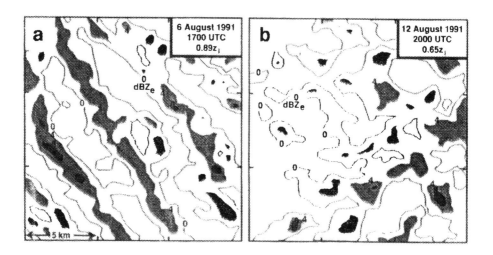

Figure 7.2: Radar-reflectivity patterns in the fair-weather boundary layer. Darker, high-reflectivity areas indicate updraft regions (bugs). (a) Roll vortices ($-h/L = 4.3$, wind speed at h 7.8 ms^{-1}), (b) cellular convection ($-h/L = 13.5$, wind speed at h 2.4 ms^{-1}). Figure from Weckwerth (1995), data from Weckwerth (personal communication).

Mixed-layer roll vortices derive much of their energy from buoyancy, but their alignment and structure are strongly related to the wind profile. Thus some combination of these mechanisms must describe roll dynamics. Several theoretical papers have successfully replicated features of rolls in neutral stratification with an Ekman profile (e.g., Lilly 1966, Faller and Kaylor 1966, Brown 1970) or more complex wind profiles (e.g. Stensrud and Shirer 1988), or have shown that shear-parallel rolls are favored for convection in the presence of vertical shear (e.g., Kuo 1963, Asai 1970). Some have combined dynamical instabilities with stratification (Etling 1971, Brown 1972). Three-dimensional numerical simulations of the Ekman layer (Coleman et al. 1990) and the planetary boundary layer (Mason and Thomson 1987, Moeng and Sullivan 1994) fail to produce rolls in neutral stratification, suggesting that shear and buoyancy mechanisms work together in producing real atmospheric rolls.

Over three decades of observational studies (e.g., Kuettner 1959,1971; LeMone 1973, LeMone and Pennell 1976, Grossman 1982, Brummer 1985, Weckwerth, 1996) and numerical modeling work (e.g., Deardorff 1972, Sykes and Henn 1989) support the theoretical prediction that roll-type convection is associated with stronger shear. A criterion for roll occurrence in the atmosphere takes the form of a 'flux' Richardson number, often expressed as

$$A < -Ri_f = -\frac{h}{L} < B, \tag{7.3}$$

where the Obukhov length $L = -\frac{\overline{T_v}}{g}\frac{u_*^3}{kw\overline{T_v}}$, u_* is the square root of the vertical flux of horizontal momentum, $[(\overline{uw})^2+(\overline{vw})^2]^{\frac{1}{2}}$, $k = 0.4$ is the von Karman constant, and the overbar indicates a horizontal average. All of these quantities are typically based on measurements between the surface and a height of 10 m. Before, I mentioned that the vertical gradients in the Richardson number are taken over the convecting fluid. In this case, although the mixed-layer depth appears in the expression, the relevant shear and thermal gradients are near the surface.[2] The previously-mentioned simulations and Deardorff (1972) suggest $A \sim 3$. Observations of rolls are typically for $A > 0$. Values of B range from 5 to 15 (LeMone, $B = 10$, Grossman, $B=5$, Weckwerth, $B=15$). Based on numerical modeling studies, Sykes and Henn (1989) find $B = 23$. Much larger values of B have been reported, by Kristovich (1993) and Christian and Wakimoto (1989). The higher values of B could reflect evolving boundary-layer flows. For B(h) increasing with time, Weckwerth et al. (1999), found rolls persisted to B~25. However, other mechanisms could produce two-dimensional boundary-layer structures at higher values of B.

7.2.1   Larger aspect-ratio mixed-layer banded structures

Many of the previous authors cite at least one case of banded mixed-layer structures with wavelength-to-height aspect ratios of six or more – much larger than the spatial scales associated with mixed-layer 'turbulence.' Although they seem to have the same type of circulation as rolls, I prefer to call these large-aspect ratio structures 'bands' rather than rolls, because they seem to have a different origin from the rolls discussed in the foregoing. Miura (1986) cites several cases with aspect ratios of six or more, with much higher ratios (up to 18) in the tropics. Figure 7.3 is a horizontal map of the thermodynamic and kinematic structure at $0.25h$, and the overhead cloud structure, for bands of cloudiness separated by 20-30 km. Bands of upwelling air are associated with higher mixing ratios and overhead cloudiness, and their location is consistent with the divergence of the horizontal wind components u and v. Aspect ratios using h reach 50:1, but setting the convective depth to cloud depth ($\sim 2-3h$) reduces the aspect ratio to less than half that.

What produces such structures? For this case and others like it documented in LeMone and Meitin (1984), the roll criterion (7.3) is not always met. Indeed, these bands are so broad that their structure is largely independent of the structure of turbulent eddies that have the aspect ratios of 1.5-3:1 that we associate with mixed-layer eddies. Indeed, the mixed-layer scale eddies on several of the days with banded structures were either random or organized into cells.

An appealing explanation for the banded structures lies in the interaction with boundary-layer convection with the troposphere overhead. Clark et al. (1986) used observations and

---

[2]Kuettner (1959) identifies a second regime for roll occurrence associated with a wind maximum in the middle of the convecting layer. Since the winds in such boundary layers are fairly strong, leading to large stress, these rolls could still satisfy the criterion (7.3).

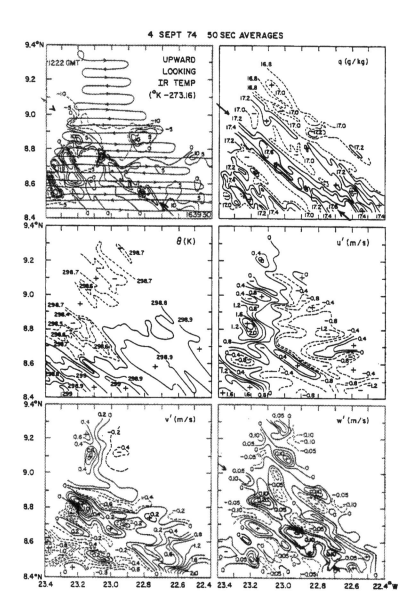

Figure 7.3: Fair-weather banded structures over the tropical ocean. Fields are of mixing ratio (q), potential temperature (θ) and the three components of wind (mean subtracted out), in a right-handed coordinate system with u pointed east. Mixed layer depth $h=600$ m; $0.2^o = 22.2$km. From LeMone and Meitin (1984). Data part of the Global Atmospheric Research Program's Atlantic Tropic Experiment, GATE.

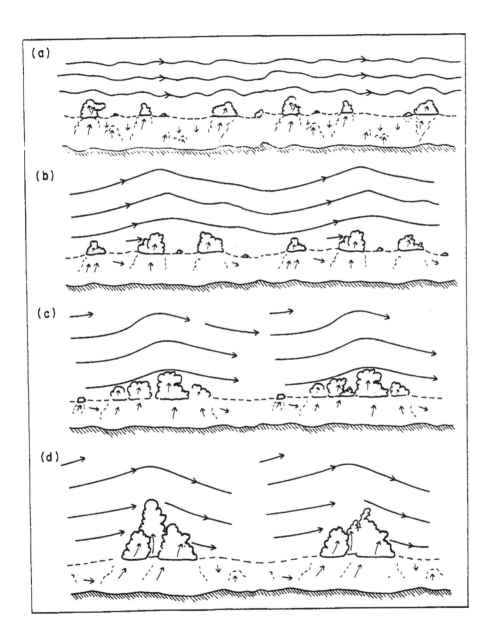

Figure 7.4: Schematic showing how mixed-layer convection induces gravity waves in the stable layer overhead, which select a scale and impose it upon the subcloud layer. After Clark, Hauf, and Kuettner (1986).

numerical simulations to infer the following plausible scenario for producing such bands, depicted in Fig. 7.4. Following the figure, (a) mixed-layer convection (like that depicted in Fig. 7.1) produces overshooting thermals and clouds that perturb the thermally stably stratified air overhead; (b) gravity waves form in the stable layer with a wavelength determined by its wind and thermal stratification; (c) the gravity waves impose their wavelength on the mixed layer beneath; and (d) the 'favored' clouds reinforce the circulation. Aside from the absence of mixed-layer scales and too large values of Ri, there are two other strands of evidence that support the Clark et al. mechanism: (1) The mechanism predicts bands that are normal to the shear across the top of the mixed layer, which matches the orientation for two of the three cases described in LeMone and Meitin (1984), and (2) Balaji et al. (1993) produced bands with the correct orientation and spacing using a numerical model initiated by the environmental sounding for the bands in Fig. 7.4. This does not mean that *all* convective banded structures are produced in this way; a summary of other proposed mechanisms appears in Etling and Brown (1993).

7.3   Precipitating Convection

7.3.1   Buoyancy

The depth scale for a Richardson number describing precipitating convection is not immediately obvious: the depth of the troposphere might seem an obvious choice, but convection frequently falls short of this depth. One good predictor of convection depth is the *equilibrium level*, the maximum height at which an air parcel lifted from near the surface has the same density as the surrounding air (Fig. 7.5). In practice, this parcel has the average properties of the lowest 500 m or so of the atmosphere (roughly the depth of the mixed layer discussed in the foregoing). The parcel is assumed to undergo a pseudoadiabatic process – cooling initially at 0.0098 Km^{-1} due to work expanding against the surrounding air, and then cooling more slowly (typically about 0.006 Km^{-1}) due to condensational heating after water-vapor saturation is reached. The area between the parcel temperature curve and the environment is a good approximation of the *Convective Available Potential Energy*, or CAPE, which often forms the numerator of the Richardson number used to define regimes of precipitating convection. More exact computations use virtual temperature, and an estimated mass of liquid water carried with the parcel is sometimes taken into account. Large values of CAPE are associated with severe thunderstorms.

Figure 7.5 shows temperature and dew point profiles for two soundings, one associated with a strong Midwestern thunderstorm (from Weisman and Klemp, 1982) and one is associated with deep convection over the tropical ocean. Observations (for 18 February) and numerical simulations (and comparison to observations in qualitatively similar environments) in both cases indicate strong vertical velocities (of the order 16-18 ms^{-1} for this tropical case, with up to 25 ms^{-1} for a different day with similar CAPE (Hildebrand et al 1996), and 20 ms^{-1} for the weakest-shear case presented in Weisman and Klemp 1984[3]. However, the *vertical distribution* of CAPE is quite different for these two cases – which leads to very different lower-level vertical velocities for the two types of soundings (Fig. 7.6). This is not a surprise if we follow an air parcel from the mixed layer. Recall the buoyancy of a parcel is roughly proportional to the difference between its temperature and that of the environment. Its maximum potential vertical velocity is the integral of the acceleration (buoyancy) felt rising

---

[3]Weisman and Klemp (1984) show how dynamic pressure effects become relatively more important in stronger shears to produce much stronger vertical velocities – up to 40 ms^{-1}

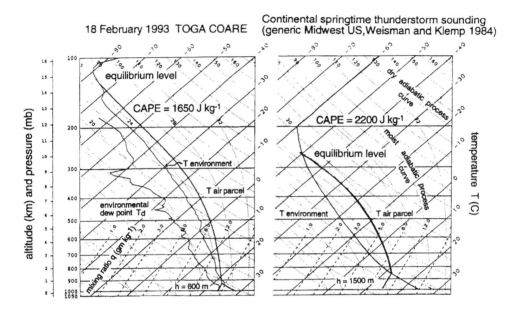

Figure 7.5: Thermodynamic profiles in the vicinity of (a) an observed band of convection over the Tropical Pacific, (b) modeled squall lines (an idealization of typical soundings over the Midwestern U.S.). The heavy line on both soundings shows the temperature of an undiluted air parcel as it rises through the atmosphere, under the influence of adiabatic cooling compensated partially by condensation of liquid water (the so-called pseudo-adiabatic process). Below the condensation level, the temperature falls dry-adiabatically. Both temperature and dew-point temperature are plotted in the sounding on the left; temperature only plotted on right. The area between the temperature curves is proportional to the convective available potential energy, CAPE.

from the mixed layer – roughly the area between the parcel and environment temperature curves below its altitude. At a pressure of 500 millibars (height $\sim$6 km), the integrated buoyancy available to the tropical parcel is less than half that available to the parcel in the Midwestern U.S. sounding. The vertical velocity at 500 mb is consequently much less for the tropical case (Fig. 7.6).[4] This has important implications. Since $T \sim 273K$ at 500 mb, strong vertical velocities at this level transport supercooled water to higher altitudes, and support growing graupel (small hail) – a mix that, with ice particles, electrifies the cloud and leads to lightning. It has been found that while storms over the tropics reach heights that can exceed 18 km, [5] where they spread out to produce areas of cloud hundreds of km in diameter, they produce lightning far less often than their counterparts over the continents (e.g. Zipser 1994).

7.3.2   Shear

The vertical shear of the horizontal wind plays an important role in the organization and evolution of atmospheric precipitating convection. Figure 7.7 shows the dissipating stage of a cloud growing in a low-shear environment. As the cloud forms, warm, moist air feeding the cloud moves straight upward. Cloud, and then rain, begin forming as the rising air cools enough for condensation to begin. Eventually, enough rain and cloud forms to neutralize the buoyancy of the rising updraft, and the rising air slows down, and then begins to sink. Once it sinks, evaporation of the rainfall contributes to relative cooling (sinking air with evaporation warms at a rate of about 0.006 K m^{-1}) and the cloud dies. The cool, rain-laden downdraft air spreads out at the surface as a density current, and can form a circle of smaller clouds around the dying mother cloud. Such a cloud pattern appears along with the environmental wind profile in Fig. 7.8.

When stronger shear occurs, two things happen. First, the updraft tilts with height. This enables rain to fall out of the updraft, so that the rain-produced downdraft is displaced horizontally. And secondly, where the horizontal vorticity of shear in the lowest few km of the environment opposes that of the shear produced in the downdraft-produced density current, convergence is strong enough and deep enough to produce strong convection (Rotunno et al. 1988, Fig. 7.9). In such a case, strong 'daughter' convection forms on the downshear side of the mother cloud's density current. The daughter clouds in turn form rain-cooled downdrafts, which merge into a common density current, and a new generation of clouds is generated along the downshear edge of the cold air, forming a band of convection, as depicted in Weisman and Klemp (1984). Because succeeding generations of clouds preferentially form on the downshear side of the storm-generated cold air, the resulting convective band or 'squall line' is normal to the low-level shear, and travels downshear. Fig. 7.10 shows the wind profile and radar pattern associated with such a system, and Fig. 7.11 is a schematic of the airflow normal to the convective band.

The structural differences between the convection in Figs. 7.8 and 7.10 have profound effects on the vertical transport of horizontal momentum. In Fig. 7.11, the main downward-moving ($w < 0$) air current moves forward ($u_s > 0$) leading to $u_s w < 0$. Similarly, the rearward-moving updraft produces $u_s w < 0$. Thus the typical momentum flux profile $-\overline{\rho}\,\overline{u_s w}$ is negative throughout storm depth, with an extremum about halfway up in convective-line

---

[4]Lucas et al. (1995) also note that the smaller diameter of tropical updrafts, which is linked to the eddy size in the boundary layer in which they occur, may also play a role in relatively weaker updrafts over the tropical oceans, since smaller eddies are less 'shielded' from mixing. The depth h of daytime mixed layers over land are 2-3 times typical fair-weather oceanic values.

[5]A severe thunderstorm in mid-latitudes might reach heights of 12-15 km

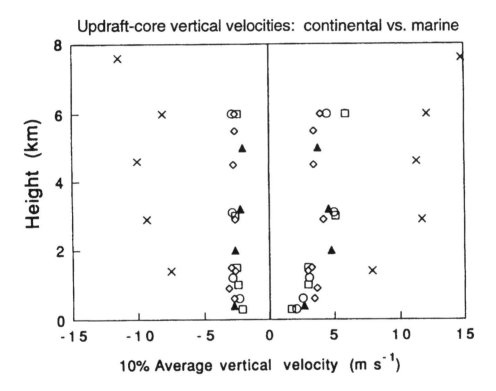

Figure 7.6: Comparison of updraft magnitudes over the tropical oceans derived from penetrating aircraft. The triangle, circle, diamond, and square symbols denote data gathered over the tropical ocean, the × symbol denotes data for continental thunderstorms. 10%-cores are stronger than 90% of the population. From Lucas et al. (1995).

NO THUNDERSTORM PROPAGATION
(low-shear case)

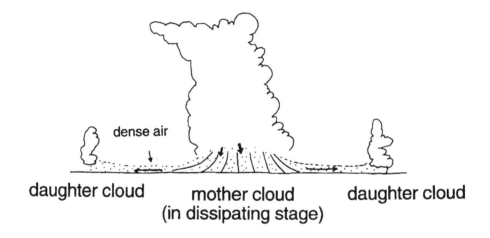

Figure 7.7: Schematic of a dissipating convective cloud. The downdraft, laden by precipitation and cooled by evaporation, has choked off the updraft feeding the cloud. The downdraft spreads out pushing up surrounding air and forming more clouds.

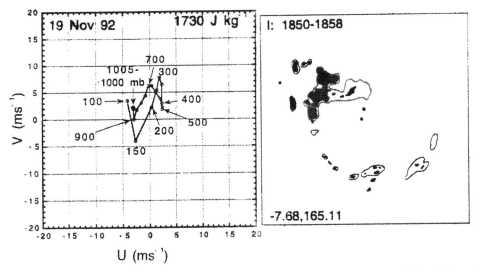

Figure 7.8: Wind profile (left) and precipitation pattern (right) for a ring of clouds formed by the spreading cold pool of a mother cloud (now dissipated). For the wind hodograph at left, the numbers indicate the pressure in millibars (hPa); the surface pressure is 1005 mb. Number at upper right is the CAPE, numbers in right frame indicate times (UTC) and the latitude and longitude of the lower left corner. Right frame is 240×240km. Data from lower-fuselage radar of the NOAA43RF P3 aircraft. Date from the Tropical Ocean-Global Atmosphere's Coupled Ocean-Atmosphere Response Experiment, TOGA COARE.

THUNDERSTORM PROPAGATION

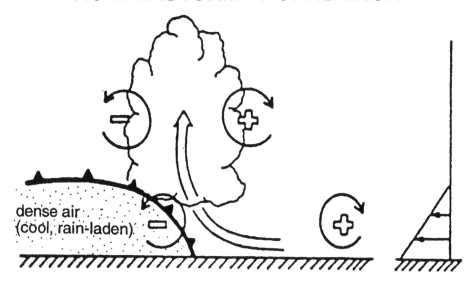

Figure 7.9: Schematic showing how the vorticity associated with environmental shear and the shear generated within a storm-generated cold outflow balance to insure a deep, strong vertical updraft. In the absence of such environmental shear, the cold pool often collapses as in Fig. 7.7, forming only small clouds. After Rotunno et al. 1988.

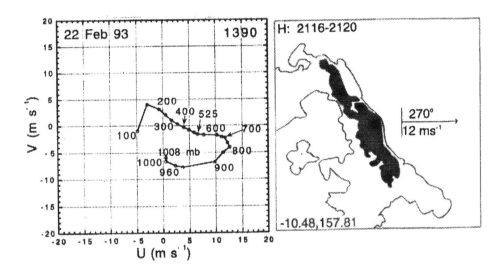

Figure 7.10: As in Fig. 7.8, but for a strong band of convection (squall line) in the presence of strong low-level shear, moving eastward (from 270°) at 12 ms⁻¹. Data from lower-fuselage radar of the NOAA42RF P3 aircraft.

relative coordinates. Since the alignment of these systems is determined by the low-level shear, the energy transport to larger scales ($\overline{uw}(\partial U/\partial z)$) at higher altitudes can be of either sign. As long as these systems remain quasi-two dimensional, their momentum transport is well-represented by an idealized three-branch dynamically-consistent flow model described in Moncrieff (1992), according to comparisons to observations in LeMone and Moncrieff (1994). In contrast, the vertical transport of horizontal momentum by cloud systems like that in Figs. 7.7 and 7.8 act to vertically mix horizontal momentum and extract mechanical energy from the larger scale. This transport is well-represented by allowing vertically-moving air parcels to preserve their horizontal momentum as described in Schneider and Lindzen (1976).

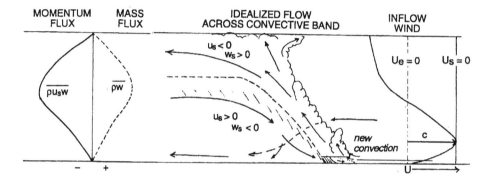

Figure 7.11: Schematic of the flow in the direction normal to convective bands like that of Fig. 7.10. Broad flow branches are separated by dashed lines; the wind profile normal to the system appears at the right. u is the band-normal wind, positive in the direction of the band motion. u_s is the relative to the band motion, u_e is relative to the earth. Note that the vertical flux of the horizontal momentum u is countergradient above the lower-level wind maximum, indicating that the momentum flux divergence is increasing the vertical shear of the horizontal wind.

It follows that a Richardson number defined to separate these regimes of convection should use shear differences in the lower part of the troposphere. Weisman and Klemp (1982), for example, found that single-celled storms were separated from those that eventually formed squall lines according to a Richardson number defined by:

$$Ri = \frac{CAPE}{\frac{1}{2}(\Delta U^2 + \Delta V^2)} \tag{7.4}$$

where ΔU and ΔV are the difference in the components of horizontal momentum averaged between 0 and 6 km, and their average in the lowest 500 m. For comparison, the equilibrium levels for both soundings in Fig. 7.5 are greater than 10 km. According to Weisman and Klemp (1982) unorganized or single-celled storms tend to occur for Ri~1000, tropical squall lines at 50-500, and 'supercells' which retain their identity for up to several hours by continuously regenerating rather than forming new cells, form at Ri~10-100. Intense supercells are are associated with severe weather, including tornadoes and large hail.

Although the low-level shear is clearly important in determining convection type, the shear at higher altitudes can also be important. For example, there are convective bands that transport momentum as in Fig. 7.11 whose alignment seems to be determined by the shear across the low-to middle troposphere (e.g. Zipser et al. 1981, Barnes and Sieckman 1984, Alexander and Young 1992). These tend to occur when the vertical shear of the horizontal wind is weak in the lowest 2 km but strong between 2 and 6-7 km (e.g. Alexander and Young 1992). We suspect some of these systems somehow locally generate 'favorable' shear so that the Rotunno et al. (1988) criterion can be satisfied for continued growth via generation of new convection. Furthermore, even for the systems whose alignment is determined by the low-level shear, observations and numerical simulations suggest that the wind shear at higher altitudes plays an important part in later evolution (Trier et al. 1996, 1997). For this reason, LeMone and Moncrieff (1994) actually normalized the wind profile using $(CAPE)^{\frac{1}{2}}$ rather than define a depth-scale for shear and computing a Richardson number in comparing behavior of linear convection in different wind-shear environments.

7.4 Conclusions

Three distinct examples of convection in Earth's atmosphere are discussed to illustrate the relative role of buoyancy and shear in determining convective structure. For all three, buoyancy is an important source of energy, but the shear seems to determine the vertical structure. Larger shear generally leads to linear organization, the exception mentioned here being supercells. Larger energy input from buoyancy tends to destroy organization. The organization is shear-parallel in non-precipitating flows driven solely by convective processes – as in the case of rolls in the convective mixed layer. At least some large aspect-ratio boundary layer 'rolls' appear to have their alignment determined by the orientation of gravity waves generated in the statically-stable air above the convective layer. In precipitating convection, the density current formed by precipitation-induced downdrafts plays a crucial role. Squall lines form normal to the low-level shear and propagate downshear, where the interaction between the density current and the ambient shear creates the best conditions for growth of new deep convection. When the low-level shear is weak, convective bands seem to align with shear at middle levels – between about 2 and 6-7 km. I have described some forms of the Richardson number than have predictive value for specific types of convection.

Bibliography

Alexander, G.D., and Young, G.S.: "The relationship between EMEX mesoscale precipitation feature properties and their environmental characteristics." *Mon. Weath. Rev.*, **120**, 554-564 (1992).

Anderson, R.K., et al.: "The use of satellite pictures in weather analysis and forecasting." *WMO Tech Note No. 75*. Geneva (1996).

Asai, T.: "Stability of a plane parallel flow with variable vertical shear and unstable stratification." *J. Meteor. Soc. Japan*, **48**, 129-139 (1970).

Balaji, V., Redelsperger, J.-L., and Klaasen, G.P.: "Mechanisms for the mesoscale organization of tropical cloud clusters in GATE Phase III. Part I: Shallow cloud bands." *J. Atmos. Sci.*, **50**, 3571-3589 (1993).

Barnes, G.M., and Sieckman, K.: "Mean inflow normal to fast and slow tropical mesoscale convective cloud lines." *Mon. Weath. Rev.*, **112**, 1782-1794 (1984).

Brown, R.A.: "A secondary flow model for the planetary boundary layer." *J. Atmos. Sci.*, **27**, 742-757 (1970).

Brown, R.A.: "On the inflection point instability of a stratified Ekman boundary layer." *J. Atmos. Sci.*, **29**, 850-859 (1972).

Brummer, B.: "Structure, dynamics, and energetics of boundary layer rolls from KonTur aircraft observations." *Beitr. Phys. Atmosph.*, **58**, 237-254 (1985).

Chandra, K.: "Instability of a fluid heated from below." *Proc. Roy. Soc. A.*, **164**, 231-242 (1983).

Christian, T.W., and Wakimoto, R.M.: "The relationship between radar reflectivities and clouds associated with horizontal roll convection on 8 August 1982." *Mon. Weath. Rev.*, **117**, 1530-1544 (1989).

Clark, T.L., Hauf, T. and Kuettner, J.P.: "Convective-forced internal gravity waves: results from two-dimensional experiments." *Quart. J. Roy. Met. Soc.*, **112**, 899-926 (1986).

Coleman, G.N., Ferzinger, J.H., and Spalart, P.R.: "A numerical study of the turbulent Ekman layer." *J. Fluid Mech.*, **213**, 313-348 (1990).

Deardorff, J.W.: "Numerical investigation of neutral and unstable boundary layers." *J. Atmos. Sci.*, **29**, 91-115 (1972).

Etling, D., and Brown, R.A.: "Roll vortices in the planetary boundary layer: A review." *Boundary-Layer Meteor.*, **65**, 215-248 (1993).

Fairall, C.W., Bradley, E.F., Rogers, D.P., Edson, J.B., and Young, G.S.: "Bulk parameterization of air-sea fluxes for the Tropical Ocean-Global Atmosphere Coupled Ocean-Atmosphere Response Experiment." *J. Geophys. Res.*, **101**, 3747-3765 (1996).

Faller, A.J., and Kaylor, R.E.: "A numerical study of the laminar Ekman layer." *J. Atmos. Sci.,*, **23**, 466-480 (1966).

Fitzjarrald, D.R. and Garstang, M.: "Vertical structure of the tropical boundary layer." *Mon. Weath. Rev.*, **109**, 1512-1526 (1981).

Grossman, R.L.: "An analysis of vertical velocity spectra obtained in the BOMEX fair-weather trade-wind boundary layer." *Boundary-Layer Meteor.*, **23**, 323-357 (1982).

Hildebrand, P.H., Lee, W.-C., Walther, C.A., Frush, C., Randall, M., Loew, E., Neitzel, R., Parsons, R., Testud, J., Baudin, F., and LeCornec, A.: "The ELDORA/ASTRAIA airborne Doppler weather radar: high-resolution observations from TOGA-COARE." *Bull. Amer. Meteor. Soc.*, **77**, 213-232 (1996).

Konrad, T.G.: "The alignment of clear air convective cells." *Proc. Intl. Conf. Cloud Phys.*, Toronto, Canada, 539-543 (1968).

Kristovich, D.A.R.: "Mean circulations of boundary-layer rolls in lake-effect snow storms." *Boundary-Layer Meteor.*, **63**, 293-315 (1993).

Kuettner, J.P.: "The band structure of the atmosphere" *Tellus*, **23**, 267-294 (1959).

Kuettner, J.P.: "Cloud bands in the atmosphere." *Tellus*, **23**, 404-425 (1971).

Kuo, H.: "Perturbations of plane Couette flow and the origins of cloud streets." *Phys. Fluids*, **6**, 195-201 (1963).

LeMone, M.A.: "The structure and dynamics of horizontal roll vortices in the planetary boundary layer." *J. Atmos. Sci.*, **30**, 1077-1091 (1973).

LeMone, M.A., and Pennell, W.T.: "The relationship of trade wind cumulus distribution to subcloud layer fluxes and structure." *Mon. Weath. Rev.*, **101**, 524-539 (1976).

LeMone, M. A., and Meitin, R. J.: "Three examples of fair-weather mesoscale boundary layer convection in the tropics." *Mon. Weather Rev.*, **112**, 1985-1997 (1984).

LeMone, M.A., and Moncrieff, M.W.: "Momentum and mass transport by convective bands: comparisons of highly idealized dynamical models to observations." *J. Atmos. Sci.*, **51**, 281-305 (1994).

Lilly, D.K: "On the instability of Ekman layer flow." *J. Atmos. Sci.*, **23**, 481-94 (1966).

Lucas, C., Zipser, E.J., and LeMone, M.A.: "Vertical velocity in oceanic convection off tropical Australia." *J.Atmos. Sci.*, **51**, 3183-3193 (1994a).

Lucas, C., Zipser, E.J., and LeMone, M.A.: "Convective available potential energy in the environment of oceanic and continental clouds." *J. Atmos. Sci.*, **51**, 3829-3830 (1994b).

Lucas, C., Zipser, E.J., and LeMone, M.A.: Reply to comments on "Convective available potential energy in the environment of oceanic and continental clouds." *J. Atmos. Sci.*, **53**, 1212-1214 (1996).

Malkus, J.S.: "On the structure of the trade wind moist layer." *Pap. Phys. Oceanogr. Meteor.*, No. 13. MIT-WHOI, 47 pp (1958).

Malkus, J.S., and Riehl, H.: *Cloud structure and distributions over the tropical Pacific Ocean.*, Univ. of Calif. Press., Berkeley and Los Angeles. 229 pp (1964).

Mason, P.J, and Thomson, D.J.: "Large-eddy simulations of the neutral-static stability planetary boundary layer." *Quart. J. Roy. Meteor. Soc.*, **113**, 801-823 (1987).

Miura, Y.: "Aspect ratios of longitudinal rolls and convection cells observed during cold air outbreaks." *J. Atmos. Sci.*, **43**, 26-39 (1986).

Moeng, C.-H., and Sullivan, P.R.: "A comparison of shear- and buoyancy-driven planetary boundary layer flows." *J. Atmos. Sci.*, **51**, 999-1022 (1994).

Moncrieff, M.W.: "Organised mesoscale convective systems: archetypal dynamical models, mass and momentum flux, theory and parametrisation." *Quart. J. Roy. Meteor. Soc.*, **118**, 819-850 (1992).

Nicholls, S., and LeMone, M.A.: "The fair weather boundary layer in GATE: The relationship of subcloud fluxes and structure to the dis- tribution and enhancement of cumulus clouds." *J. Atmos. Sci.*, **37**, 2051-2067 (1980).

Pennell, W.T., and LeMone, M.A.: "An experimental study of turbulence structure in the fair-weather trade wind boundary layer." *J. Atmos. Sci.*, **31**, 1308-1323 (1974).

Rotunno, R., Klemp, J.B., and Weisman, M.L.: "A theory for strong, long-lived squall lines." *J. Atmos. Sci.*, **45**, 463-3485 (1988).

Schneider, E.K, and Lindzen, R.S.: "A discussion of the parameterization of momentum exchange by cumulus convection." *J. Geophys. Res.*, **81**, 3158-3160 (1976).

Stensrud, D.J., and Shirer, H.N: "Development of boundary layer rolls from dynamic instabilities." *J. Atmos. Sci.*, **45**, 1007-1019 (1988).

Sykes, R.I, and Henn, D.S.: "Large-eddy simulation of turbulent sheared convection." *J. Atmos. Sci.*, **46**, 1106-1118 (1989).

Trier, S.B., Skamarock, W.C., LeMone, M.A., Parsons, D.B., and Jorgensen, D.P.: "Structure and evolution of the 22 February 1993 TOGA-COARE squall line: Numerical simulations." *J. Atmos. Sci.*, **53**, 2861-2886 (1996).

Trier, S.B., Skamarock, W.C., and LeMone, M.A.: "Structure and evolution of the 22 February 1993 TOGA-COARE squall line: Organization mechanisms inferred from numerical simulation." *J. Atmos. Sci.*, in press (1997).

Weckwerth, T.M.: *A Study of Horizontal Convective Rolls Occurring within Clear-Air Convective Boundary Layers.* NCAR Cooperative Thesis No. 160, UCLA/NCAR, 179 pp (1995).

Weckwerth, T.M., Horst, T.W. and Wilson, J.W.: *An observational study of the evolution of horizontal convective rolls.* Mon. Wea. Rev., **56**, in press (1999).

Weisman, M.L., and Klemp, J.B.: "The dependence of numerically simulated convective storms on vertical wind shear and buoyancy." *Mon. Weath. Rev.*, **110**, 504-520 (1982).

Weisman, M.L., and Klemp, J.B.: "The structure and classification of numerically simulated convective storms in directionally varying wind shears." *Mon. Weath. Rev.*, **112**, 2479-2498 (1984).

Woodcock, A.: "Soaring over the open sea." *The Scientific Monthly*, **25**, 226-232 (1941).

Zipser, E.J.: "Deep cumulonimbus cloud systems in the tropics with and without lightning." *Mon. Weath. Rev.*, **122**, 1837-1851 (1994).

Zipser, E.J., Meitin, R.J., and LeMone, M.A.: "Mesoscale motion fields associated with a slowly moving GATE convective band." *J. Atmos. Sci.*, **38**, 1725-1750 (1981).

Chapter 8

IMAGES OF HARD TURBULENCE: BUOYANT PLUMES IN A CROSSWIND

ANDREW BELMONTE[1] and ALBERT LIBCHABER[2]

[1] *Institut Non-Linéaire de Nice*
1361 route des Lucioles
06560 Valbonne, France
[2] *Center for Studies in Physics and Biology*
The Rockefeller University
1230 York Avenue, New York, NY 10021

We present a visualization study of turbulent convection in gas under pressure for Rayleigh numbers (Ra) from 5×10^5 to 1×10^{11}, with Prandtl number $Pr = 0.7$. Measurement of the thermal boundary layer thickness indicates that it scales as $Ra^{-2/7}$ for $Ra > 2 \times 10^7$. Here we present shadowgraph images of the flow near the boundary layers for a cubic cell with $L = 17$ cm. We study the plumes and other structures which form above the thermal layer and are advected horizontally in the large scale circulation which persists in the cell. Measurements of the two-point intensity correlation function of these images provide quantitative information: the average size and penetration length of the plumes into the bulk of the cell (the development height z_c). The dependence of these new lengths on Ra indicates that the high Reynolds shear of the large scale circulation is modifying the dynamics near the plates.

8.1 Introduction

It is a common aspect of geophysical and astrophysical convection that the plumes or other buoyant structures at the boundaries rise into regions where strong winds or vortices transport them horizontally. This is a natural result of other intervening physical forces such as rotational effects, waves, or larger scale flow systems; the reader will find ample illustrations in this volume, especially contributions by Kerr, Grabowski, and LeMone. The overall convective transport can thus be either free (buoyancy-dominated) or forced (wind-dominated), or, most likely, somewhere in between. It is somewhat surprising, however, that in the laboratory one finds the same competition of buoyancy and mean flow in a closed stationary box of fluid heated from below.

165

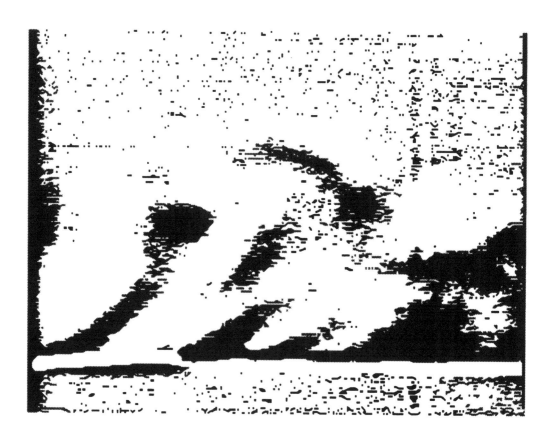

Figure 8.1: Shadowgraph image (1.5 cm across) of a rising plume, at $Ra = 7 \times 10^8$.

We are referring here to the hard turbulent state of Rayleigh-Bénard convection, which is observed in gases for $Ra > 4 \times 10^7$. This state was first observed in helium gas at 5 K for Ra up to 10^{14} (Heslot $et\ al.$, 1987; Castaing $et\ al.$, 1989; Sano $et\ al.$, 1989; Wu and Libchaber, 1992), and similar states have now also been seen in other experiments (Xin $et\ al.$, 1996 ; Takeshita $et\ al.$, 1996), and numerical simulations (Werne, 1993; Kerr, 1996). Hard turbulence is characterized by a number of different aspects: the thermal boundary layer thickness which scales as $Ra^{-2/7}$ (or equivalently the heat flux or Nusselt number $Nu \sim Ra^{2/7}$), a persistent large scale circulation in the cell with a speed $U \sim Ra^{1/2}$, non-Gaussian histograms of temperature fluctuations, and a power law in the temperature spectrum. Despite the large number of theoretical and experimental studies, there is to date no unequivocal understanding of hard turbulence.

It may be that no direct analogue of this scaling state exists in stellar or atmospheric convection. On the other hand, experiments on side heated convection (Belmonte $et\ al.$, 1995), and numerical simulation of rotating convection (Julien $et\ al.$, 1996), suggest that the 2/7 scaling of hard turbulence is more generic. Putting this question aside for the moment, we focus on the interplay of the buoyant instabilities at the thermal boundary layer with the crosswind of the large scale convection. What does the flow look like in the region near the plate? What are the coherent structures at high Ra? What are their characteristics and dynamics? Ultimately, what is the role of plumes in average "thermodynamic" quantities of the convective state, for example the heat flux and large scale circulation?

In most fluids, the fact that the index of refraction is temperature dependent allows visual images of the flow to be obtained by nonintrusive optical techniques such as shadowgraph and schlieren. Here we present a study of turbulent convection in room temperature gas performed at the Physics Department of Princeton University (Belmonte, 1994). We combine the shadowgraph technique with modern methods of video image processing to study the objects which leave the thermal boundary layer and are swept along the plate by the coherent large scale circulation (an example is shown in Fig. 8.1). This complements our previous measurements of thermal and velocity boundary layers in water and gas (Tilgner $et\ al.$, 1993, Belmonte $et\ al.$, 1994).

8.2   Hard Turbulence

The earliest prediction for the scaling of the thermal boundary layer thickness λ_{th} in turbulent convection (Malkus, 1954; Priestley, 1954; Howard, 1963) can be summarized by the statement that the Rayleigh number of the thermal boundary layer ($Ra_\lambda \sim \lambda_{th}^3$) is always at a constant critical value. This implies that λ_{th} does not depend on the height of the cell L. The assumption $Ra_\lambda \simeq 1000$ leads to the scaling $Nu = 0.05 \times Ra^{1/3}$, known as the classical model. But the scaling observed in hard turbulence is $Nu = 0.2 \times Ra^{2/7}$. This implies that the thermal layer $does$ depend on the height of the cell. Note that although the heat flux in hard turbulence increases more gradually with Ra than the classical model predicts, the actual magnitude of the heat flux is larger. It is generally suspected that these differences are due to the coherent large scale circulation (which is an eddy of size L), and a model taking this into account derives the observed scaling exponents for the heat flux and velocity (Shraiman and Siggia, 1990; Siggia, 1994). This model, however, neglects plumes entirely, and assumes a constant shear rate across the plates. Recently, a more general version which includes the plume contribution and a position-dependent shear has been made by Ching (1996); the same scaling exponents are derived, dependent on the third derivative

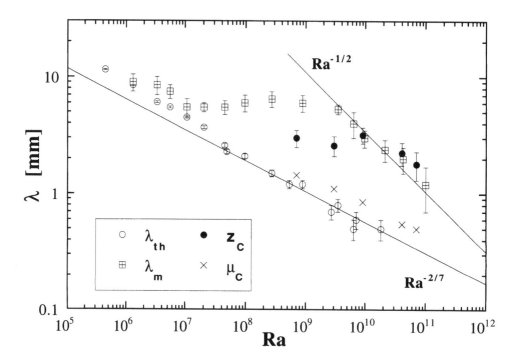

Figure 8.2: Length scales of turbulent convection. The thermal boundary layer thickness λ_{th}, the position of maximum cutoff frequency λ_m, the development height z_c, and the bulk size μ_c are plotted vs. the Rayleigh number Ra.

of the temperature profile at the plates. Other scaling models have also been proposed (see contribution by Zaleski).

The first experiments on hard turbulence used helium gas at 5 K to change Ra over many orders of magnitude, while using the same cell (Threlfall, 1975). However, the use of moving probes and flow visualization is difficult under cryogenic conditions, so we chose to perform our experiments at room temperature. Part of the motivation of these visualization studies is to look for further evidence of a transition observed at $Ra \simeq 2 \times 10^9$ (Belmonte et al., 1994). We measured the thermal layer thickness λ_{th} with moveable thermistors, and found that $\lambda_{th} \sim Ra^{-2/7}$. We also measured the highest frequency in the temperature power spectrum, defined by the cutoff frequency f_c, which we found to vary with distance from the plate. We observed that f_c reaches a maximum value at a distance λ_m larger than λ_{th}, as shown in Fig. 8.2. For Ra from 2×10^7 to 2×10^9, this length has a fixed value of $\lambda_m/L \simeq 0.039$. It decreases from this value to 0.007 between Ra from 2×10^9 to 1×10^{11}. What is the nature of this transition? If it indeed corresponds to the Reynolds number of the flow along the plate reaching a turbulent value (Belmonte et al., 1994), one should see a change in the mixing. By using the two point correlation function of the intensity in our images, we measure a characteristic size for the thermal objects. The dependence of this average size on height appears to change when it intersects λ_m, indicating that the large scale circulation is affecting the mixing of temperature in the cell.

The questions thus raised by the observations of hard turbulence in the laboratory are closely related to similar questions in atmospheric convection. Among these are the effects of mean circulation due to horizontal convection rolls on the planetary thermal boundary layer (LeMone, 1976), and the estimation of buoyant and velocity scales for the case of moist convection (Emanuel and Bister, 1996). Simple models of the atmosphere also rely on the same basic physics faced in the lab (see for example Rennó and Ingersoll, 1996). Among all other complicating factors, the most fundamental difference between laboratory and atmospheric convection is that the typical lengths over which gravity stratifies a gas are much larger than any laboratory cell; the scale height plays no role in the experiments reported here. Nonetheless, an increase in dialogue between the two disciplines, such as this conferences entails, should benefit all concerned.

8.3   Experimental Techniques

8.3.1   The Convection Cell

Our visual studies were performed in the same pressure vessel as the temperature measurements (Belmonte et al., 1994). We use a cubic cell with a height $L = 17.0$ cm; the top plate (brass) is held at a constant temperature by a refrigerated water circulation system, and the bottom plate is heated, thus imposing a temperature difference Δ which is measured by thermistors embedded in the plates. The convection cell reaches a steady state for which Δ is approximately constant, with fluctuations of less than 1%. To permit visual observation, a clear path between the cell and the endcap windows is established through the cotton batting surrounding the cell (Belmonte, 1994).

To scan Ra at room temperature we change the density of the gas, varying the pressure between 0.6 and 18 atmospheres for three different gases: helium, nitrogen, and sulfur hexafluoride (SF$_6$). *For each gas we stay far from the critical point.* The resulting span of Rayleigh number is almost 6 decades, from 5×10^5 to 1×10^{11}, while the Prandtl number remains within 5% of 0.7. Experimental values of Ra are calculated using the pressure and temperature dependence of the material properties of each gas (Belmonte, 1996) evaluated at the midway temperature between the plates. Because these properties are not strongly dependent on temperature, the precision in Ra is about 10%. Our experiments are therefore within the Boussinesq approximation. We measure only in the upper half of the cell, while verifying that in the bulk that the mean temperature is halfway between the top and bottom plate temperatures (Wu and Libchaber, 1991).

8.3.2   Visualization

To obtain images of the turbulent flow in our convection cell, we use the standard shadow-graph technique (Merzkirch, 1987); this is the phenomenon responsible for the shimmering one often sees above some objects on a hot day. A parallel beam of light passing through the fluid is focused or defocused by differences in the index of refraction. In our case these differences come from temperature fluctuations of the fluid. For gases, the index of refraction n has a simple dependence on the density, known as the Gladstone-Dale relation:

$$n - 1 = K\rho. \tag{8.1}$$

The value of K is about $0.1 - 0.2$ for the gases in our experiment, but as the density of SF_6 is the largest, it produces the clearest shadowgraph image (Belmonte, 1996). The intensity variations in the shadowgraph image are caused by the second spatial derivative of the temperature: a constant gradient in index would deflect light uniformly, but a varying gradient can focus the light. Though extracting quantitative information about the temperature is difficult from such a technique, for our purpose the shadowgraph provides a convenient way to visually follow thermal objects in the flow.

One of the aspects of a shadowgraph image is that it is an integration along the beam direction; there is no focal plane. Although this is often cited as a disadvantage, this collapse of the third dimension is actually advantageous to our experiment. The large scale circulation in a cubic cell is along the diagonal (Zocchi et al., 1990; Zocchi, 1990), and any focal plane would be parallel to the walls, in which case the objects advected by the flow would only be in focus for a short time. We did attempt to define such a focal plane using a technique called focused schlieren (Kantrowitz and Trimpi, 1950; Mortensen, 1950; Weinstein, 1993). However, the schlieren technique, which measures the first spatial derivative of the temperature (Merzkirch, 1987), was not sensitive enough for our experiment with gas.

Our optical setup consists of two main parts. The illuminator provides a collimated beam of monochromatic light which passes through the cell; on the other side of the cell, the imaging system projects the shadowgraph image into the camera, from which each image is computer processed. A diagram is shown in Fig. 8.3. The illuminator provides uniform lighting of the boundary layer region using a low intensity sodium vapor lamp as a light source; our design is based on Kohler illumination (see Lacey, 1989). The collecting lens L1 ($f_1 = 4.5$ cm) projects an image of the source onto the aperture A1 (0.6 cm diameter). The condenser L2 ($f_2 = 5.0$ cm) reproduces this image, reducing it by about a factor of 5, onto a $50\mu m$ pinhole. The pinhole is located at the focal point of L3 ($f_3 = 10.0$ cm), so that a beam of parallel light exits the objective aperture A2, with a diameter of about 2.5 cm.

Once the beam has passed through the convection cell, it enters the lens L4 ($f_3 = 12.7$cm), the compound magnifying lens L5, and the CCD camera (NEC TI-24A). The compound lens is an old photocopying lens, with one focal point 8.8 cm from its front face (towards the cell), and the other one 13.1 cm from its back face (towards the camera). The total numerical

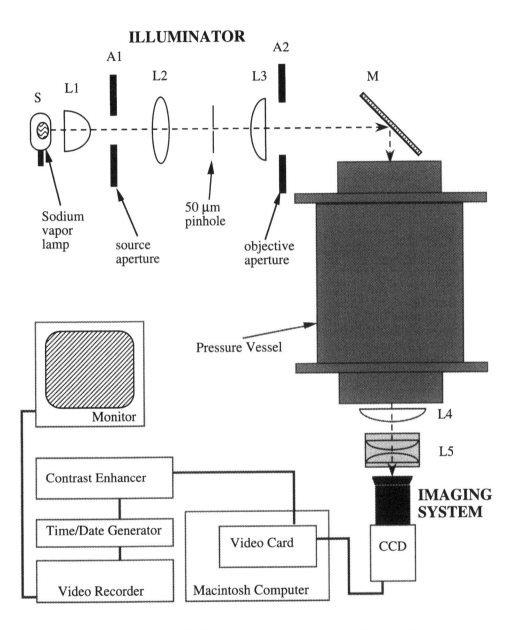

Figure 8.3: Shadowgraph visualization setup. Source S is imaged by lens L1 onto aperture A1, then reduced by lens L2 onto a pinhole which sits at the focal point of lens L3. The beam is stopped down by aperture A2, reflected by mirror M through the pressure vessel and cell, and imaged by lenses L4 and L5 into the CCD camera. The video signal path is described in Section 8.3.3.

aperture of this imaging system (image size / object size) ranges from 0.9 to 1.6, thus the boundary layer is essentially mapped one-to-one onto the camera element.

Our visualization technique is limited to Ra high enough that the shadowgraph is visible; for SF_6 this means in practice $Ra > 10^8$. We have made many hours of videotape from our shadowgraph visualization, for Ra ranging from 3×10^8 to 7×10^{10}, which encompasses the transition at $Ra \simeq 10^9$.

8.3.3   Image Processing

The essential aspect of our image processing system is background subtraction. Working with gas under pressure requires that the plexiglas windows in the pressure vessel be 2" thick. These windows are the main source of noise in the shadowgraph images. Because this noise does not change with time, a long enough time average of the images contains only the noise (the "background"); the structures in the flow will average out. To "background subtract" our images, we first make a time average and store it in the memory of the video processing hardware in our computer. We use a video frame grabber (PixelPipeline, Perceptics), installed in a Macintosh computer. This hardware allows us to subtract this averaged image in real time from the incoming video images. We then adjust the contrast and gray level before recording it onto video tape. The resulting image reveals structures in the flow with greatly improved clarity.

8.4   Shadowgraph Images

In water, the visualization of turbulent convection has led to the detailed study of coherent structures which come from the boundary layer (Zocchi et al., 1990; Moses et al., 1993). Although we have not performed similar studies in gas, we have observed both swirls and plumes above the thermal boundary layer at every Ra. Examples are shown in Fig. 8.4; in both pictures the motion is from left to right. These are the two prototypical structures of a buoyant, sheared boundary layer (Shelley and Vinson, 1992), although experimentally they are not as common as one might suppose. Typically we observe what might be called thermal "stems" moving above the thermal layer.

Sending almost any light through the convection cell permits one to see something of the flow. We visualize the overall flow of the turbulent convection by projecting a slightly diverging laser beam through the cell onto a screen. For all Ra up to 1×10^{11}, we observe a single circulating roll in the cell, confirming the original observation made in low temperature helium gas (Sano et al., 1989; Wu 1991).

To study the coherent thermal structures in detail, we focus on the boundary layer region. In Fig. 8.5 we show images of the flow near the center of the bottom plate at four different Ra; the mean flow is from left to right. These images give some idea of the moving filaments or stems that are so evident in the videotape. These long structures extend from the boundary layer upwards in a slanting curve, as seen most clearly in Fig. 8.5A for $Ra = 7 \times 10^8$. These filaments are similar to the sheets seen peeling off the thermal layer in our convection experiment in water (Tilgner et al., 1993).

Qualitatively these images are becoming more complicated with increasing Ra; one sees an increased activity, and a shortening of the thermal filaments which extend away from the plate. What is clear in the video tape is that the long filaments become much less frequent somewhere between Ra of 2×10^9 (Fig. 8.5B) and 9×10^9 (Fig. 8.5C). In Fig. 8.5C-D, many of these filaments do not remain attached, but appear to disconnect into the mean flow.

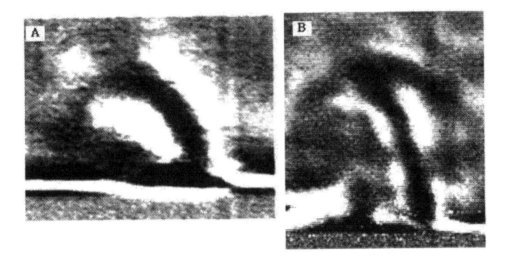

Figure 8.4: Swirl at $Ra = 7 \times 10^8$ (a), and plume at $Ra = 4 \times 10^9$ (b). Each image is taken near the center of the bottom plate, and is 0.8 cm and 0.65 cm across respectively.

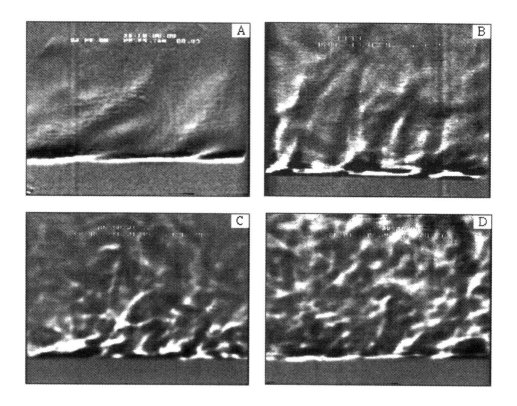

Figure 8.5: Typical shadowgraph images taken near the center of the bottom plate for Ra = (A) 7×10^8; (B) 2×10^9; (C) 9×10^9; (D) 4×10^{10}. Each frame is 3 cm across.

All of these qualitative visual observations may be indications of a transition to turbulent mixing (large Reynolds number) in the large scale circulation for $Ra > 2 \times 10^9$, as also suggested by measurement of λ_m (see Fig. 8.2) and the skewness of the derivative (Belmonte and Libchaber, 1996).

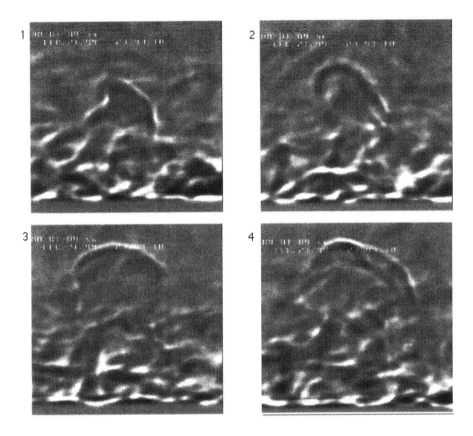

Figure 8.6: Ascending plume cap near the center of the hot plate, at $Ra = 9 \times 10^9$. Each frame is separated by 1/15 second, and is 2 cm across.

By timing the transit of passing thermal objects, we estimate the speed of the large scale circulation U_L: it ranges from 3 to 5 cm/s, and does not change as Ra is varied. This observation is in agreement with previous observation in low temperature helium (Sano *et al.*, 1989). Occasionally one sees a plume develop and move upwards, undistorted, for several video frames. From this we can estimate the ascending speed w_P for plumes. Typically $w_P \sim 3 - 4$ cm/s, also independent of Ra; thus $w_P \simeq U_L$. This suggests that the velocity of the plumes is driving the mean circulation, and that the plumes exist in a balance of buoyant (vertical) and advected (horizontal) motion.

For $Ra > 10^{10}$, it becomes increasingly difficult to find any object which can be followed for even a few frames; one has the impression of droplets being flung from the boundary layer and evaporating. The plume cap which we catch in its ascent in Fig. 8.6 for $Ra = 9 \times 10^9$ is a rare find. This cap is ascending at about 3 cm/s. Most of the time the motion is not easy to follow, an indication that mixing is increasing with Ra. Viewing these images at high Ra is an absorbing if a bit overwhelming experience, and one is rapidly led to consider a statistical approach.

Visual studies of hard turbulence have often described the bulk of the cell as dominated by thermals, or plume caps (Solomon and Gollub, 1990, 1991; Zocchi et $al.$, 1990). Most of our observations in the boundary layer region are of the long filaments discussed above. The conclusion is that the caps have detached at an earlier time (as shown in Fig. 8.6), leaving only plume stems behind.

After a few minutes of watching the mean flow along the plate, one is struck by the realization that there are regular pauses in the mean flow; the large scale circulation is not steady. An increase of activity from the boundary layer occurs during these momentary pauses, which we observe for all Ra in the hard turbulence regime. These bursts of local activity indicate that, in the sudden absence of mean flow, the heat is building up. As more plumes are emitted, the thermal layer must locally be approaching a marginally stable state, the one described in the classical scaling law $\lambda_{th} \sim Ra^{-1/3}$ (Malkus, 1954; Priestley, 1954; Howard, 1963). However, after pausing locally the mean flow starts up again. The interactions with other parts of the cell, including the cold boundary layer above, reassert the mean flow in the section we observe; it has not stopped entirely, but just locally. This bears some similarity to a proposed model for the large scale flow in terms of nonlinear delay equations (Villermaux, 1995). The fact that the thermal layer cannot achieve a state of local stability (during the pause) before it is interrupted by the recurring mean flow, is direct evidence that the thermal boundary layer is not determined locally. In other words, different regions of the cell are interconnected so that λ_{th} cannot scale as $Ra^{-1/3}$. The local oscillation of the large scale flow remains a question for further study.

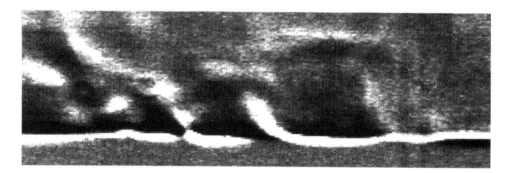

Figure 8.7: Plumes emerging from the thermal boundary layer into the mean flow, at $Ra = 4 \times 10^9$. The image is 1.6 cm across.

Before turning to statistical measurements of the intensity fluctuations of these images, we leave you with a beautiful picture of the dynamics at the boundary, a "cascade of plumes on the horizon", in Fig. 8.7. The large scale circulation is from right to left, and the plumes appear to be growing in the breeze - or are they pushing it along? This picture sums up our present view of the thermal boundary layer in hard turbulence: buoyant plumes in a crosswind.

8.5   Intensity Correlation Measurements

In our shadowgraph images, a plume or other coherent object appears as a light or dark shape on the screen. To extract information from the image in the simplest way possible, we study the two point correlation function of the intensity at a given point and at some fixed separation; the intensity should be correlated over the size of the thermal structures in the flow. By restricting the separation to the horizontal direction x, we obtain correlation curves at different distances z from the bottom plate. From each correlation function we estimate the mean correlation length, and thereby measure the average size of the plumes, stems, swirls, and other coherent structures in the flow in the image at that height, using an in-house program written at Princeton (see Belmonte, 1994).

We define the intensity fluctuation at the point (x, z) in the image as

$$i(x, z, t) \equiv I(x, z, t) - \langle I(x, z) \rangle, \tag{8.2}$$

where x is the distance *along* the plate, and z the perpendicular distance *from* the plate. For a horizontal separation d, the two point correlation function is then defined as

$$C(d, z) \equiv \frac{\langle i(x_0, z) \cdot i(x_0 + d, z) \rangle}{I_{RMS}(x_0, z) \cdot I_{RMS}(x_0 + d, z)}. \tag{8.3}$$

This function is equal to 1 at $d = 0$; for large enough d we expect C to approach 0. In practice we define $I(x, z)$ as the average intensity in a 3 pixel by 3 pixel box, and vary the distance d between two boxes. We compute the curve $C(d, z)$ at each height by averaging from 3 to 10 minutes of data from video tape (Belmonte, 1994). Several such curves are shown in Fig. 8.8.

The general trend apparent in Fig. 8.8 is that the intensity is correlated over shorter lengths at larger z (further from the boundary layer). To quantify this we extract a length scale from $C(d, z)$. There are many possibilities, but perhaps the most natural one is the mean correlation length, defined by

$$L_{corr}(z) \equiv \frac{\sum\limits_{i} d_i \cdot C(d_i, z)}{\sum\limits_{i} C(d_i, z)} \tag{8.4}$$

This is the average of separations d weighted by the correlation at that separation. In practice we use the following approximation to L_{corr}: for small d the correlation decreases roughly linearly in d; we define the length μ as the separation at which $C(\mu) = 0.5$. Note that μ would be equal to L_{corr} if the correlation function $C(d)$ were in fact linear. Our definition ignores the tails of the correlation function, which are a small correction. We therefore consider μ to be the approximate average size of thermal objects in our images. For the $z = 9.9$ mm curve in Fig. 8.8, $\mu = 0.80$ mm.

We thus estimate the typical size of the thermal objects at each height z, where $z = 0$ is the position of the bottom plate. In Fig. 8.9 we plot μ vs. z for several different Ra. We observe first that μ is rapidly decreasing with z in the region close to the plate, then varies much more slowly, not significantly different from a saturation value μ_c beyond the height

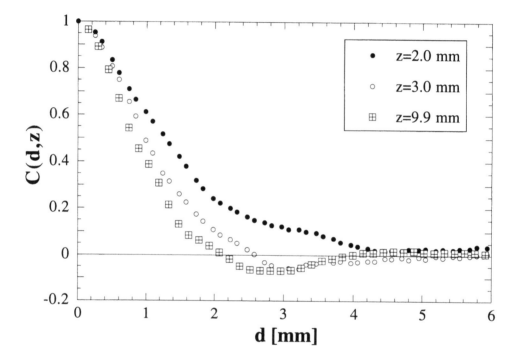

Figure 8.8: Correlation function $C(d, z)$ at three different distances z from the bottom plate, for $Ra = 9 \times 10^9$.

z_c. We define these lengths by approximating the two dependences of $\mu(z)$ as straight lines, and finding their intersection. We call the height z_c the *development height*, since the size of the thermal objects μ is developing within this distance of the plate, and has reached some stable value μ_c (the bulk size) outside of it. This development height is similar to the mixing zone postulated in one of the proposed models for the 2/7 scaling (Castaing *et al.*, 1989). We plot μ_c and z_c against Ra in Fig. 8.2, along with the thermal boundary layer thickness λ_{th} and the position of the maximum velocity λ_m.

8.6 Discussion

We begin with a brief discussion of the large scale circulation. A fluid parcel which has a temperature difference δT with its surroundings feels a buoyant acceleration equal to $g\alpha\delta T$. The associated free fall velocity is the speed this parcel would achieve if it were able to accelerate through a height L unimpeded by viscosity:

$$v_{ff} = \sqrt{2gL\alpha\delta T}. \tag{8.5}$$

The temperature drop across each thermal layer is $\Delta/2$, so we expect the temperature difference of detachments from the thermal layer δT to be some portion of $\Delta/2$. Previous measurement has shown that the maximum RMS is about $\Theta_{MAX} \sim 0.06\Delta$ in the hard turbulence regime (Belmonte *et al.*, 1994). Using $\delta T \simeq \Theta_{MAX}$ gives $v_{ff} \sim$ 8cm/s, for $\Delta = 20°$C. Since Θ_{MAX} is due to the fluctuations of the boundary layer itself, it is entirely reasonable that the circulation velocity, driven by boundary layer detachments, would be less than this. Further, we can make an estimate of the temperature difference δT of these detachments, from our estimate $v_{ff} = 4$ cm/s, which implies that δT is 1.4% of Δ.

Thus the free fall velocity v_{ff} depends only on Δ and L: in an experiment which changes Ra but keeps Δ and L constant, v_{ff} will be constant. Because we have $\Delta \sim 20°$C for all Ra, and we observe that our velocities elsewhere do not change with Ra, we associate them with the free fall velocity. The measurement of the scaling $U_L L/\kappa \sim Ra^{1/2}$ in low temperature helium by Sano *et al.* (1989) means that the actual speed U_L in cm/s is independent of Ra, so this is also related to the free fall velocity.

Close to the bottom plate $\mu(z)$ is decreasing rapidly with z. Thus the plume stems or other detachments are thickest as they begin to leave the thermal layer. The fluid flow is incompressible (except for buoyancy), which means that the density is approximately constant. Thus if a stem is narrowing with height, its velocity must increase. We therefore associate the decrease of $\mu(z)$ with the acceleration of plumes. Indeed the region close to the plate is where we expect high shear from the large scale circulation, and from Fig. 8.2 we see that this rapidly decreasing portion of $\mu(z)$ is within the position λ_m of the maximum mean velocity (or where mixing effects become dominant (Shraiman and Siggia, 1996)). Based on this reasoning we would expect $z_c < \lambda_m$, since beyond this point the fluid is well mixed.

For positions outside of the development height z_c, the size of thermal objects μ appears to have stopped evolving and reached a stable size μ_c. Fig. 8.2 shows that μ_c follows the thermal boundary layer thickness λ_{th}, in agreement with the observations made in water, that beyond the thermal layer the plumes are still characterized by the size λ_{th} (Zocchi *et al.*, 1990). This was also an assumption used in the interpretation of the cutoff frequency (Belmonte *et al.*, 1994). From our measurements of $\mu(z)$ we conclude that this occurs not because the objects start with this size, but because they become stable at the size λ_{th}. Visual observations of images like those in Fig. 8.5 confirm that the stems are thicker at their point of attachment to the thermal layer.

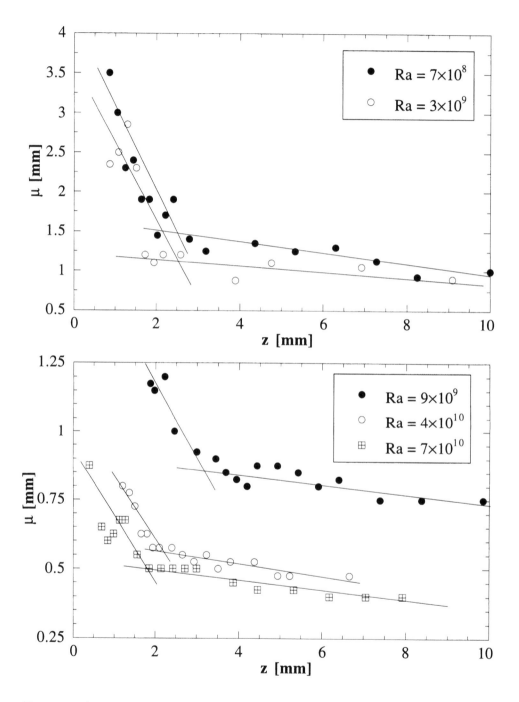

Figure 8.9: Average size μ vs. height z for different Ra; the values of μ_c and z_c at each Ra are given by the intersection points of the drawn straight lines.

As Fig. 8.9 and Fig. 8.2 show, the development height z_c, and the size of thermal objects beyond this distance, μ_c, vary with Ra; there seems to be a change in z_c at $Ra \sim 10^{10}$. The development height for the thermal detachments is being encroached upon by the decreasing λ_m, characterized by a turbulent Reynolds number. This confirms our previous interpretation: a turbulent flow at the plates occurs for $Ra > 2 \times 10^9$. We have seen further evidence that this is restricting the structure of the flow outside of the thermal layer. We suggest that this restriction of the development distance results in the reduction in the skewness of the temperature derivative at high Ra (Belmonte and Libchaber, 1996). It may even be that the occurrence of the development distance z_c is a characteristic of hard turbulence; optical difficulties at lower Ra have so far kept us from investigating the boundary layer visually for $Ra < 10^7$. The technique of intensity correlation has been surprisingly fruitful in connecting the length scale from temperature measurements with our thinking about the plumes and other objects which depart from the thermal layer into the large scale flow.

Acknowledgements

We thank L. P. Faucheux for help with the correlation programs, and M. Shelley, B. Shraiman, E. Siggia, and A. Tilgner for discussions during the course of these experiments. A. B. thanks R. S. Belmonte for proofreading, and for help in preparing the manuscript.

Bibliography

Belmonte, A., and Libchaber, A. "Thermal signature of plumes in turbulent convection: The skewness of the derivative." *Phys. Rev. E* **53**, 4893–4898 (1996).

Belmonte, A., "Buoyant plumes and internal waves: two experiments in turbulent convection", preprint (1996).

Belmonte, A., Tilgner, A., and Libchaber, A., "Turbulence and internal waves in side-heated convection." *Phys. Rev. E* **51**, 5681–5687 (1995).

Belmonte, A., "Boundary layer measurements in turbulent convection: vertical plumes in a horizontal wind." Ph. D. Thesis, Princeton University, 1994 (unpublished).

Belmonte, A., Tilgner, A., and Libchaber, A., "Temperature and velocity boundary layers in turbulent convection." *Phys. Rev. E* **50**, 269–279 (1994).

Castaing, B., Gunaratne, G., Heslot, F., Kadanoff, L., Libchaber, A, Thomae, S., Wu, X. Z., Zaleski, S., and Zanetti, G., "Scaling of hard thermal turbulence in Rayleigh-Bénard convection." *J. Fluid Mech.* **204**, 1 (1989).

Ching, E. S. C., "Heat flux and shear rate in turbulent convection", preprint (1996).

Emanuel, K. A., and Bister, M., "Moist convective velocity and buoyancy scales." *J. Atmos. Sci.* **53**, 3276–3285 (1996).

Heslot, F., Castaing, B., and Libchaber, A, "Transitions to turbulence in helium gas." *Phys. Rev. A* **36**, 5870 (1987).

Howard, L. N., "Heat transport by turbulent convection." *J. Fluid Mech.* **17**, 405–432 (1963).

Julien, K., Legg, S., McWilliams, J., and Werne, J. "Hard turbulence in rotating Rayleigh-Bénard convection." *Phys. Rev. E* **53**, 5557–5560 (1996).

Kantrowitz, A., and Trimpi, R., "A sharp-focusing Schlieren system." *J. Aeronaut. Sci.* **17**, 311–319 (1950).

Kerr, R.M, "Rayleigh number scaling in numerical convection", *J. Fluid Mech.* **310**, 139–179 (1996).

Lacey, A. J., *Light Microscopy in Biology: A Practical Approach* (IRL Press, Oxford, 1989).

LeMone, M., "Modulation of turbulence energy by longitudinal rolls in an unstable planetary boundary layer." *J. Atmos. Sci.* **33**, 1308–1320 (1976).

Malkus, W. V. R., "Heat transfer and spectrum of thermal turbulence." *Proc. R. Soc. Lond.* **A225**, 196 (1954).

Merzkirch, W. *Flow Visualization*, (Academic Press, New York, 1987).

Mortensen, T., "An improved schlieren apparatus employing multiple slit-gratings." *Rev. Sci. Instr.* **21**, 3-6 (1950).

Moses, E., Zocchi, G., and Libchaber, A., "An experimental study of laminar plumes." *J. Fluid Mech.* **251**, 581 (1993).

Priestley, C. H. B., "Convection from a large horizontal surface." *Aust. J. Phys.* **7**, 176–201 (1954).

Rennó, N. O., and Ingersoll, A. P., "Natural convection as a heat engine: a theory for CAPE." *J. Atmos. Sci.* **53**, 572–585 (1996).

Sano, M., Wu, X. Z., and Libchaber, A, "Turbulence in helium-gas free convection." *Phys. Rev. A* **40**, 6421–6430 (1989).

Shelley, M., and Vinson, M., "Coherent structures on a boundary layer in Rayleigh-Bénard turbulence." *Nonlinearity* **5**, 323 (1992).

Shraiman, B., and Siggia, E., "High Rayleigh number convection and passive scalar mixing." *Physica D* **97**, 286–290 (1996).

Shraiman, B., and Siggia, E., "Heat transport in high Rayleigh number convection." *Phys. Rev. A* **42**, 3650–3653 (1990).

Siggia, E., "High Rayleigh number comvection." *Ann. Rev. Fluid Mech.* **26**, 137–168 (1994).

Solomon, T. H., and Gollub, J. P., "Sheared boundary layers in turbulent Rayleigh-Bénard convection." *Phys. Rev. Lett.* **64**, 2382–2385 (1990); "Thermal boundary layers and heat flux in turbulent convection: The role of recirculating flows." *Phys. Rev. A* **43**, 6683–6693 (1991).

Takeshita, T., Segawa, T., Glazier, J. A., and Sano, M., "Thermal turbulence in mercury", *Phys. Rev. Lett.* **76**, 1465 (1996).

Threlfall, D. C., "Free convection in low-temperature gaseous helium." *J. Fluid Mech.* **67**, 17–28 (1975).

Tilgner, A., Belmonte, A., and Libchaber, A., "Temperature and velocity profiles in turbulent convection in water." *Phys. Rev. E* **47**, 2253–2256 (1993).

Villermaux, E., "Memory-induced low frequency oscillations in closed convection boxes", *Phys. Rev. Lett.* **75**, 4618–4621 (1995).

Weinstein, L. M., "Large-field high-brightness focusing schlieren system", *AIAA Journal* **31**, 1250–1255 (1993).

Werne, J., "Structure of hard-turbulent convection in two dimensions: Numerical evidence." *Phys. Rev. E* **48**, 1020–1035 (1993).

Wu, X. Z. and Libchaber, A., "Non-Boussinesq effects in free thermal convection." *Phys. Rev. A* **43**, 2833 (1991).

Wu, X. Z. and Libchaber, A., "Scaling relations in thermal turbulence: The aspect-ratio dependence." *Phys. Rev. A* **45**, 842–845 (1992).

Wu, X. Z., "Along a road to developed turbulence: free thermal convection in low temperature helium gas." Ph. D. Thesis, University of Chicago, 1991 (unpublished).

Xin, Y.-B., Xia, K.-Q., and Tong, P., "Measured velocity boundary layers in turbulent convection." *Phys. Rev. Lett.* **77**, 1266–1269 (1996).

Zocchi, G., "Flow structures in turbulent convection." Ph. D. Thesis, University of Chicago, 1990 (unpublished).

Zocchi, G., Moses, E., and Libchaber, A., "Coherent structures in turbulent convection, an experimental study." *Physica A* **166**, 387–407 (1990).

Chapter 9

CONVECTION IN CLOUD-TOPPED ATMOSPHERIC BOUNDARY LAYERS

CHRISTOPHER S. BRETHERTON[1]

[1] *Department of Atmospheric Sciences*
Box 351640, University of Washington
Seattle, WA 98195-1640

Convecting cloud-topped boundary layers a few hundred meters to 2 km deep typically overlay more than half of Earth's surface. They play a vital role in Earth's radiation balance and the exchange of moisture, heat and momentum between the atmosphere and ocean and land surfaces. Boundary layers capped by layer clouds, or stratocumulus, are common over cold surfaces, while shallow cumuli are seen over warmer surfaces. The boundary layer and its cloud are maintained by an intimate set of feedbacks between the cloud layer, radiative cooling, surface fluxes, precipitation, turbulence, and entrainment of air from above the boundary layer. Using the subtropical marine boundary layer as an example, we use observations and models to discuss these feedbacks and how changing surface and above-boundary layer conditions bring about changes in boundary layer and cloud structure.

9.1   Introduction

The dynamics of atmospheric convection are greatly enriched by the phase changes of water that can occur as air circulates. These phase changes are vital to the development of severe thunderstorms and many other forms of deep moist convection. They are equally important in the development of convecting cloud-topped boundary layers (CTBLs). CTBLs are less spectacular than deep moist convection, but they are ubiquitous and hence climatically important, covering the majority of Earth's oceans and much of its land surface.

CTBLs extend from the surface to a typical depth of 500-2000 m. The cloud may be a solid stratus layer a few hundred meters thick, or shallow cumuli up to 1500 m thick. The convection is maintained by an intimate set of feedbacks involving condensation into cloud, radiative cooling, surface fluxes, precipitation, and entrainment of air from above the boundary layer. These feedbacks have proven challenging to represent in numerical global circulation models for climate simulation, because the convective circulations are not resolved in such models, and their effects must be parameterized.

In this chapter, we examine the global distribution and climatic importance of boundary layer cloud, then discuss the fascinating dynamics that produce different types of CTBLs. While the details of the arguments are probably relevant to Earth's atmosphere alone, some aspects of CTBL convection, such the selection of the lapse rate of temperature with height and the development of vertically separated layers of convection within a single boundary layer, could conceivably be relevant to stellar convection as well.

9.2 Global distribution and importance of boundary layer cloud

Perhaps our most reliable climatology of boundary layer cloud comes from routine hourly weather reports from surface observers at weather observing sites and on commercial ships. They classify the cloud type and estimate the fraction of the sky (in eighths) covered by cloud. Although individual observations are imprecise, millions of routine surface cloud observations have now been archived. Satellite cloud climatologies are also useful, but usually cannot detect low-lying clouds underneath a higher cloud layer.

Figure 9.1 shows the annually averaged cloud amount (the product of cloud fraction and frequency of occurrence) for boundary layer stratus (or layer) clouds. These cloud layers are typically 100-500 m thick, with a cloud base anywhere from the surface to 1500 m, and tend to be nonprecipitating. Over much of the midlatitude oceans and parts of the eastern subtropical oceans, stratus cloud cover exceeds 50%. Klein et al. showed that the cloud cover in these regions is highest when the sea-surface is coldest compared to the air above the boundary layer, which tends to occur in the summertime. In some parts of the Aleutian Islands, the average stratus cloud cover in June, July and August is 90%...a dreary sky indeed. Over land, there is much less stratus cloud due to the lesser availability of surface water. In the tropical and subtropical oceans, stratus clouds are rare. Instead, shallow 'trade' cumulus clouds are seen almost all the time. Even in these regions, trade cumulus cloud amounts do not exceed 10-20%, because they only cover a small part of the sky.

Both of these cloud types are important to global climate. Trade cumulus clouds help convect moist air upward from the ocean surface, greatly enhancing the amount of evaporation that occurs in the subtropics. The moistened air is drawn into a zone of persistent deep convection, the intertropical convergence zone, where much of the moisture is precipitated out. The resulting latent heating drives the entire tropical circulation. Tiedtke et al. (1988) found that a better parameterization of boundary layer cumulus convection in a weather forecast model considerably improved their representation of the strength of the mean tropical circulation.

Stratus clouds are also usually associated with convection (in which case they are often called stratocumulus clouds, but we will not concern ourselves with this technical distinction), but their biggest climate impact is on Earth's radiation budget. Liquid water clouds as little as 100 m thick are almost opaque to infrared radiation. Therefore, they have a 'greenhouse' effect, absorbing upwelling infrared blackbody radiation emitted by the underlying surface, while radiating less infrared radiation upward because they are typically at a colder temperature than the surface. This effect is not that large for boundary layer clouds, which are very low in the atmosphere and hence nearly as warm as the surface, but is considerable for high cirrus clouds. Clouds also have an 'albedo' effect – they usually reflect solar radiation more efficiently than the underlying surface, so they increase the amount of

Annual Stratus Cloud Amount

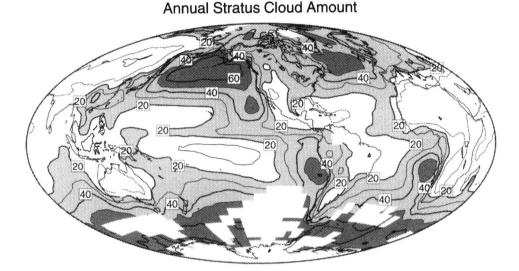

Figure 9.1: Annual average boundary layer stratus, stratocumulus and fog cloud amount in percent as seen by surface observers, from Klein and Hartmann (1993).

solar radiation reflected back to space.

Both of these effects have been measured from satellites, which can measure the 'cloud radiative forcing', which is defined as the difference between the combined upwelling infrared and visible radiance at times when a given location is cloud-free and its average at that location over all times. Figure 9.2 shows the annually averaged cloud radiative forcing. It bears a strong resemblance to the annually averaged stratus cloud amount. Because of their global coverage, boundary layer stratus clouds have a large albedo effect which dominates their greenhouse effect to create negative cloud radiative forcing, i.e. increased radiation of energy back to space. This helps cool the Earth. Hence, boundary layer stratus clouds can be thought of giant natural refrigerators for Earth and particularly its colder oceans. Cirrus clouds, by contrast, have almost cancelling albedo and greenhouse effects and have little net effect on Earth's current radiation balance.

In the rest of this chapter, we will focus on marine CTBLs because they cover a much larger area than CTBLs over land and involve a more limited set of feedbacks which are more likely relevant to astrophysical convection. We have mentioned that stratus clouds are prevalent over the colder oceans and shallow cumulus clouds over the warmer oceans. How do boundary layer dynamics help to produce this pattern? The answers will prove quite subtle, but we can learn more by looking at how the vertical structure of marine CTBLs varies between different locations.

Figure 9.3 shows composite soundings from four field experiments that studied marine subtropical and tropical CTBLs (Albrecht et al. 1995). The experiments were conducted over locations with very different sea-surface temperature (SST). The first variable plotted is the virtual potential temperature θ_v. This is a slight modification of the potential temperature θ,

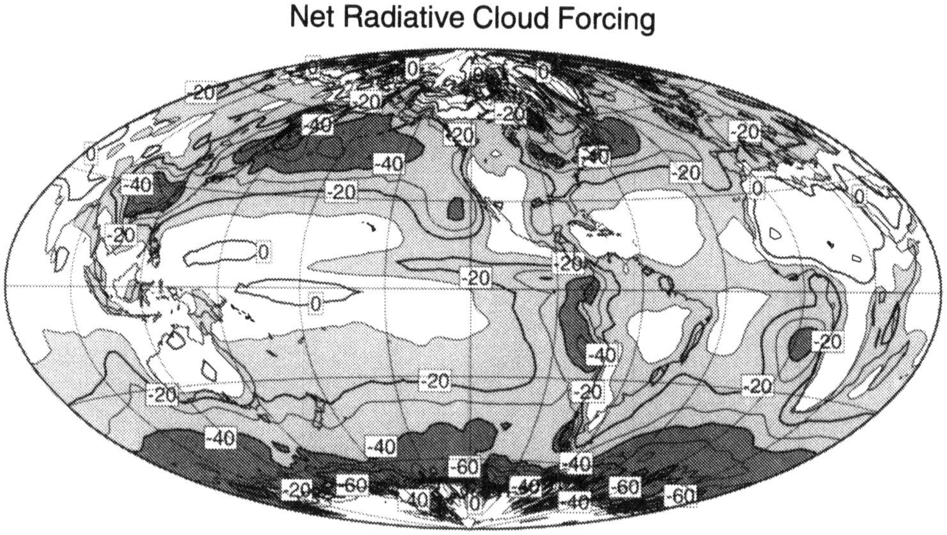

Net Radiative Cloud Forcing

Figure 9.2: Satellite derived annual average cloud radiative forcing (W m^{-2}) for 1985-1986, from Klein and Hartmann (1993).

which is the temperature multiplied by a pressure-dependent factor that corrects for cooling due to adiabatic expansion on temperature. For dry air, θ is inversely proportional to air density. For moist air, the density also depends slightly on the water vapor content of the air, which is accounted for by adding a small water vapor correction to θ to get θ_v. For moist air, θ_v is inversely proportional to density and is conserved in adiabatic motions of an unsaturated air parcel. Vertical gradients in θ_v indicate static stability to unsaturated overturning.

The vertical virtual temperature profile in an unsaturated and turbulently well-mixed convecting layer will approximately follow a 'dry' adiabat in which θ_v is constant. In saturated (cloudy) air, latent heat is released by condensation as the air rises to lower pressure and expands. Thus a turbulently well-mixed and fully saturated layer will have a 'moist-adiabatic' temperature profile in which θ_v increases with height at 3-6 K km^{-1}.

The second variable plotted is the total water mixing ratio $q_t = q_v + q_l$, which is the ratio of the mass of water in both vapor (q_v) and liquid (q_l) phases to the mass of dry air. This is conserved following all adiabatic motions of an air parcel, even including phase changes, as long as water is not lost to precipitation. In boundary layer clouds, most liquid water is in cloud droplets approximately 10 microns in diameter, which fall negligibly slowly relative to the air motions, so to a first approximation many boundary layer clouds can be regarded as nonprecipitating, i.e., all condensed water remains with the air parcel in which it condensed.

In all four locations, the CTBL is capped by a thin stable layer or 'inversion' in which θ increases in a nearly steplike fashion, and above which the air is much drier. The boundary layer above the coldest water (SNI) has the lowest and strongest inversion and the simplest internal structure of the shallowest boundary layer (SNI) is simplest. The SNI q_t profile

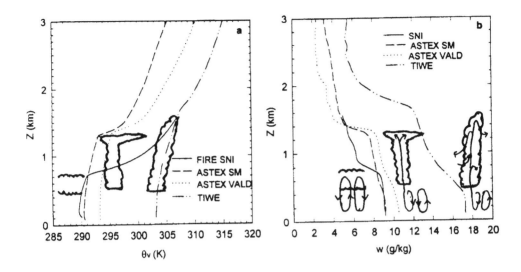

Figure 9.3: Composite soundings of (a) θ_v and (b) q_t from four CTBL experiments from Albrecht et al. (1995). Sketches of the typical boundary layer cloud structure observed in (left to right) FIRE, ASTEX and TIWE are overlaid. In (b), the air motions that accompany the clouds are also sketched.

is nearly uniform with height. In individual profiles used to build up this composite, θ_v is dry-adiabatic below cloud base and moist-adiabatic above cloud base (this is somewhat smeared out by the compositing method). This is consistent with a turbulently well-mixed CTBL.

The CTBLs above warmer water are deeper and have lower cloud amount. Observations show that the intermediate regimes (ASTEX SM and VALD) are associated with cumulus clouds rising into a thin and patchy stratus layer just beneath the inversion, while the warmest CTBL (TIWE) supports only trade cumulus clouds. The ASTEX and TIWE CTBLs are well-mixed only below the cumulus cloud base 400-500 m above the sea surface. In the next 100-200 m (around this cloud base) there is a 'transition layer' in which moisture decreases with height and individual temperature profiles often show a thin layer of static stability. Within the cumulus cloud layer, there is a continued decrease of q_t and increase of θ_v with height which is small in the intermediate profiles but considerable (but still less than moist-adiabatic) in the TIWE trade cumulus profile.

We conclude that the phase change of water has considerably complicated the dynamics of these deeper CTBLs such that neither dry nor moist convective adjustment gives a reasonable approximation to their vertical structure. It is also at first sight puzzling that the shallowest boundary layers, with the shallowest clouds, should tend to have the highest cloud cover. To probe these enigmas, we must develop a better understanding of CTBL dynamics.

9.3   Convective dynamics of CTBLs

The turbulence in CTBLs is more often than not convective, driven by radiative cooling of air in the upper part of the boundary layer and heat fluxes from the surface. We start by examining the maintenance of a shallow, well mixed subtropical boundary layer such as the SNI case above. Figure 9.4 shows the physical processes that control the evolution of such a CTBL. The most important driving mechanism for convection is infrared radiative cooling at the cloud top. The air within 50 m of the cloud top is rapidly cooled by blackbody radiation, as it emits considerably more infrared radiation than it absorbs from downwelling radiation emitted by water vapor in the overlying atmosphere. This cooled air sinks in convective cells that extend down to the ocean surface, where it is moistened. It is usually slightly warmed by surface heat fluxes as well. In subtropical boundary layers these are typically 10% or less of the radiative cooling that is the principal thermal driving for the convection. In mid-latitudes, particularly when colder air flows out over warmer water, surface fluxes can become dominant. When the moistened air rises, it condenses to form the cloud.

The convective downdrafts and updraft speeds are around 1 m s^{-1}. Near the cloud-top, turbulent entrainment by the convective eddies brings some warm, dry above cloud air into the convective circulation. This would slowly deepen the boundary layer, but these CTBLs are generally found in regions of large-scale subsidence. The subsidence can counteract the entrainment deepening to permit a nearly steady state convecting layer to form in which surface moistening balances entrainment drying, radiative cooling balances entrainment and surface warming, and entrainment balances large-scale subsidence.

Lilly (1968), Schubert et al. (1979) and many others have used mixed layer models to explore well-mixed CTBLs. These models assume that thermodynamic variables which are conserved in adiabatic motions with phase change will be uniform throughout the mixed layer up to the inversion height z_i. One such variable is q_t. A second is the 'liquid water' potential temperature $\theta_l = \theta - cq_l$, where c = 2500 K is a thermodynamic constant that reflects the increase of θ by latent heating due to condensation of the liquid water q_l in an air parcel. The evolution of the mixed layer is determined by equations for the rates of change of θ_l, q_t and z_i. θ_l and q_t change due to surface fluxes, radiative cooling and entrainment. z_i evolves due to specified large scale subsidence and entrainment.

A key part of a mixed layer model is an 'entrainment closure' that determines the entrainment rate w_e in terms of known quantities, usually considering the budget of turbulent kinetic energy. The appropriate entrainment closure for a CTBL is controversial. In dry boundary layers heated from below, considerable evidence from observations, laboratory experiments and computer modelling shows that the entrainment rate is such that there is a downward entrainment flux of warm air from above into the boundary layer that is 20% as large as the rate of surface heating (Stull 1976). In CTBLs, the entrainment rates are small and very difficult to measure, and the above-cloud air can evaporate cloud droplets if it mixes with CTBL air, reducing the buoyancy of mixtures above the cloud base. A variety of disparate approaches have been proposed, all of which generalize the dry, surface heated case in different ways. Stage and Businger (1981) and Nicholls and Turton (1986) have compared these closures against the very limited observational data. Nicholls and Turton's closure, which is perhaps the most appealing closure that is consistent with the data, takes the form

$$\frac{w_e}{w_*} = \frac{A}{Ri}. \tag{9.1}$$

Here the 'entrainment efficiency' A is a parameter which increases rapidly with the maximum amount of evaporative cooling that can be produced by the mixing of cloudy and above-cloud

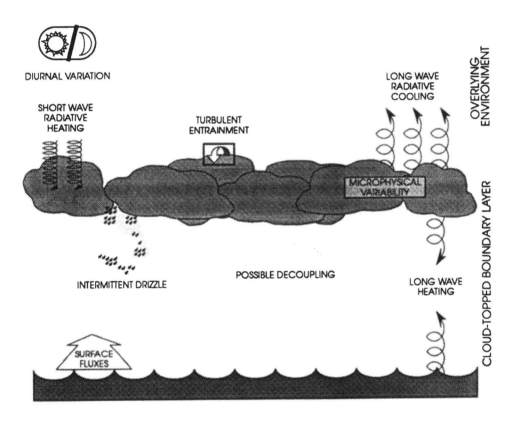

Figure 9.4: Physical processes that affect a subtropical cloud-topped mixed layer, from Siems (1991)

air. For typical subtropical CTBLs, $A \approx 2$, about ten times as large as for dry surface-heated boundary layers. w_* is a convective velocity scale given by

$$w_*^3 = 2.5 \int_0^H (g/\theta_0)\overline{w'\theta'}dz, \qquad (9.2)$$

where g is gravity, θ_0 is a reference potential temperature, and $(g/\theta_0)\overline{w'\theta'_v}$ is the vertical buoyancy flux, which is proportional to the vertical flux of virtual potential temperature. w_* can be determined from the mixed layer parameters using parameterizations of the radiative cooling rate, the surface fluxes, and w_e itself. Ri is the interfacial Richardson number

$$Ri = \frac{gz_i\Delta\theta_{vi}/\theta_0}{w_*^2}, \qquad (9.3)$$

where $\Delta\theta_{vi}$ is the virtual potential temperature jump across the inversion.

Lilly (1968) and others found that convective mixed layer models make good predictions of the depth, cloud thickness, and turbulence levels observed in shallow stratus-capped CTBLs off the coast of California. However, problems arose when researchers tried to apply these models on a more global scale. Wakefield and Schubert (1981) pioneered a Lagrangian approach to mixed layer evolution, treating the boundary layer air as a column moving in the wind from one location to another, thereby experiencing changing boundary conditions. In the summer, air columns move in the trade winds from locations slightly off the California coast to Hawaii in roughly a week, and the cloud is observed to change from a stratus layer to isolated cumulus clouds. The mixed layer model predicts a continuous thickening of the stratus layer instead.

Bretherton and Wyant (1997) resolved this inconsistency by showing that the well-mixedness of the convective layer breaks down as the CTBL warms and deepens. Figure 9.5 shows the profile of buoyancy fluxes in a mixed layer model at three stages in the evolution of a mixed layer moving from colder to warmer sea-surface temperature. As the mixed layer deepens, it develops an expanding region of negative buoyancy fluxes below cloud base. While the mixed layer model is forced to keep the CTBL well-mixed through this process, the negative buoyancy fluxes (less dense air being forced down) tend to damp out convective motions in this region and cause 'decoupling' of the turbulent circulations into separate layers near the surface and within the cloud. Typically this occurs by the time the CTBL is 1 km deep; most convective CTBLs more than a few hundred km away from a coast (i.e. the majority of the ocean areas covered by boundary layer cloud) are probably decoupled.

Among convective systems, the CTBL is uniquely vulnerable to decoupling because of the unusual profile of buoyancy flux that is created by the phase change of water (Fig. 9.6). Updrafts are slightly moister and have a lower condensation level than downdrafts (which have been diluted by slight mixing with entrained dry above-cloud air). Even if an updraft and downdraft have nearly the same temperature below the cloud, the updraft will have a higher temperature in the cloud, because it has more condensed liquid water, which has released more latent heat. This effect causes the buoyancy flux to rise considerably in the cloud. In fact, in subtropical cloud-topped mixed layers, the convection is typically driven mainly from within the cloud, with comparatively small buoyancy fluxes below cloud base. As the water warms, the updrafts are moistened more and more, and the difference between updraft and downdraft condensation levels increases until decoupling occurs.

The dynamics that follow decoupling have been documented in observations of deepening, warming CTBLs from the subtropical North Atlantic (Bretherton and Pincus 1995;

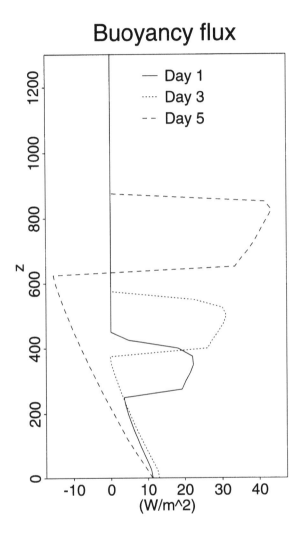

Figure 9.5: The evolution of the buoyancy flux profile in a mixed layer model of a subtropical CTBL (Bretherton and Wyant 1997) under which sea-surface temperature is raised at 1.5 K day^{-1} from 285 K at Day 0 to 294 K at Day 6, while other external parameters (wind speed, mean subsidence, above-inversion temperature, mixing ratio, and downwelling solar and infrared radiation) are held constant at typical subtropical values. Numerical simulations and observations of CTBLs suggest that decoupling (i. e. the breakdown of well-mixedness in the CTBL) occurs when considerable negative buoyancy fluxes develop below cloud base, which occurs between days 3 and 5.

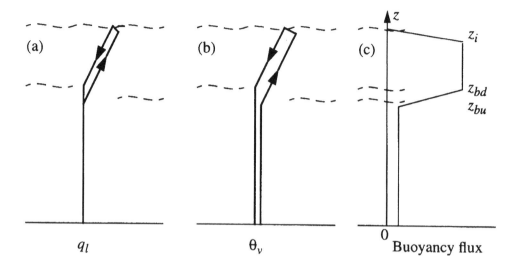

Figure 9.6: Air parcel properties and buoyancy fluxes in an convective eddy in a well-mixed CTBL. Dashed lines indicate heights of updraft condensation level z_{bu}, downdraft condensation level z_{bd} and inversion z_i. Updraft-downdraft differences in (a) q_l and (b) θ_v are shown, along with (c) the resulting buoyancy flux profile.

Bretherton et al. 1995), and eddy-resolving numerical models of boundary layer air columns (Krueger et al. 1995a, b; Wyant et al. 1997). As the cloud base of the updrafts and downdrafts gradually separates, the updrafts become more intense and widely spaced, resembling small cumulus clouds rising into an overlying cloud layer, while the downdraft areas below the downdraft cloud base broaden. The updrafts originate in the subcloud layer, which has become considerably moister than the cumulus layer above (see Fig. 9.3b). When the inversion reaches 1500 m, the stratuscloud layer below the inversion has started to break into broad, thin patches. It is sustained by intermittent injection of liquid water by the cumulus updrafts rising from the moister subcloud layer. These updrafts rise until they encounter the inversion, where they become negatively buoyant and detrain their air.

The stratification in the cumulus layer is very weak (see Fig. 9.3a). This is due to the balance between convection and radiation. Extensive cloud at the inversion causes considerable infrared cooling at the inversion, but provides a greenhouse effect that makes cooling beneath the inversion very weak. Thus the air that subsides around the cumulus clouds experiences only slight radiative cooling as it sinks through the cumulus layer, producing a very weak stratification around the cumulus clouds.

The weak stratification allows cloudy air parcels to accelerate quite rapidly as they ascend through this layer. Hence cumulus clouds overshoot their equilibrium level of neutral buoyancy at the inversion. Wyant et al. (1997) present evidence from their numerical model that this overshoot causes the ultimate breakup of the upper cloud layer to break up, to leave just a cumulus cloud field as is observed over the warmer subtropical waters. The overshooting cumuli entrain some of the overlying drier, warmer air that they penetrate into. This air mixes with the cloudy updraft air, evaporating its liquid water. As the CTBL deepens, the cumulus updrafts have more distance to accelerate, so they can penetrate further into and entrain more from above the inversion. This dries the updraft air before it is detrained below and into the inversion, so that this air no longer supports a stratus cloud layer. A conceptual model of the entire transition from subtropical stratus to cumulus capped CTBLs is presented in Fig. 9.7.

9.4   Further observations and conclusions

Many other processes affect the evolution of marine CTBLs. While boundary layer clouds typically do not precipitate heavily, they can precipitate enough to affect both the water and energy balance of the boundary layer. Precipitation processes are tightly coupled to the cloud 'microphysics', i.e. the distribution of droplet sizes within the cloud. Droplets about 5-10 microns in radius condense on small submicron diameter aerosol particles, then coalesce to form larger precipitation-size droplets 50-2000 microns in radius. The less condensation nucleii there are, the larger the initial condensed droplets are and the more readily a few of them can grow to precipitation size. In pristine marine airmasses, there are typically 50 condensation nucleii per cubic centimeter. Under these conditions, stratus clouds more than 200 m thick can drizzle. This promotes decoupling by causing water to evaporate much lower than it condenses, which causes latent heat release in the cloud and cooling below the cloud, a distribution unfavorable to convection. Cumulus clouds as little as 1 km deep in pristine airmasses can also produce showers. This depletes the cloud liquid water in cumulus updrafts so they do not detrain as much liquid water, which can considerable reduce the stratus cloud cover in the intermediate decoupled regime. In airmasses that have flowed off of polluted continents, condensation nucleus concentrations can be as much as 500 cm^{-3}, which almost entirely suppresses precipitation in boundary layer clouds.

If the liquid water in the clouds is subdivided into many small droplets instead of a few larger ones, this also increases the surface area of the droplets and hence their effectiveness in scattering solar radiation. While this appears to have little direct effect on the CTBL dynamics, it can considerably increase the cloud albedo. A dramatic demonstration of this is seen in Fig. 9.8. This figure shows a satellite image of 'ship tracks', lines of brightening several hundred km long and 5-10 km wide often visible in shallow marine stratus cloud layers. These have been traced to aerosols in stack effluent that act as condensation nucleii, increasing cloud droplet concentrations by 10-500% above background values (Radke et al. 1987). Ship tracks are much more rarely detected in deeper boundary layers, in which decoupling makes the mixing of effluent near the surface into the main cloud layer, which must occur in cumulus updrafts, much more intermittent and patchy. The concomitant reduction in precipitation in polluted boundary layer clouds may also increase their liquid water content and their cloud cover, further increasing their areally average albedo (Albrecht 1989). Anthropogenic sources of cloud condensation nucleii feedbacks on boundary layer clouds and their convective dynamics may be helping to significantly increase the typical albedo of CTBLs over the oceans and counteract greenhouse gas induced global warming of our current climate (Boucher and Lohmann 1995).

Boundary layer convection is surprisingly different from even deep moist convection in the atmosphere, let alone laboratory analogues. Phase change, precipitation, and the interaction of clouds and radiation considerably affect the dynamics of the convection, the vertical thermodynamic structure of the convecting layer, and the degree of horizontal homogeneity. However, numerical models, guided by observations, have been used with considerable success

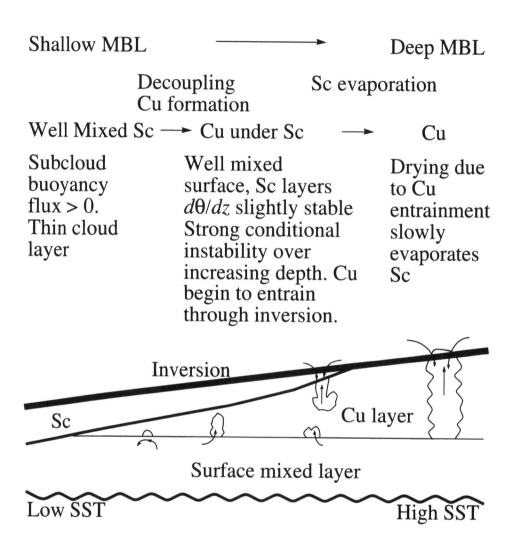

Figure 9.7: A conceptual model of the entire transition from subtropical stratus to cumulus capped CTBLs, from Wyant et al. (1997)

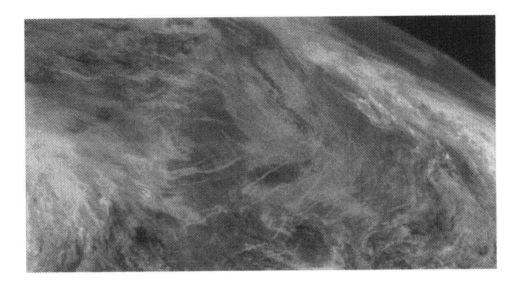

Figure 9.8: Geostationary satellite image (in visible light) of ship tracks in stratus clouds over the summertime northeast Pacific Ocean

to understand the most important feedbacks. The moral for the astrophysical context, where processes akin to phase change are important and radiation is also a principal driving force for convection, is simple. Do not assume that convection or the convectively induced temperature profile will be in accord with a priori expectations, and use whatever tools are available (mainly detailed numerical simulation) to try to understand the likely structure of the convection.

Acknowledgements

Research described in this review was supported by ONR grant N00014-90-J-1136 and NASA grant NAG1-1711.

Bibliography

Albrecht, 1989: Aerosols, cloud microphysics and fractional cloudiness. *Science*, **245**, 1227-1230.

Albrecht, B. A., M. P. Jensen and W. J. Syrett, 1995: Marine boundary layer structure and fractional cloudiness. *J. Geophys. Res.*, **100**, 14209-14222.

Bretherton, C. S., and R. Pincus, 1995: Cloudiness and marine boundary layer dynamics in the ASTEX Lagrangian experiments. Part I: Synoptic setting and vertical structure. *J. Atmos. Sci.*, **52**, 2707-2723.

Bretherton, C. S., Austin, P., and S. T. Siems, 1995: Cloudiness and marine boundary layer dynamics in the ASTEX Lagrangian experiments. Part II: Cloudiness, drizzle, surface fluxes and entrainment. *J. Atmos. Sci.*, **52**, 2724-2735.

Bretherton, C. S., and M. C. Wyant, 1997: Moisture transport, lower tropospheric stability and decoupling of cloud-topped boundary layers. *J. Atmos. Sci.*, **54**, 148-167.

Boucher, O., and U. Lohmann, 1995: The sulfate-CCN-cloud albedo effect: A sensitivity study with two general circulation models. *Tellus*, **47**, 281-300.

Klein, S. A, and Hartmann, D. L., 1993: The seasonal cycle of low stratiform cloud, *J. Climate*, **6**, 1587–1606.

Krueger, S. K., G. T. McLean, and Q. Fu, 1995a: Numerical simulations of the stratus to cumulus transition in the subtropical marine boundary layer. Part 1: Boundary layer structure. *J. Atmos. Sci.*, **52**, 2839-2850.

Krueger, S. K., G. T. McLean, and Q. Fu, 1995b: Numerical simulations of the stratus to cumulus transition in the subtropical marine boundary layer. Part 2: Boundary layer circulation. *J. Atmos. Sci.*, **52**, 2851-2868.

Lilly, D. K., 1968: Models of cloud-topped mixed layers under a strong inversion. *Quart. J. Roy. Meteor. Soc.*, **94**, 292-309.

Nicholls, S., and J. D. Turton, 1986: An observational study of the structure of stratiform cloud layers: Part II. Entrainment. *Quart. J. Roy. Meteor. Soc.*, **112**, 461-480.

Radke, L. F., J. A. Coakley, Jr., and M. D. King, 1989: Direct and remote sensing observations of the effects of ships on clouds. *Science*, **246**, 1146-1149.

Schubert, W. H., J. S. Wakefield, E. J. Steiner and S. K. Cox, 1979: Marine stratocumulus convection. Part I: Governing equations and horizontally homogeneous solutions. *J. Atmos. Sci.*, **36**, 1286-1307.

Siems, S. T., 1991: *A Numerical Investigation of Cloud Top Entrainment Instability and Related Experiments.* Ph. D. Dissertation, Department of Applied Mathematics, University of Washington, 116 pp.

Stage, S. A., and J. A. Businger, 1981: A model for entrainment into a cloud-topped marine boundary layer. Part II: Discussion of model behavior and comparison with other mod els. *J. Atmos. Sci.*, **38**, 2230-2242.

Stull, R.B. 1976: The energetics of entrainment across a density interface. *J. Atmos. Sci.*, **33**, 1260-1267.

Tiedtke, M., W. A. Heckley and J. Slingo, 1988: Tropical forecasting at ECMWF: The in fluence of physical parametrization on the mean structure of forecasts and analyses. *Quart. J. Royal. Meteor. Soc.*, **114**, 639-664.

Wakefield, J. S., and W. H. Schubert, 1981: Mixed-layer model simulation of Eastern North Pacific stratocumulus. *Mon. Wea. Rev.*, **109**, 1952-1968.

Wyant, M. C., C. S. Bretherton, H. A. Rand, and D. E. Stevens, 1996: Numerical simulations and a conceptual model of the subtropical marine stratocumulus to trade cumulus transition. *J. Atmos. Sci.*, **54**, 168-192.

Chapter 10

SOLAR GRANULATION: A SURFACE PHENOMENON

Advanced Study Program*
National Center for Atmospheric Research[1]
PO Box 3000, Boulder, CO 80307-3000 USA

Convection in the solar envelope globally transports nearly all the incident radiative flux from the interior, yet the dynamics of solar granulation as observed at the photosphere appears dominated by local radiative cooling and new downflow formation. In this paper we discuss how the details of granule fragmentation and evolution may be understood in these terms. We also examine the role of downflow plumes in convective heat transport and assess the stability of compressible starting plumes with depth.

Regions of maximum granular upflow lie not in the center of granules, as may be expected if granulation were a manifestation of overshooting upflowing thermals or simple cellular convection, but immediately adjacent to the intergranular downflow lanes. Fragmentation results when granular expansion further diminishes the already weak central upflow, radiative losses exceed the advected heat supply from below, and a new central downflow plume initiates. New lanes then form in those directions predisposed to weak upflow by the strength and shape of the downflows defining the granule boundary.

Ideal gas studies have indicated that the vigorous downflows found in compressible convection may play little role in net heat transport. This is because the enthalpy and kinetic energy carried by such flows nearly cancel. This is likely not the case in the outer solar envelope. The fluid there is partially ionized and two-thirds of the enthalpy transported is carried as latent heat of ionization. Buoyancy and enthalpy fluctuations are thus less tightly coupled and the role of downflow plumes in convective energy transport is elevated.

Given the importance of photospherically driven downflow plumes to the dynamics of and heat transport by granular flows, it is of interest to assess the depth to which they might penetrate. We do so here by examining individual cool plumes descending through an adiabatically stratified layer of increasing density with depth. We find that such plumes are subject to vigorous secondary instabilities even in a quiescent background medium, and in contrast to upflow plumes in the same environment *de*trainment of fluid from the plume region is suggested.

---

[1] NCAR is sponsored by the National Science Foundation. Current address: Joint Institute for Laboratory Astrophysics, University of Colorado, Boulder, CO 80309-0440 USA

10.1 Introduction

The Sun is a middle-aged moderate-mass main-sequence star and as such burns hydrogen to helium in its core (Bethe 1939). The energy produced radiatively diffuses outward through much of the solar interior, is carried almost entirely by convection in the outer envelope, and finally escapes at the solar photosphere. Convective transport in the outer layers is initiated because of the elevated radiative opacity of the gas due primarily to the the partial ionization of its principal constituents, hydrogen and helium. The convection is very efficient. Current helioseismic observations estimate that the mean thermal stratification of the solar envelope is nearly adiabatic to a depth of about $0.71R$, where R is the solar photospheric radius (Christensen-Dalsgaard et al. 1991, Gough et al. 1996). This implies a convection zone depth of 2×10^5 km, over which the temperature of the gas increases about 400 times, its density nearly 10^6 times, and its pressure over 10^8 times (Model S of Christensen-Dalsgaard et al. 1996). The fluid stratification is thus enormous, most of it occurring in the upper layers where the pressure and density scale heights are about 150 km and 350 km respectively.

Observations of the solar surface reveal motions of multiple spatial and temporal scales. These include contributions from differential rotation, meridional circulation, acoustic oscillations, and convection. The convective contributions appear to be at somewhat discrete scales, termed supergranulation, mesogranulation, and granulation. Supergranulation is observed either as horizontal flows of $0.3 - 0.5$ km s^{-1} or as outlined by magnetic network. It has a spatial scale of about $20 - 50,000$ km and a lifetime of $20 - 40$ h. Mesogranulation is observed primarily by its vertical velocity signature of amplitude $0.05 - 0.1$ km s^{-1}. It has a horizontal scale of about $5 - 10,000$ km and a lifetime of about about 2 h. The smallest scale of convective motion on the Sun is granulation. Both velocity components are observed with amplitudes of $0.5 - 2$ km s^{-1}. It has a horizontal scale of $500 - 2500$ km, a lifetime of about $5 - 15$ m, and the appearance of bright irregularly shaped polygons separated and surrounded by dark lanes. The correlation of this observed intensity pattern with fluid velocity is well established (e.g., Mattig et al. 1969, Nesis et al. 1992), with the bright granular regions generally upflowing and the darker intergranular lanes generally downflowing gas. Most granules evolve by expansion and fragmentation, with essentially all granules forming from fragments of previous granules (Mehltretter 1978). In this paper we focus on the fragmentation process, how it can be understood by considering the dynamics of strongly driven downflow plumes, and what such plumes imply for heat transport and morphology with depth.

In our discussion we reference the results of three separate sets of numerical experiments: two-dimensional pure hydrogen convection (Rast & Toomre 1993a,b), three-dimensional radiative hydrodynamic convection (Nordlund 1985, Stein & Nordlund 1989, Nordlund & Stein 1990, Rast et al. 1993, Nordlund 1996), and compressible thermal plume convection (Rast 1995, 1998). The two-dimensional pure hydrogen experiments study compressible convection in a plane-parallel layer of single-atomic-level hydrogen undergoing collisional ionization and recombination and in thermodynamic equilibrium. The fluid is confined to a horizontally periodic domain with stress free and impenetrable horizontal boundaries. The temperature is held constant along the upper boundary, and a constant heat flux is applied at the bottom. The ionization formulation focuses on equation of state effects, incorporating particle number, latent heat, and specific heat changes, but omitting radiative opacity variations by employing a constant thermal conductivity. The three-dimensional radiative hydrodynamic code employs greater realism, using a solar mixture of elements, a more complete equation of state, and an accurate treatment of radiative energy exchange. The domain boundaries are horizontally periodic and open in the vertical, with flows, waves, and shocks being transmitted vertically with as little reflection as possible. A constant heat flux is maintained by specifying the entropy of the incoming fluid at the bottom, with radiative losses determining the temperature at the top. The final set of experiments employed, the thermal plume experiments, are designed for maximal physical simplicity and correspondingly maximal numerical resolution. They study isolated compressible thermal plumes descending through an adiabatically stratified layer of ideal gas spanning four density (about seven pressure) scale heights. The fluid is confined within a two-dimensional domain, horizontally periodic and with stress free, impenetrable, and constant temperature horizontal boundaries. A cool Gaussian temperature perturbation centered along the upper boundary is applied and maintained, driving a cool plume downward through the layer. Given the simplified flow geometry, extremely high resolution on nonuniform grids is achieved in the plume region. Secondary instabilities of the flow are resolved to better than 30 grid points per Taylor micro-scale while still minimizing the lateral boundary influences by considering large domains. Each of these three sets of simulations offers insight into the dynamics of solar granulation. The three-dimensional model is physically most complete, whereas the simple models isolate and address specific issues of equation of state and flow stability.

10.2   Granular Dynamics

The photoionization cross section and thus radiative properties of the partially ionized gas in the surface layers of the Sun are extremely temperature and density sensitive. In Fig. 10.1 the radiative conductivity of pure hydrogen

$$K_r = \frac{16\sigma T^3}{3\bar{\kappa}\rho} , \qquad (10.1)$$

where σ is the Stefan-Boltzman constant and $\bar{\kappa}$ is the Rosseland mean opacity, is plotted as a function of temperature T and density ρ. For reference the value of K_r and extent of the convection zone in a modern solar model (Model S of Christensen-Dalsgaard $et\ al.$ 1996) are also indicated. The rapid increase of conductivity with decreasing temperature at low temperatures and with increasing temperature at high temperatures is responsible for the finite extent of the convective envelope. Above the photosphere (T less than ~ 6000K) and below the base of the convection zone (T greater than $\sim 10^6$K) the fluid is able to carry the heat flux by radiation while still possessing a subadiabatic stratification. Between these two depths, where the conductivity of the gas is low, convective motions are essential to heat transport. The changes in transport properties of the fluid at low temperatures are even more dramatic than Fig. 10.1 indicates. Radiative diffusion is a poor approximation to heat transport in the surface layers of the Sun. Instead, there is a rapid (over a depth of $\sim 50 - 100$ km) photospheric transition from nearly adiabatic convective heat transport to nearly unimpeded free streaming of radiant energy, a transition from an optically thick to an optically thin medium. A rising parcel of fluid experiences an opacity drop of several orders of magnitude over a distance of just a few hundred kilometers (Nordlund & Stein 1991, Rast $et\ al.$ 1993).

   While the radiative properties of the fluid in the solar photosphere are complex, the essential surface energy balance maintained by the convection is that between advection of heat (both thermal and ionization energy) from below and radiative cooling from above. A

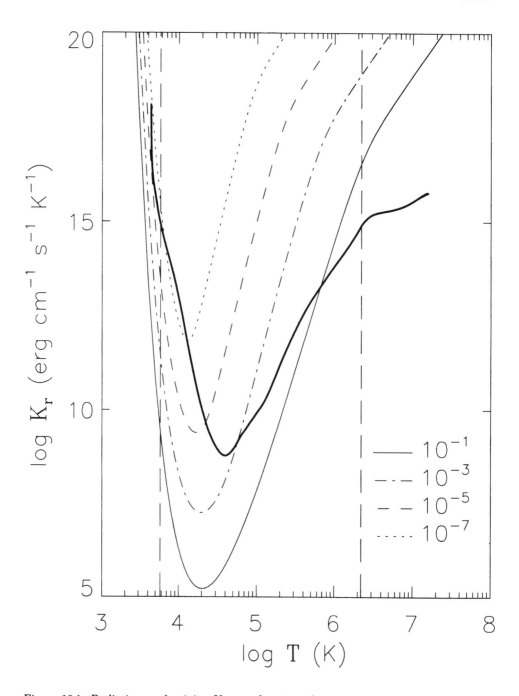

Figure 10.1: Radiative conductivity K_r as a function of temperature T and density (indicated by linestyle in g cm^{-3}). Thick solid line indicates values of K_r in solar Model S of Christensen-Dalsgaard *et al.* (1996). The top and base of the solar convection zone are indicated by vertical dashed lines.

simple balancing of the solar luminosity by the vertical advection of enthalpy

$$\rho w \Delta h \approx \sigma T^4 \tag{10.2}$$

yields an estimated ascent velocity of about 2 km s^{-1} to sustain radiative losses (Nordlund 1985, Nordlund and Stein 1991). Local reduction of vertical advection can lead to heat loss rates in excess of supply, consequent temperature reduction, and new downflow plume formation. It is this sensitivity to radiative loss which governs the dynamics of granule fragmentation. Note that it is only through the release of latent heat of hydrogen ionization that solar convection is able to sustain the radiative losses incurred at the photosphere. If the only available energy source were thermal (as for an ideal gas) much larger vertical velocities (on the order of 15 km s^{-1}) or significantly higher upflow/downflow temperature contrasts (on the order of 10 times greater) would be required. The availability of ionization energy allows significant energy loss with comparably small advective velocities or temperature changes in the fluid.

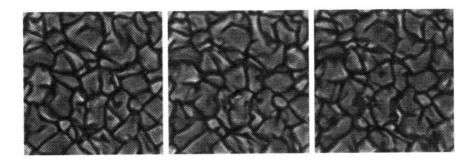

Figure 10.2: Horizontal planforms of the vertical velocity at successive 2 minute intervals in a numerical simulation of solar granulation by Nordlund & Stein (references cited in text). A 12×12 Mm area at a nominally photospheric depth and of numerical resolution 128×128 is shown. Fiducial marks indicate location of the expanding granule through which horizontal cuts at each successive time are plotted in Fig. 10.4. Note that simulations of four times greater resolution have been performed, with the results displaying additional intricate flow structure not seen here (cf. Nordlund's URL http://www.astro.ku.dk/ãake/convection).

Often solar granulation is interpreted as roughly cellular convection, with overshooting upflowing thermal plumes being surrounded by intergranular downflow lanes. Recent observational and numerical studies indicate however that regions of maximum granular upflow lie not in the centers of granules, as may then be expected, but immediately adjacent to the intergranular downflow lanes. Figure 10.2 shows horizontal planforms of the vertical velocity at a nominal photospheric level in the three-dimensional granulation simulations of Nordlund and Stein (Nordlund 1985, Stein & Nordlund 1989, Nordlund & Stein 1990, Rast *et al.* 1993). Clearly apparent in these images is that maximal upflow velocities (bright regions) usually occur immediately adjacent to downflow lanes (dark regions) and only rarely in the center of the granules. High spatial resolution observations of both intensity (de Boer *et al.* 1992) and Doppler velocity (Nesis *et al.* 1992, Rast 1995) indicate that the same is also true on the real Sun. Moreover, the dominant granule fragmentation scenario observed in these simulations is also that observed on the Sun. Many granules undergo initial expansion, followed by central darkening, and finally splitting by the formation of dark interior radially-directed lanes (Rösch 1960, Carlier *et al.* 1968, Namba & Diemel 1969, Mehltretter 1978 , Kitai & Kawaguchi 1979, Bray *et al.* 1984, Namba 1986). When the fragmentation process occurs particularly vigorously (expansion to 3-5 arc-seconds within ∼ 8 minutes) the phenomenon has been called an 'exploding' granule, but the scenario is not uncommon to many less vigorous fragmentations.

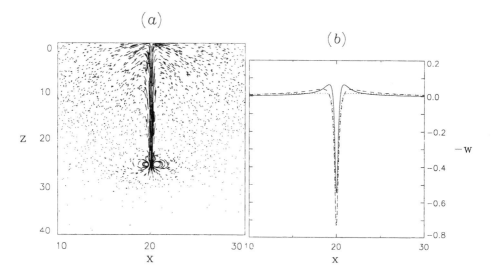

Figure 10.3: (*a*) The instantaneous vector flow field of a cool thermal plume buoyantly driven through an (initially) adiabatically stratified layer, and (*b*) horizontal cuts of the vertical velocity, $-w$ (upward plotted positive), as a function of horizontal position x at depths $z = 0.75$, $z = 3.80$, and $z = 17.2$ corresponding to 0.25, 1.0, and 2.5 density scale heights downward from the top of the domain. Distance is scaled by the width of the applied temperature perturbation and velocity by the initial isothermal sound speed at the top of the layer. In both figures only one half of the full horizontal extent of the computational domain is shown.

Compressible plume simulations suggest that both this process of granular fragmentation and the observed granular upflow topology may be understood in terms of the properties of strongly driven compressible downflow plumes (Rast 1995). Figure 10.3*a* displays the instantaneous vector flow field of a cool downflowing plume. Horizontal cross sections of vertical velocity through the plume at three depths, $N_\rho = 0.25$, $N_\rho = 1.0$, and $N_\rho = 2.5$, where N_ρ is the number of scale heights down from the top of the domain and corresponding to $z = 0.75$, $z = 3.80$, and $z = 17.2$ respectively, are plotted in Fig. 10.3*b*. Note that at all depths, but particularly in the upper layers, the upflow velocity peaks immediately adjacent to the downflow plume. This is in contrast to motion exhibited by either compressible or incompressible cellular Rayleigh-Bénard convection in which predominantly horizontal motions are found adjacent to the downflows. In the single plume simulations the fluid is accelerated upward and inward before descending the plume stem. The influence of the lower boundary is dynamically weak, and the upflows are driven locally by buoyancy forces and large pressure gradients induced in the vicinity of the downflow plume (cf. Rast 1998), not by buoyancy production on a lower heated boundary as in Rayleigh-Bénard convection. The three-dimensional granulation simulations similarly favor downflow-dominated flows, with the strong radiative cooling at the surface, an open lower boundary which allows for unimpeded downflow penetration, and upward lower-boundary inflow whose properties are determined solely by the mass and entropy of the outgoing fluid. On the Sun granular flows are similarly strongly surface driven and are likely only weakly dynamically coupled to the lower boundary located many pressure scale heights away. In all these cases the upflow topology is determined by a local response to downflow plume formation, with maximum upflow velocities immediately adjacent to the site of downflow.

Given such a flow structure granule fragmentation can be interpreted as follows. Darkening and new downflow plume formation in the center of granules is a consequence of the dynamical link between the intergranular downflow lanes and the adjacent upflows. As a granule expands the upflows move with the downflows. Figure 10.4a shows the vertical velocity at $z = 0.5$ in a domain similar to that of Fig. 10.3 but now through which two downflow plumes of equal strength are driven. Three realizations, differing only in the distance between the cool boundary perturbations applied, are shown. As the separation between the plumes increases, the vertical velocity between them decreases, leaving a central region of weak flow. While the flow in the granulation simulations is much more complex, with many downflow plumes present in a three-dimensional geometry, simple horizontal cuts of the surface vertical velocity at successive times through an expanding granule (granule location indicated in Fig. 10.2 and horizontal cuts plotted in Fig. 10.4b) shows a remarkably similar structure. As the granule expands the region of maximal upflow moves with the downflow and the central velocity weakens. The fluid in the center of an expanding granule is thus subject to radiative cooling, and ever diminishing vertical advection of heat to sustain those losses. As the central upflow velocity drops below about 2 km s^{-1}, the fluids temperature and its opacity decreases, and the depth of the radiating surface increases. Overlying layers continue to cool until buoyancy forces become sufficient to initiate a new central plume. Neighboring flow stagnate or reverse in response, and the cooling instability propagates outward from the granule center (Rast et al. 1993). Fragmentation occurs by new lane formation in those directions already predisposed to weak upflow, a predisposition determined by the distance to, the strength, and the shape of the intergranular flow boundaries (cf. Rast 1995).

Such a scenario suggests that the fundamental scale of solar granulation is linked not to a cellular size or vertical scale height, but instead to the structure and horizontal extent of the upflows induced in the vicinity of very strongly buoyantly driven downflow plumes. Unfortunately, the details of that structure and its extent depend on the downflow induced pressure fluctuations which drive the upflows, and these in turn are likely sensitive to the fluid stratification and viscosity, the overall plume strength, and the domain boundary conditions. The granular scale is consequently not immediately determinable from the simple plume simulations discussed. However, qualitative support for this interpretation comes from the more realistic three-dimensional simulations whose mean granular scale does compare well with observation and is reduced by a reduction in numerical viscosity (Lites et al. 1989) and

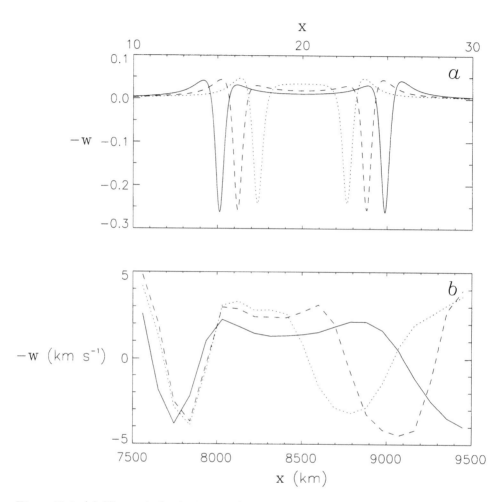

Figure 10.4: (*a*) The vertical velocity, $-w$ (upward plotted positive), as a function of horizontal position x at a depth $z = 0.5$ for three realizations of increasing plume separation and equal plume strength. (*b*) Horizontal cuts of the photospheric vertical velocity displayed in Fig. 10.2. Cuts are through each image successively (dotted line for the earliest image and solid for the last) at the location indicated in Fig. 10.2, and illustrate the reduction in upflow velocity which accompanies granule expansion.

the consequent sharpening of the shear layers and plume structures.

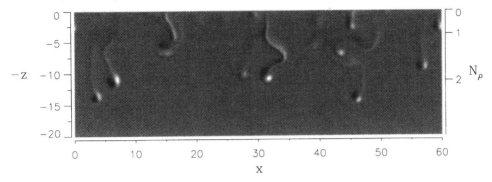

Figure 10.5: Instantaneous snapshot of the horizontal vorticity in a periodic domain subject to plume initiation by randomly located upper boundary temperature perturbations. Note both the small scale separation between weak plumes and the sites of stronger downflow which are maintained by the advection of neighboring perturbations.

Viewing granulation as a manifestation of radiatively-driven cool-pluming surface convection also suggests a model for multiple scale production. Consider, as previously, a two-dimensional domain of adiabatically stratified fluid, but now instead of applying an maintaining a centrally located source, position cool sources of finite duration at locations chosen at random along the upper boundary. If the interval between applied perturbations is also randomly distributed about some mean, then the separation between source locations is given approximately by $\Delta \approx Td/t$, where T is the mean interval between events, t is their mean duration, and d is the width of the domain. We conceptually associate this scale with granulation. Figure 10.5 shows the vorticity at one instant in time as a function of depth in such an experiment, and the small scale separation between weak plumes in the uppermost layers reflects this driving scale. A larger more slowly evolving horizontal scale also emerges in the experiment. It characterizes the separation between sites of strong and relatively persistent downflow. While initially associated with the random occurrence of large plumes, or the merger of multiple plumes in close proximity, such sites are dynamically maintained by the advection and consumption of subsequent adjacent smaller plumes. They persist longer than any single source lifetime as a local convergence of the flow. While this simple experiment is very preliminary, it suggests a kinematic model for the multiple scales of solar convection, similar in spirit to that of Simon and Weiss (Simon & Weiss 1989, Simon et al. 1991) but one in which downflow plumes rather than sources of radial outflow play the dominant dynamical role. One complication which must be address in further work is the nonrandom nature of granulation, since as we have seen new granular downflows form preferentially near the centers of an existing expanding granules.

10.3   Heat Transport

As we have seen the transport of ionization energy to the surface of the Sun is critical to the maintenance of radiative losses there. It is also critical to heat transport in the interior. In this section we contrast the heat transport properties found in ideal gas models of turbulent compressible convection with those in nonideal ionizing convection.

Numerical simulations of compressible convection over the past decade have revealed the importance of strong downdrafts to the dynamics of thermal convection in a stratified medium (e.g., Hurlburt *et al.* 1984, Stein & Nordlund 1989, Cattaneo *et al.* 1991). The narrow downflows originate in the upper thermal boundary layer and often extend to the bottom of the computational domain. Upflows, in contrast, are broad, weak, and more local in character. This fundamental asymmetry arises because positive pressure perturbations in both the upflow and downflow regions contribute positively to density perturbations, enhancing buoyancy driving in the downflows but contributing to buoyancy braking of the upflows (Massaguer & Zahn 1980, Hurlburt *et al.* 1984). Additionally, mean stratification of fluid necessitates compression of the downflowing and expansion and overturning of the upflowing material.

The net energy flux carried by mass motions in a convecting fluid is the sum of two parts: the enthalpy flux F_h and the kinetic energy flux F_k. The vertical convective flux F_c, defined to be positive when directed upward (in the negative z-direction), can thus be expressed explicitly as

$$F_c = F_h + F_k = -\rho w h - \rho w (u^2 + v^2 + w^2) , \tag{10.3}$$

where h is the enthalpy of the fluid, ρ is its density, and u, v, and w are the three fluid velocity components. Since ρw is positive for downflowing and negative for upflowing fluid, and since cool fluid has lower enthalpy than the mean while warm fluid has higher, the two contributions to F_c oppose each other in the downflows while reinforcing in the upflows. In other words, the upflows transport both kinetic energy and enthalpy upward, while the downflows transport enthalpy upward (cool fluid moving down) but kinetic energy downward. Flux cancellation within the downflows is thus possible if the magnitude of the kinetic energy carried by the flow equals that of the enthalpy it transports. This is indeed what is found for strong downflow plumes in the turbulent ideal-gas convection simulations of Cattaneo *et al.* (1991).

A simple argument can be advanced to understand this. The specific enthalpy of an ideal gas scales simply with its temperature;

$$h = \frac{5}{2} N_A k T , \tag{10.4}$$

where N_A is Avogadro's number ank k is Boltzmann's constant. For small pressure perturbations

$$\frac{\rho'}{\bar{\rho}} \approx \frac{T'}{\bar{T}} , \tag{10.5}$$

where ρ' and T' are fluctuations about the horizontal mean states $\bar{\rho}$ and $\bar{T}$, and together with Eq. 4 this implies that

$$\rho' \sim \frac{\bar{\rho}}{\bar{T}} h' . \tag{10.6}$$

Thus buoyant acceleration of an ideal gas is tightly coupled to enthalpy transport, with changes in the enthalpy of the fluid directly affecting its density. In fact, if the downflows flows are nearly steady with vertical advection and gravitational acceleration providing the dominant momentum balance, and if the mean state is nearly adiabatic so that $\bar{T} \sim z$, then

$$w^2 \sim \frac{\rho'}{\bar{\rho}} z \sim h' \tag{10.7}$$

and the kinetic energy and enthalpy fluxes of the downflows must balance pointwise precisely.

While the actual cancellation measured in the simulations appears more complex, occurring for horizontal averages of the fluxes but not pointwise (Brummell this volume), the net consequence is that the strong coherent downflows, so prominent in the large scale dynamics of compressible convection, play little role in energy transport in an ideal gas. Instead the more local small-scale turbulent motions between these coherent flows carry the vast majority of the convective flux. It is important to note that, while the transport of kinetic and thermal energy by the downflows in these simulations nearly cancel, their dissipation in the lower layers are quite different. Enthalpy transported downward is dissipated by thermal diffusion in the lower thermal boundary layer. Kinetic energy transported downward, on the other hand, is viscously dissipated in the small scale motions induced by the splashing impact of the strong downflows on the lower boundary. Small scale turbulent motions build up in the lower layers until the mean viscous dissipation there balances the influx of kinetic energy.

The remarkable cancellation between the downflow kinetic energy and enthalpy fluxes found in the ideal gas simulations of turbulent convection is not likely to occur if the fluid is partially ionized. Such a fluid is nonideal, its enthalpy is not a simple function of its temperature but depends also on its density. For pure single-atomic-level hydrogen the specific enthalpy is given by

$$h = \frac{5}{2} N_A k (1 + y) T + y N_A \chi_H \ , \tag{10.8}$$

where χ_H is the hydrogen ionization potential and y, the ionized fraction, is given in collisional equilibrium by the Saha equation,

$$\frac{y^2}{1 - y} = \left(\frac{2\pi m_e k}{h^2} \right)^{3/2} \frac{T^{3/2}}{\rho N_A} e^{-\chi_H / kT} \ . \tag{10.9}$$

Since the ionization potential of hydrogen is large (13.6 eV per atom), small fluctuations in the ionization state of the fluid correspond to large enthalpy perturbations. Moreover, since the ionization state of hydrogen is quite temperature sensitive and less density sensitive, such large fluctuations in enthalpy can be associated with quite small density perturbations, and the close link between buoyant acceleration and enthalpy transport found to hold for an ideal gas is weakened. The advection of latent heat of hydrogen ionization thus provides a relatively buoyancy insensitive means of heat transport.

This mode of energy transport dominates the convective flux in a partially ionized fluid. Figure 10.6 plots the horizontally averaged enthalpy and kinetic energy fluxes as a function of depth in a two-dimensional layer of ionizing convection. Also plotted are the average ionized fraction y (dash-dot line), latent heat contribution to the enthalpy flux $F_{lh} = \rho w y N_A \chi_H$ (dotted line), and enthalpy and kinetic energy fluxes in a comparable layer of fully ionized ideal gas (dashed lines). All fluxes are normalized to the input radiative flux imposed at the bottom of the computational domain, $z = 1$. Note that in the region of partial ionization about 2/3 of the enthalpy is carried as latent heat. This follows directly from the relative contributions of thermal and ionization energy to the enthalpy as expressed by Eq. 8 and evaluated at solar pressures and temperatures, and is found to also hold true for the more elaborate equation of state used in the granulation simulations (Rast et al. 1993). The latent heat contribution to the enthalpy flux diminishes to zero when the fluid is either fully ionized of fully neutral because fluctuations about the mean ionization state then vanish. Thus the mean enthalpy flux carried by ionizing convection shows a double peak, a peak in the region of 50% ionization where ionization state perturbations are maximum, not seen

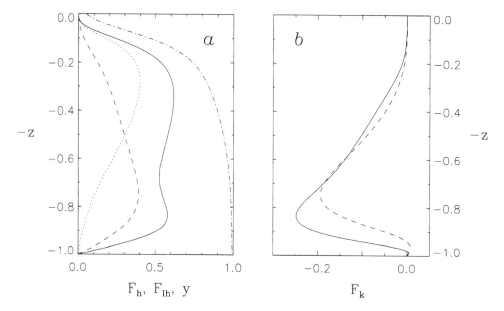

Figure 10.6: Variation with depth of the horizontally averaged (a) enthalpy flux F_h and (b) kinetic energy flux F_k, in a partially ionized convecting layer of pure hydrogen (solid lines) and a fully ionized ideal gas comparison (dashed lines). Also plotted in (a) are the latent heat flux F_{lh} (dotted line) and ionized fraction y (dash-dot line) in the partially ionized layer.

in the ideal gas case, and a second peak deeper in the layer where maximum density and maximum velocities occur. The kinetic energy flux, while slightly elevated over the ideal gas value in the region of partial ionization (Fig. 10.6b) shows only the second peak, as the buoyancy effects of partial ionization are small. Thus the kinetic energy and enthalpy fluxes do not share the same vertical profiles, and cancellation between the two in the downflows is likely to be incomplete even if their magnitudes in the lower layers under more vigorous flow conditions were to become comparable.

It thus seems likely that convective flux asymmetries in a turbulent ionizing fluid will depend strongly on depth. If the ionization zone is located in the upper portion of the domain, as it is in the case of the solar convection zone, then energy transport by convection in that region will be dominated by strong narrow downdrafts carrying latent heat enhanced enthalpy. Deeper down, as the fluid becomes more fully ionized, the latent heat contribution to the enthalpy fluctuations will diminish and the downward-directed kinetic energy flux will increase. Flux cancellation and upflow dominance of the convective energy transport may then occur as it does in an ideal gas. There are hints that such a division into regions of downflow dominance at the top and upflow dominance at greater depths actually occurs in the three-dimensional simulations of solar granulation (Rast et al. 1993). The results are however tentative, as those simulations are neither deep enough nor turbulent enough fully address the issue. A three-dimensional version of the simplified pure-hydrogen model presented here is currently under development to further study the transport properties of a turbulently convecting partially ionized medium.

10.4   Flow Stability

Given the importance of downflow plumes to both flow dynamics and heat transport in these models of solar convection, it is of interest to address the issue of their stability. This is of particular interest because the apparent propensity toward downflow plumes in compressible convection simulations has inspired a number of authors to consider them as an essential ingredient in understanding the thermal stratification of the overshoot region below the solar convection zone and possible magnetic dynamo mechanisms operating there (Schmitt et al. 1984, Simon & Weiss 1991, Zahn 1991, Nordlund et al. 1992, Ginet 1994). But do the surface driven downflow plumes actually descend through the entire depth of the solar convection zone and penetrate below? Previous authors (Schmitt et al. 1984, Simon & Weiss 1991, Rieutord & Zahn 1995) have answered this in the affirmative based on simplified plume models which included the effects of stratification but not compressibility (small pressure fluctuations) and employed velocity proportional entrainment prescriptions. Here we examine the results of direct numerical simulations of fully compressible plumes, and show that such plumes are subject to vigorous secondary instabilities leading to disruption and an apparent detrainment of fluid from the plume region.

Compressible downflow starting plumes are subject to both varicose and sinuous instability (Fig. 10.7). The varicose mode is compressive and results when dynamic pressure fluctuations behind the plume head initiate secondary head formation, successively pinching off vortex pairs which continue to travel downward through the domain. Because the fluid in the vortices is spinning they resist compression and find themselves less dense at depth than the adiabatically stratified surroundings. In fact, somewhat surprisingly the total integrated buoyancy force on these detached pairs is actually upward, an example of what Parker (1991) has called dynamical buoyancy. What keeps them propagating downward is the front to back

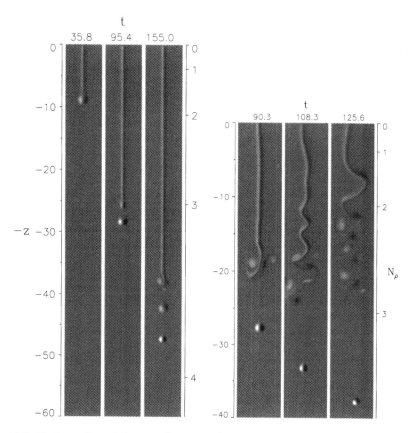

Figure 10.7: Horizontal vorticity as a function of time for two separate single plume experiments, illustrating the varicose (left) and sinuous (right) instabilities. Each frame encompasses only the very central portion of a much larger [60 × 60 (left) and 40 × 40 (right)] computational domain. Depth z and number of density scale heights spanned N_ρ label the vertical axes.

pressure gradient induced across them as they were spun up (cf. Rast 1998). Since each successive pair is weaker and travels slightly slower than the one ahead, the separation between them increases with time. This process continues until weak left/right asymmetries develop in the flow, initiating the onset of the sinuous stem instability. Such flow asymmetries can be enhanced by low amplitude thermal fluctuations in the background or by multiple plume interactions, in which case the onset of the sinuous mode closely follows the first vortex pair detachment.

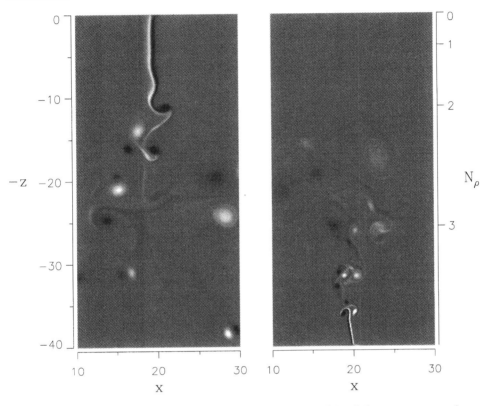

Figure 10.8: Instantaneous horizontal vorticity as a function of depth for two separate plume experiments showing the evolved state of a downflow plume (left) and an upflow plume (right) of comparable magnitude. One half the horizontal extent and the full vertical extent of the computational domain is shown. Extreme values of vorticity are saturated to show the details of the weaker vorticity regions.

The sinuous mode develops as a shear instability of the underlying plume stem flow. It usually initiates immediately behind the plume head and propagates upward along the stem to a depth of about one or two density scale heights. Subsequently the flow behaves quasiperiodically, shedding vortices alternately to either side at that depth. Coherent penetration into the deeper layers of the domain never again occurs, and the vortices shed travel along complex trajectories as they interact with those previously detached. The resulting structure is quite different from the classical conical turbulent plume one might imagine. Figure 10.8 contrasts the evolved flow of an upflowing and a downflowing compressible plume. The plumes are of comparable strength and are embedded in initially identical adiabatically stratified media. Flow development in the two cases is quite distinct. The upflow starting plume is not subject to the varicose instability. Pressure fluctuations behind the head are weak due to expansion of the rising vortex pair, and no pinch is realized. Since expansion tends to smooth fluctuations, no sinuous instability is observed either if the background is quiescent. That mode can be realized, however, by inducing asymmetries in the flow with low amplitude random background fluctuations. The stem flow then becomes unstable above a certain height and sheds vortices into the domain above (Fig. 10.8). The vortices expand and slow as they rise, with successive pairs traveling faster, penetrating and disrupting the previous one as in the familiar leapfrogging of two vortex rings seen in laboratory experiments (eg. Yamada & Matsui 1978). The fluid, rather than moving away from the plume region as a series of detached vortex pairs, as in the compressible downflowing plume (Fig. 10.8), is instead wrapped by this interaction back into the plume region. The whole region expands with height, within a roughly conical shaped envelope reminiscent of incompressible turbulent plumes. These two-dimensional plume studies thus suggest that while an entrainment hypothesis may prove useful in describing compressible upflowing plumes in a stratified medium, as it already has for incompressible plumes in a variety of stratified settings (eg. Turner 1986 and references within), such an hypothesis is likely not applicable to the compressible downflowing plume case. The dynamics of the flow following instability of the downflowing plumes in these experiments is better described by the interactions between detached vortex pairs, with the plume quickly losing its identity with depth.

The structure of the laminar plume stem flow, even before instability, is also strongly influenced by the fluids mean stratification. Figure 10.9 plots the stem diameter d as a function of depth for a series of low Prandtl number plume experiments of differing fluid viscosity. The diameter is measure as the distance between oppositely signed vorticity maxima on either side of the plume. Thick vertical bars in the figure are composed of individual data points, and their separation reflects twice the horizontal grid spacing in the domain (the positions of the vorticity maxima each shift by one grid point at a time as the plume diameter changes). Thin solid lines plot the best power-law fit to the data. The plume stems increase in diameter in the top density scale height, with the maximum diameter scaling empirically with viscosity to the one-half power. With depth below this maximum, the plume diameters decrease. This can be understood as follows.

The primary vertical force balance in the stem flow of these solutions is between buoyancy, horizontal viscous dissipation, and horizontal advection. At the locations of vorticity amplitude maxima, which determine our measure of plume diameter, all three of these terms in the momentum equation are of comparable magnitude,

$$\rho u \frac{\partial w}{\partial x} \sim \rho' g \sim \frac{1}{Re} \frac{\partial^2 w}{\partial x^2} \ . \tag{10.10}$$

This implies a plume stem diameter which scales as

$$d^4 \sim \frac{z}{\rho Re^2 \rho' g} \ , \tag{10.11}$$

where ρ' measures density fluctuations about the background value ρ. For an adiabatic descent of fluid and weak pressure fluctuations, ρ'/ρ is approximately constant, implying that, since $\rho \sim z^{3/2}$ in the background, the plume stem diameter should scale with depth as

$$d \sim z^{-\frac{1}{2}} \ . \tag{10.12}$$

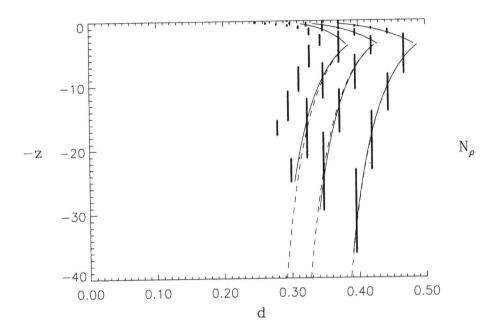

Figure 10.9: Plume stem diameter d, measured as the distance between peak oppositely signed vorticity, as a function of depth for six experiments of differing fluid viscosity. Thick vertical bars are composed of individual data points, their separation reflects twice the horizontal grid spacing. Thin solid lines plot the best power-law fit to the data and dashed lines indicate scaling theory behavior, which in the absence of thermal diffusion predicts $d \sim z^{-1/2}$ (dash-dot line).

That scaling is plotted in Fig. 10.9 with a dash-dot line, and is steeper than the results found for the experiments. The actual thermodynamics of the fluid descent in the experiments is, however, not adiabatic. Formed on a diffusive scale, the plumes suffer horizontal heat losses even at very low values of thermal conductivity. This is particularly true in the upper layers where the temperature gradients are sharpest and fluid density is the lowest. The broadening of the plume by thermal diffusion can be accounted for (cf. Rast 1998), and the dashed lines in Fig. 10.9 indicated how well the scaling theory then fits the data.

Such scalings imply significant reduction in plume diameter with depth. For example, a plume descending through the solar envelope should shrink, if the adiabatic scaling holds, from 300 km in the solar photosphere (typical dimensions of a granular downflow lane) to about 10 km at the base of the solar convection zone. This presumes that no secondary instabilities where to affect it. Such instabilities are, however, very likely. The secondary instabilities observed in our plume simulations occur despite unrealistically high (although low for numerical simulations) nonsolar values of fluid viscosity. Moreover, the instabilities are significantly enhanced in the experiments when the background fluid stratification is increased, more than one plume is present, or a nonquiescent background exists. All three of these conditions apply to the solar case. While the stratification of our domain is significant, it is substantially less than that in the Sun where the photospheric density scale height is approximately equal to one-half the downflow lane diameter. Additionally, the solar photosphere generates many new plumes which interact, and the background state through which they travel is far from quiescent. All this suggests to the author that the survival of photospherically driven downflows to the base of the solar convection zone is unlikely.

10.5   Conclusion

Solar granulation is a surface phenomenon. Its dynamics is governed by strong radiative cooling of the fluid in the solar photosphere, and we argue that its structure and evolutionary behavior can be understood in terms of the dynamics of compressible downflowing thermal plumes whose ultimate depth of penetration is limited. Issues that remain outstanding include the relationship of granulation to the other scales of observed solar convective motion. While it has been proposed that mesogranulation and supergranulation are components of the same convective phenomenon (November 1989, 1994) and also that a dynamical link between granulation and mesogranulation exists (Title et al. 1989, Simon et al. 1991, Rast 1995, §2 of this paper), such ideas require further development and examination before we understand precisely how the different scales arise and what their interrelationships are.

The importance of strong downflows to heat transport in a partially ionized convecting medium was also discussed. While studies to date indicate that the transport of latent heat of ionization by the downflows is the dominant energy transport mechanism in a partially ionized fluid, it is important to verify that this remains so even when the motions become fully turbulent. Confirmation that stellar envelope convection is divided, as has been suggested, into domains of upflow and downflow dominance, and thus regions of local and nonlocal heat transport, depending on the ionization state of the fluid, would be very significant to efforts aimed at parameterizations of convection useful to stellar modeling.

Finally, the stability of photospherically driven plumes is critical to our understanding of the deeper convective properties of the solar envelope. Simulation of the entire convective envelope is well beyond current computational capabilities, and one of the greatest difficulties in modeling the deeper layers is the inappropriateness of the applied upper boundary conditions. How to link the dynamics of the surface flow with that of the interior depends critically on the stability of the former. As discussed, our studies indicate that compressible downflowing plumes, when computed at extremely high numerical resolution in two dimensions, are subject to vigorous secondary instabilities which disrupt the coherence of the plume with depth. These studies should, and will, be extended to three dimensions, but it is difficult to imagine how three dimensionalization would serve to stabilize the flow. It is also possible that single thermal-plume dynamics is significantly different from downflow dynamics in other convecting environments, although we have argued that the study of single plumes dynamics is appropriate in light of the properties of observed and simulated solar granulation. Unequivocal demonstration, however, awaits a significant increase in computational capabilities which would allow for comparable resolution to that achieved here under more realistic flow conditions.

Bibliography

Bethe, H.A., "Energy production in stars," *Phys. Rev.* **55**, 434-??? (1939).

Bray, R.J., Loughhead, R.E., and Durrant, C.J., *The Solar Granulation*, Cambridge University Press, Cambridge (1984).

Carlier, A., Chaveau, F., Hugon, M., and Rösch, J., "Cinématographie à haute résolution spatiale de la granulation photosphérique," *Compt. Rend. Acad. Sci. Paris* **266**, 199-201 (1968).

Cattaneo, F., Brummell, N.H., Toomre, J., Malagoli, A., and Hurlburt, N.E., "Turbulent compressible convection," *Astrophys. J.* **370**, 282-294 (1991).

Christensen-Dalsgaard, J., Gough, D.O., and Thompson, M.J., "The depth of the solar convection zone," *Astrophys. J.* **378**, 413-437 (1991).

Christensen-Dalsgaard, J. *et al.*, "The current state of solar modeling," *Science* **272**, 1286-1292 (1996).

de Boer, C.R., Kneer, F., and Nesis, A., "Speckle observations of solar granulation," *Astron. Astrophys.* **257**, L4-L6 (1992).

Ginet, G.P., "Downflow plumes and entropy balance in deep convection zones," *Astrophys. J.* **429** 899-908 (1994).

Gough, D.O. *et al.*, "The seismic structure of the sun," *Science* **272**, 1296-1300 (1996).

Hurlburt, N.E., Toomre, J., and Massaguer, J.M., "Two-dimensional compressible convection extending over multiple scale heights," *Astrophys. J.* **282**, 557-573 (1984).

Kitai, R., and Kawaguchi, I., "Morphological study of the solar granulation I: Dark dot formation in the cell," *Sol. Phys* **64**, 3-12 (1979).

Lites, B.W., Nordlund, Å., and Scharmer, G.N., "Constraints imposed by very high resolution spectra and images on theoretical simulations of granular convection," *Solar and Stellar Granulation*, eds. R.J. Rutten and G. Severino (Dordrecht: Kluwer Academic), 349-357 (1989).

Massaguer, J.M., and Zahn, J.-P., "Cellular convection in a stratified atmosphere," *Astron. Astrophys.* **87**, 315-327 (1980).

Mattig, W., Mehltretter, J.P., and Nesis, A., "Studies of granular velocities I: Granular Doppler shifts and convective motion," *Sol. Phys.* **10**, 254-261 (1969).

Mehltretter, J.P., "Balloon-borne imagery of the solar granulation II. The lifetime of solar granulation," *Astron. Astrophys.* **62**, 311-316 (1978).

Namba, O., "Evolution of 'exploding' granules," *Astron. Astrophys.* **161**, 31-38 (1986).

Namba, O., and Diemel, W.E., "A morphological study of the solar granulation," *Sol. Phys.* **7**, 167-177 (1969).

Nesis, A., Hanslmeier, A., Hammer, R., Komm, R., Mattig, W., and Staiger, J., "Dynamics of the solar granulation I. A phenomenological approach, " *Astron. Astrophys.* **253**, 561-566 (1992).

Nordlund, Å., "Solar Convection," *Sol. Phys.* **100**, 209-235 (1985).

Nordlund, Å., URL http://www.astro.ku.dk/ãake/convection (1996).

Nordlund, Å., and Stein. R.F., "3-D simulations of solar and stellar convection and magneto-convection," *Comp. Phys. Comm.* **59**, 119-125 (1990).

Nordlund, Å., and Stein. R.F., "Dynamics of and radiative transfer in inhomogeneous media," *Stellar Atmospheres: Beyond Classical Models*, eds. L. Crivellari, I. Hubeny, and D.G. Hummer (Dordrecht: Kluwer Academic), 263-279 (1991).

Nordlund, Å., Brandenburg, A., Jennings, R., Rieutord, M., Ruokolainen, J., Stein, R.F., and Tuominen, I., "Dynamo action in stratified convection with overshoot," *Astrophys. J.* **392** 647-652 (1992).

November, L.J., "The vertical component of the supergranular convection," *Astrophys. J.* **344**, 494-503 (1989).

November, L.J., "Inferring the depth extent of the horizontal supergranular flow," it Sol. Phys. **154**, 1-??? (1994).

Parker, E.N., "Dynamical buoyancy of hydrodynamic eddies," *Astrophys. J.* **380**, 251-255 (1991).

Rast, M.P., "On the nature of 'exploding' granules and granule fragmentation," *Astrophys. J.* **443**, 863-868 (1995).

Rast, M.P., "Compressible plume dynamics and stability," *J. Fluid Mech.* **369**, 125-149 (1998).

Rast, M.P., Nordlund, Å, Stein. R.F., and Toomre, J., "Ionization effects in three-dimensional solar granulation simulations," *Astrophys. J.* **408**, L53-L56 (1993).

Rast, M.P., and Toomre, J., "Compressible convection with ionization. I. Stability, flow asymmetries, and energy transport," *Astrophys. J.* **419**, 224-239 (1993).

Rast, M.P., and Toomre, J., "Compressible convection with ionization. II. Thermal boundary-layer instability," *Astrophys. J.* **419**, 240-254 (1993).

Rieutord, M., and Zahn, J.-P., "Turbulent plumes in stellar convective envelopes," *Astron. Astrophys.* **296**, 127-138 (1995).

Rösch, J., "Observations from Pic du Midi," *IAU Symp. 12, Aerodynamic Phenomena in Stellar Atmospheres* ed. R.N. Thomas (Bologna: Nicola Zanichelli), 313-319 (1960).

Schmidt, J.H.M.M., Rosner, R., and Bohn, H.U.,"The overshoot region at the bottom of the solar convection zone," *Astrophys. J.* **282**, 316-329 (1984).

Simon, G.W., and Weiss, N.O., "Simulation of large-scale flows at the solar surface," *Astrophys. J.* **345**, 1060-1078 (1989).

Simon, G.W., and Weiss, N.O., "Convective structures in the Sun," *Mon. Not. R. astr. Soc.* **252**, 1P-5P (1991).

Simon, G.W., Title, A.M., and Weiss, N.O., "Modeling mesogranules and exploders on the solar surface," *Astrophys. J.* **375**, 775-788 (1991).

Stein, R.F., and Nordlund, Å., "Topology of convection beneath the solar surface," *Astrophys. J.* **342**, L95-L98 (1989).

Title, A.M., Tarbell, T.D., Topka, K.P., Ferguson, S.H., Shine, R.A., and the SOUP team, "Statistical properties of solar granulation derived from the SOUP instrument on Spacelab 2," *Astrophys. J.* **336**, 475-494 (1989).

Turner, J.S., "Turbulent entrainment: the development of the entrainment assumption, and its application to geophysical flows," *J. Fluid Mech.* **173**, 431-471 (1986).

Yamada, H., and Matsui, T., "Preliminary study of multiple slip-through of a pair of vortices," *Phys. Fluids* **21**, 292-294 (1978).

Zahn, J.-P., "Convective penetration in stellar interiors," *Astron. Astrophys.* **252**, 179-188 (1991).

Chapter 11

TURBULENT CONVECTION: WHAT HAS ROTATION TAUGHT US?

JOSEPH WERNE[1,2]

[1] *National Center for Atmospheric Research*
P.O. Box 3000, Boulder, CO 80307-3000

[2] *Joint Institute for Laboratory Astrophysics &*
Laboratory for Atmospheric and Space Physics
University of Colorado, Boulder, CO 80309-0440

Rotation alters turbulent convection in fundamental ways. Thermal plumes forming on domain boundaries spin-up and rotate faster than the external rotation rate. Their shapes differ from plumes in nonrotating convection, as do their basic motions. They participate in vigorous lateral mixing with neighboring plumes as they mutually advect. Unstable mean density stratification results. Despite these changes, other aspects of nonrotating convection remain (*e.g.*, the nonclassical heat-transport efficiency $Nu \sim Ra^{2/7}$ is seen in both cases). The similarities shared by and differences between rotating and nonrotating convection teach us much about turbulent transport in general.

11.1   Introduction

There are examples in nature where turbulent fluid motion is induced by buoyancy, then molded by rotation. These include many astrophysical and geophysical flows, from solar and giant planetary convection to oceanic deep convection. In each case, differential heating initiates convective motion while spinning of the star or planet provides the basic rotation of the fluid. Ordered structures and flow coherence often result and can dominate the dynamics, features not anticipated by classical stochastic theories of turbulence. Plumes and vortices, overturning circulations, zonal jets, and ocean gyres are seen.

Attempts to characterize some of the fundamental aspects of buoyancy induced turbulence have proceeded through laboratory and numerical studies of incompressible convection. The highly nonlinear nature of the flows make purely theoretical inquiry extremely difficult (or impossible) without guidance from empirical sources. In this paper I highlight some of the important discoveries provided by recent laboratory experiments on turbulent Rayleigh-Bénard convection, both in rotating tanks and in stationary vessels, and the theoretical ideas they have inspired. I then describe new numerical solutions for rotating convection (computed by Keith Julien, Sonya Legg, James McWilliams and me) which help evaluate some of the theoretical assumptions.

11.2   Nonrotating Rayleigh-Bénard convection

Rayleigh-Bénard convection is a classic experiment in which a fluid layer is heated from below. The magnitude of the buoyancy force which results is measured by the non-dimensional Rayleigh number $Ra = g\alpha\Delta L^3/\nu\kappa$, where Δ is the temperature drop maintained across the layer's depth L. g is the local acceleration due to gravity. α, ν, and κ are the fluid's thermal expansion coefficient, kinematic viscosity, and thermal diffusivity, respectively. For values of Ra larger than $O(10^3)$, buoyancy overcomes diffusive processes, and convective motion ensues (Chandrasekhar, 1961). For much larger values of Ra, $O(10^7)$, only very sharp gradients in the flow fields may elevate viscous drag and thermal diffusion sufficiently to balance the excessive buoyancy force. As a result, at these large Rayleigh numbers thin shear layers and thin thermal plumes participate in the fluid dynamics (Heslot et al., 1987; Castaing et al., 1989; see also the Adrian's contribution in this volume). The Rayleigh numbers appropriate to geophysical and astrophysical fluid layers, such as the earth's planetary boundary layer, the oceanic mixed layer, and the solar convection zone, are much larger, attaining values of the order 10^{18}. The flows that result at such high Rayleigh numbers possess 1) many thin plumes which punctuate the buoyancy and motion fields at small scales, but also 2) sufficient dynamic range for plumes to act collectively to organize larger scales of motion.

In 1989, Sano, Wu & Libchaber hinted at the ability of plumes to organize larger scales of convective motion in their low-temperature Helium-gas experiments (Sano et al., 1989). A single persistent turbulent roll filled their 8.7 cm tall, 8.7 cm diameter vessel for $Ra > 4 \times 10^7$. Though such a unit aspect (width-to-height) ratio roll does not evoke an image of 'large scale', for Ra below 4×10^7 (but $> 3 \times 10^5$), single rolls are unstable, frequently reversing orientation.[1] More dramatic evidence for large-scale coherence in laboratory convection is offered by Krishnamurti's $40 \times 40 \times 1$ experiments with water (Krishnamurti, 1995). Even in this very wide container, a domain-filling pattern develops in which plumes cluster together and march in 'soldier-row' formation across the cell.

The organized motions in these experiments affect the transport properties of the fluid layer. For example, at Rayleigh numbers above which plumes spontaneously appear, the heat-transport law for flux through the layer becomes less efficient. Heslot et al. report $Nu \sim Ra^{1/3}$ when no plumes are present, but $Nu \sim Ra^{2/7}$ for $Ra > 4 \times 10^7$ (Heslot et al., 1987). They christened the 2/7 state 'hard turbulence' and dubbed the 1/3 regime 'soft'. Similarly, Krishnamurti sees inefficient heat transport in her organized plumey flows, actually consistent with an even smaller exponent: 1/4 (Krishnamurti, 1995). Here the Nusselt number $Nu = H/(\kappa\Delta/L)$ is the normalized heat flux H. Though the small difference between these exponents can appear academic when the data are plotted on a logarithmic axis (which is typical), when a linear axis is used the difference is clearly sizeable. For example, at $Ra = 10^{18}$ the 1/3 law predicts more than three times the heat transport than does the 2/7

---

[1] Despite the axisymmetric geometry of the container, the rolls in these experiments flowed only in a single plane, pinned by the few thermal probes used to record fluid temperatures (Sano et al., 1989).

law (for laws that coincide at $Ra = 4 \times 10^7$).

The experiments by Libchaber and coworkers have drawn a good deal of attention from theorists for one important reason: their impressive range in Ra. By using Helium gas near its triple point (~ 5 K), Ra up to 10^{15} has been obtained (Wu & Libchaber, 1992). Over seven decades in Ra means the exponent can be determined to at least three significant digits. Also, just three decades shy of geophysical and astrophysical values, the experiments promise relevance to realms of turbulence occurring in nature. With these motivations, theorists have attempted explanation of the 2/7 exponent.

11.3   Turbulent convection theories

11.3.1   Priestley's theory

Before reviewing modern theories for turbulent convection, it is instructive to review the classic theory for the 1/3 law; it is due to Priestley (Priestley, 1954, 1959; see also Howard, 1966; Malkus, 1963; Spiegel, 1962). Priestley suggested that all of the temperature drop imposed across a fluid layer should concentrate in thin thermal layers adjacent to the top and bottom boundaries. The reason for this partitioning of the flow is that while turbulent motion efficiently transports heat in the interior of the layer, only diffusive processes act at the domain boundaries where the normal velocity is zero. The heat flux H through the entire layer can therefore be estimated by measuring the thickness λ of these thermal boundary layers: $H \approx \Delta/2\lambda$, from which we have $Nu \approx L/2\lambda$. Priestley further argued that for sufficiently large values of L (or equivalently Ra), the thin boundary layers should be incapable of communicating to each other across the turbulent layer separating them; hence, $\lambda \neq \lambda(L)$. With these two expressions and the definition for Ra, it is readily shown that $Nu \sim Ra^{1/3}$.

11.3.2   Kadanoff, Zaleski & Zanetti's theory

Because thermal plumes appeared simultaneously with the 2/7 law in the experiments of Heslot *et al.*, the first theoretical attempt by Kadanoff, Zaleski & Zanetti to derive the 2/7 law concentrated on plumes (Castaing *et al.*, 1989). They conjectured a 'mixing zone' where plumes grow out of the boundary layers, then mix through turbulent agitation before merging with the bulk fluid. The crux of their argument is that the dynamics of plumes are dominated by two competing effects: buoyancy and drag. Their picture of plumes consists of fragmented pieces of highly unstable boundary layers with temperatures $\Delta/2$ and thicknesses λ. As these plumes move under the buoyancy force $g\alpha\Delta/2$, viscous drag $\nu u/\lambda^2$ acts to slow them. Here u is the plume velocity. Balancing these two, we obtain an expression relating u and λ:

$$\frac{uL}{\kappa} \sim Ra\left(\frac{\lambda}{L}\right)^2 . \tag{11.1}$$

In addition to Eq. (11.1), Kadanoff, Zaleski & Zanetti made use of earlier attributions to turbulent convection, namely Priestley's relation for λ and Nu

$$Nu \sim \frac{L}{\lambda} , \tag{11.2}$$

and Prandtl's view of the turbulent bulk (Prandtl, 1932) in which the heat flux is carried by eddy motion, $H \sim u_c \Delta_c$, or

$$Nu \sim \left(\frac{u_c L}{\kappa}\right)\left(\frac{\Delta_c}{\Delta}\right) , \qquad (11.3)$$

and the kinetic energy of the eddies is gained at the expense of buoyancy work as they traverse the layer, $u_c^2 \sim g\alpha\Delta_c L$, or

$$\left(\frac{u_c L}{\kappa}\right)^2 \sim \sigma Ra\frac{\Delta_c}{\Delta} . \qquad (11.4)$$

Here Δ_c and u_c are measures of the temperature and velocity of the dominant eddies in the turbulent interior. The system is closed with the argument that as plumes merge with the bulk turbulence, they attain velocities consistent with the interior: $u \sim u_c$. Combining the four expressions Eqs. (11.1)–(11.4) for the four unknowns Nu, u_c, Δ_c, and λ, the 2/7 law is derived (along with other laws for u_c and Δ_c). All of the resulting laws agree with experimental observation (Castaing et $al.$, 1989 for Nu and Δ_c; Shen et $al.$, 1995 for u_c).

11.3.3   Shraiman & Siggia's theory

In 1990, Shraiman & Siggia proposed an alternate theory (Shriaman & Siggia, 1990). This time plumes would not be the central players, but rather only the intermediaries necessary to organize the large-scale circulation. In fact, Shraiman & Siggia propose the 2/7 law emerges precisely when the dominant circulation is fast enough to produce shear-flow turbulence near the boundaries. They estimate the dissipation in the layer as $\varepsilon \approx u_*^3/L$, where $u_* = \sqrt{\nu(\partial U/\partial z)_{wall}}$ is the friction velocity, U being the mean velocity parallel to the boundary (wall) and z the normal direction ($e.g.$, Tennekes & Lumley, 1972). For such wall-bounded turbulent-shear layers, very near the wall (in the so-called viscous sub-layer) the velocity profile is linear $U = z(\partial U/\partial z)_{wall}$ ($e.g.$, Tennekes & Lumley, 1972). By assuming a thermal boundary layer λ so thin that it nestles within the viscous sub-layer, Shraiman & Siggia simplify the heat equation and obtain the result

$$Nu \sim (\partial U/\partial z)_{wall}^{1/3} . \qquad (11.5)$$

Then, employing the exact relation (Shraiman & Siggia, 1990)

$$\frac{\varepsilon L^4}{\kappa^3} = \sigma Ra(Nu - 1) , \qquad (11.6)$$

the asymptotic ($Nu \gg 1$) scaling laws $Nu \sim Ra^{2/7}$ and $(\partial U/\partial z)_{wall} \sim Ra^{6/7}$ are obtained. The new prediction for the wall stress has been directly verified with 2D numerical simulation (Werne, 1993).

11.3.4   Cautionary comment on scaling theories

The stark contrast between these two approaches points to the caution with which one must digest scaling theories for turbulent fluids: derivation of a known result does not guarantee the underlying assumptions sufficient to realize the behavior in nature. For theoretical approaches to laboratory convection to be useful to those who model natural flows at $Ra \sim O(10^{18})$, the necessary and key dynamical ingredients responsible for the observed transport laws must be identified. Is the mere presence of plumes enough to restrict the heat flux, as Kadanoff, Zaleski & Zanetti suggest? Or do boundary conditions play a crucial role as suggested by Shraiman & Siggia? Or is another theoretical description in fact correct?

11.3.5 She's theory

She offered an alternate view to Kadanoff, Zaleski & Zanetti in the same year, 1989 (She, 1989). He suggests a simple modification to Kolmogorov's 1941 description of homogeneous and isotropic turbulence (Kolmogorov, 1941), one that includes buoyancy. Two major changes accompany the inclusion of buoyancy: 1) anisotropic forcing and 2) non-conservative cascades of energy to dissipative scales. She's theory incorporates only the latter. He proposes Kolmogorov's well-known scaling law relating the velocity of an eddy v_ℓ to its size ℓ, $v_\ell \sim \varepsilon^{1/3}\ell^{1/3}$, be modified to include injection of energy from the buoyancy force: He suggests

$$v_\ell \sim \varepsilon^{1/3}\ell^{1/3}\left(\frac{\ell}{\delta}\right)^{\beta} . \tag{11.7}$$

Here δ is the thermal dissipation scale of the flow, and β is a constant exponent. Like Kadanoff, Zaleski & Zanetti, She also employs Prandtl's arguments for turbulent convection,

$$H_\ell \sim v_\ell \Delta_\ell \qquad \text{and} \qquad v_\ell^2 \sim g\alpha\Delta_c\ell , \tag{11.8}$$

only here She considers quantities appropriate to the scale ℓ of individual eddies. A noteworthy exception to this rule, however, is the suggestion that eddies of all scales gain kinetic energy at the expense of buoyancy work from the *largest scale*, possessing the temperature Δ_c. By examining Eqs. (11.7) and (11.8) at both scale L and δ, and by suggesting the equivalence of δ and λ, She obtains $\beta = 1/6$ as well as the 2/7 law. She's computed spectrum for turbulent convection derived in this theory has not been seen in experiment or simulation.

11.3.6 Yakhot's theory

The fourth and final attempt to derive the 2/7 law is due to Yakhot (Yakhot, 1992). Like She, Yakhot concentrates on spectral properties and modifies one of Kolmogorov's turbulence formulas to include buoyancy effects. The starting point here is Kolmogorov's so-called 4/5 law for velocity differences in decaying homogeneous and isotropic turbulence (Kolmogorov, 1941). Yakhot suggests that on sufficiently small scales, the turbulent pressure field reorients and isotropizes motions initiated by buoyancy, eventually leading to isotropic fields on these small scales. He suggests, therefore, that Kolmogorov's 4/5 relation applies to small scales, with only the minor addition of a randomly oriented buoyancy-work term. Yakhot then proceeds to suggest a scaling theory by examining velocity differences at large separations of order L. A few errors are made by Yakhot, such as proposing an eddy diffusivity inconsistent with the exact result $Nu = N$ where N is the total thermal dissipation in the layer. Nevertheless, these technical errors can be corrected (Werne, 1996) and, with She's assumption ($\delta \approx \lambda$), the 2/7 law can in fact be derived.[2] Despite this achievement, a major conceptual flaw in the theory is the employment of isotropic gravity to deduce scaling behavior. It is certain that on the largest scales of motion, gravity is anisotropic; since these scales dominate the transport characteristics, randomly oriented gravity is a questionable concept on which to base a global transport theory.

---

[2]Without correcting the error $Nu \neq N$, She's assumption leads to $Nu \sim Ra^{5/19}$ in Yakhot's paper. Only when one takes $Nu \sim N$ does the 2/7 exponent result.

11.4   Rotating Rayleigh-Bénard convection

When a convecting fluid layer is rotated, basic changes in the dynamics take place. Both the detailed motions of plumes and their collective interaction are affected. One might expect, therefore, that the global transport properties (e.g., the total heat flux) should also exhibit dependence on rotation.

In 1969, Rossby was the first to conduct an extensive study on rotating Rayleigh-Bénard convection (Rossby, 1969). With two parameters to vary, Ra and Ta, instead of only one, data collection can be overwhelming. Nevertheless, Rossby presents an impressive series of Nu curves in Ra-Ta space. The non-dimensional Taylor number $Ta = (2\Omega L^2/\nu)^2$ is proportional to the square of the rotation rate Ω. His data span $0 < Ta < 10^8$ and $0 < Ra < 10^7$, with individual experiments sufficiently numerous to individually plot smooth contours of $Nu \in [1, 1.5, 2, 2.5, 3, 4, 5, 6, 7, 8, 9, 10, 11, 12]$. Though quantitative conclusions are difficult to draw from Rossby's data, most importantly because the aspect ratio is varied between some experimental series,[3] it appears that Nu is relatively less sensitive to Ta (than to Ra) for sufficiently large Ra.

More recently, Zhong et al. report new experiments in rotating Rayleigh-Bénard convection (Zhong et al., 1993). Like Rossby, Zhong et al. use a cylindrical vessel; theirs has aspect ratio 2. Visualizations of their flow at $Ra = 7 \times 10^6$ and $Ta = 1.8 \times 10^7$ reveal motions of interacting vortices which swirl cyclonically (in the direction of rotation $\vec{\Omega}$) before spiraling toward each other and finally merging. Spinning plumes have been seen in convection experiments before (Nakagawa & Frenzen, 1955; Rossby, 1969; Boubnov & Golitsyn, 1986); however, Zhong et al. were the first to quantify the co-rotation rate and merger dynamics.

In addition to these studies on Rayleigh-Bénard convection between rigid, constant-temperature boundaries, other experiments target configurations more appropriate to geo-physical flows. For example, Boubnov & Golitsyn (1986, 1990) study steady-state convection bounded by a rigid, constant-heat-flux surface below and a free surface above; Fernando et al. (1991) use a similar configuration to study the transient dynamics when bottom heating is turned on instantaneously; finally, Maxworthy & Narimousa (1994) initiate a surface buoy-ancy flux through salt release, either through the entire upper surface or confined to a smaller patch. Though the results from these studies are interesting, particularly as they address the role rotation plays in determining characteristic length- and velocity scales, here we concentrate only on statistically steady Rayleigh-Bénard convection to focus our discussion.

11.5   Numerical simulation of rotating convection

To illustrate some of the aspects of rotating convection discovered experimentally (and mentioned above), as well as some new results, I now present numerical solutions com-puted by Julien, Legg, McWilliams & Werne. The algorithm used is a pseudo-spectral Fourier/Chebyshev-'tau' method developed by Keith Julien and me. Side boundaries are periodic, while the top and bottom surfaces are held at fixed temperatures. Separate series of calculations are performed with no-slip and stress-free velocity conditions. All simulations resolve the dissipation scale. For more details on the simulations, see Julien et al., 1996b.

---

[3]Rossby's experiments include aspect ratios ranging from 6.4 to 31. He does not report the value of the aspect ratio for individual data points on his $Nu(Ra, Ta)$ diagram.

11.5.1   Intermittent flow fields

Figure 11.1 presents the temperature and vertical vorticity on slices through the three-dimensional domain at increasing values of Ra and Ta. For all simulations, the ratio Ra/Ta is held fixed at $(0.75)^2$ so as to keep rotation and buoyancy time-scales comparable. The Prandtl number $\sigma = \nu/\kappa$ is 1.0. The associated color tables for the figure are included in Fig. 11.2. From the figure we see that the horizontal scale of convection decreases as Ta is increased; this is predicted by linear stability calculations. The horizontal line at the bottom of each column in the figure indicates the wavelength of the most unstable linear mode. At low Ra, convection takes on the pattern of chaotic rolls (Fig. 11.1, $Ra = 3.1 \times 10^4$). For increasing Ra, however, the fields become punctuated by more closely spaced plumes; see Fig. 11.1, $Ra \geq 2.5 \times 10^6$. By $Ra = 1.1 \times 10^8$, the temperature field is filled with sharp thermal structures; see Fig. 11.2.

11.5.2   Cyclonic plumes

The vorticity is even more intermittent than the temperature. This is evident in both Fig. 11.1 and the probability density functions (PDFs) for the two fields (Julien *et al.*, 1996a). Strong correlation exists between cyclonic vorticity and thermal plumes, as was observed in the experiments (Nakagawa & Frenzen, 1955; Rossby, 1969; Boubnov & Golitsyn, 1986; Zhong *et al.*, 1993). The cyclonic relative spinning of plumes results from angular momentum constraints: as plumes grow from perturbations on the boundary-layer, they form tubes which extend vertically away from the boundary; their angular velocity amplifies as they contract in the horizontal direction, much like an ice skater pulling in her arms.

Figure 11.4 illustrates the cyclonic skewing of plume vorticity through the profile of vertical-vorticity skewness, $\langle \omega_3^3 \rangle / \omega_{3\ rms}^3$. Here, averages over horizontal planes and time are indicated by $\langle \ \rangle$. PDFs at three points within the layer are also plotted. Near the boundary where plumes form, the skewness is high, and the associated PDF reflects large numbers of intense cyclonic events. As plumes are pulled into the interior by buoyancy, their cyclonicity is dramatically reduced. Values close to zero occur at mid-layer where turbulent interactions between plumes mix and reorient them.

11.5.3   Ekman pumping

A consequence of intense cyclonic rotation for plumes can include Ekman pumping (Ekman, 1905; Gill, 1982), depending on the boundary conditions. Consider a vortex tube that terminates at a plane boundary. If the tube is in geostrophic balance (with outward centrifugal acceleration $(2\vec{\Omega} + \vec{\omega}) \times \vec{v}$ balanced by an inward radial pressure-gradient force $\vec{\nabla} P$), then the nature of the velocity conditions determines the circulation pattern around the tube. For no-slip boundaries, the centrifugal acceleration drops to zero at the surface (because $\vec{v} = 0$). Hence, the pressure gradient prevails there, resulting in a radial in-flow of fluid. In order to conserve mass flow, the radial in-flow must be compensated by an expulsion away from the boundary along the tube axis in z. Stress-free conditions do not produce this kind of shear-induce Ekman pumping because the velocity need not be zero at the boundary.

To illustrate the role Ekman pumping plays in rotating convection, Fig. 11.5 shows the vertical-velocity skewness $\langle W^3 \rangle / W_{rms}^3$ versus depth z for three different simulations. Profile (a) contains the skewness for nonrotating convection between no-slip boundaries. As was found by Moeng & Rotunno, near the boundaries, motion is skewed toward incoming fluid, the dominant dynamics being that of plume bombardment from the opposite boundary

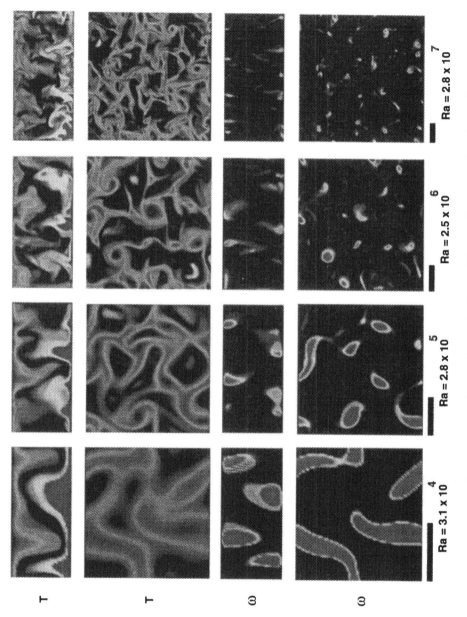

Figure 11.1: Temperature and vertical vorticity for rapidly rotating convection. Views from side and above

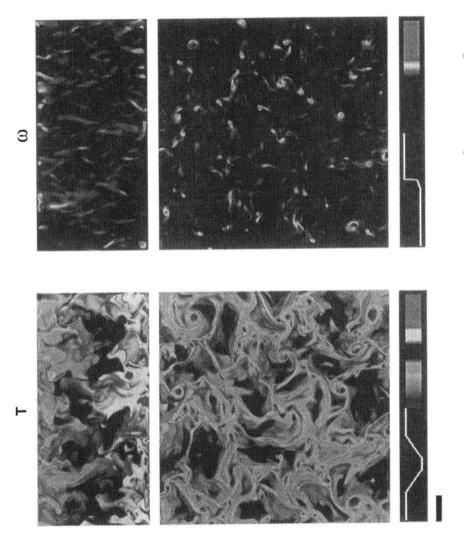

Figure 11.2: Temperature and vertical vorticity for $Ra = 1.1 \times 10^8$ and $Ta = 2 \times 10^8$.

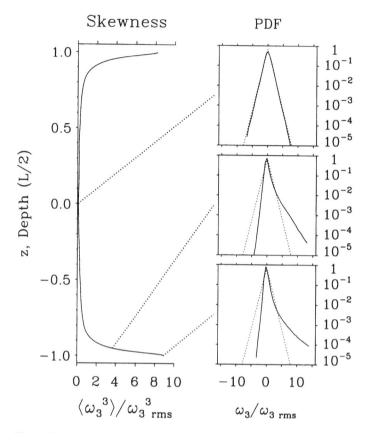

Figure 11.4: Vertical-vorticity skewness and PDFs for rotating Rayleigh-Bénard convection between no-slip boundaries. Exponential (dotted) PDFs are also included for references.

(Moeng & Rotunno, 1990). Profile (c) shows results for rotating convection. Here Ekman effects are clearly evident. Strong pumping away from the boundaries is indicated by the outward skewing of the velocity field. This is exactly opposite to the nonrotating result (a). Profile (b) presents results for rotating convection between stress-free boundaries. Here Ekman pumping is absent, as expected. Furthermore, the bombardment from incoming plumes is also absent, which might not have been expected. As we shall see, rotation dramatically increases the mixing dynamics between plumes, making traversal of the entire layer (and bombardment of the opposite boundary) unlikely.

11.5.4 Linear thermal Ekman layer

A feature of plumes in rotating convection is their distinctive shape when compared to nonrotating convection. Figure 11.6 shows some typical features exhibited by plumes when viewed from above. In particular, in rotating flows, plumes tend to have ring-shaped cross-sections, especially for the divergence and vertical vorticity fields. These features are present in both stress-free and no-slip simulations. A simple linear model of plume formation in the

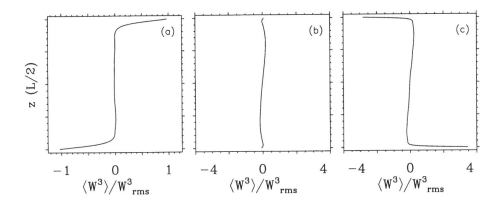

Figure 11.5: Vertical-velocity-skewness profiles. (a) $Ra = 10^7$, $Ta = 0$, no-slip boundaries. (b) $Ra = 1.8 \times 10^8$, $Ta = 3.2 \times 10^8$, stress-free. (c) $Ra = 1.1 \times 10^8$, $Ta = 2 \times 10^8$, no-slip.

boundary layers predicts this ring shape.

Consider a linear thermal Ekman layer whose pressure and horizontal velocity fields are described by the boundary-layer approximation:

$$fv \approx \nu \partial_{zz} u - \partial_x P \, , \tag{11.9}$$

$$-fu \approx \nu \partial_{zz} v - \partial_y P \, , \tag{11.10}$$

$$\partial_z P \approx g\alpha T \, . \tag{11.11}$$

T is a prescribed temperature field, normally taken in the classic Ekman-layer analysis to be zero (Ekman, 1905; Gill, 1982). Here let us consider the simple boundary-layer model

$$T = \begin{cases} 1 - z/\lambda & z < \lambda \\ 0 & z > \lambda \end{cases} \tag{11.12}$$

with $\lambda(x, y)$ a function of the horizontal directions x and y. u and v are the horizontal velocity components, P is the pressure divided by density, and f is the Coriolis parameter 2Ω. Because the system (11.9)–(11.11) is linear, solutions are straightforward to compute; one must take care to match the velocity fields and their derivatives at $z = \lambda$. Solutions are of the form $X = X_E + X_{th}$ where X is the horizontal velocity field in complex notation $X = u + iv$. X_E is the classic Ekman-spiral solution obtained for $T = 0$; i.e., $X_E = q^{-1}\partial_z X_0(1 - e^{-qz})$ for no-slip boundaries, and $X_E = constant$ for stress-free. Here q is the inverse Ekman layer thickness $q = \sqrt{if/\nu}$, and $\partial_z X_0$ denotes the wall stress at $z = 0$. X_{th} has a more complicated form $X_{th} = (\Lambda/q^4\lambda^2) R(q\lambda, qz)$ where $\Lambda = \partial_x \lambda + i\partial_y \lambda$ depends on horizontal gradients in λ, and R is a function of $q\lambda$ and qz.[4] Note that for λ independent of x and y, $X_{th} = 0$. For a Gaussian-shaped bump, $\lambda = e^{-(x^2+y^2)/2}$, X_{th} results in the horizontal divergence and vertical vorticity shown in the top row of Fig. 11.6. Interestingly, local Ekman pumping can in fact act near a stress-free boundary, as is evident in the computed horizontal divergence field (see also Hide, 1964). Nevertheless, the positive pumping in the ring circling the Gaussian

---

[4]For the exact form of $R(q\lambda, qz)$, see Julien et al., 1996b.

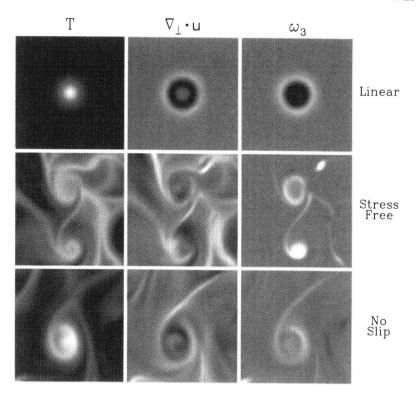

Figure 11.6: Temperature, horizontal divergence, and vertical vorticity on a horizontal plane through developing plumes. The top row depicts fields for the thermal Ekman layer result-ing from a Gaussian-shaped 'plume' on the thermal boundary layer; see text. The middle (bottom) row displays fields for numerical simulations with stress-free (no-slip) boundaries within (slightly above) the mean hot thermal boundary layer. The shading is lighter for more intense temperatures, horizontal convergence, and cyclonic flow.

bump is compensated (exactly) by the negative pumping within the ring; therefore, no net thermal Ekman pumping results. Though details in the computed fields differ from this simple analysis (presumably because of an imperfect prescription for T), general features are shared by both.

11.5.5   Nonlinear Ekman spirals

Though the linear analysis just presented correctly predicts the qualitative shapes of plumes as they emerge from the boundary layers, its quantitative value is limited because of the high degree of nonlinearity in the plumes. To demonstrate, Fig. 11.7a compares the predicted linear Ekman spiral (dotted curve) with three hodographs — plots of $v(x_i, y_i, z)$ versus $u(x_i, y_i, z)$ — at slightly different horizontal positions near a plume. Nonlinear effects are evident in the numerical solutions by their more tightly spiraling velocity fields (Carrier, 1971; McWilliams, 1971). The locations (x_i, y_i) at which the hodographs are produced are

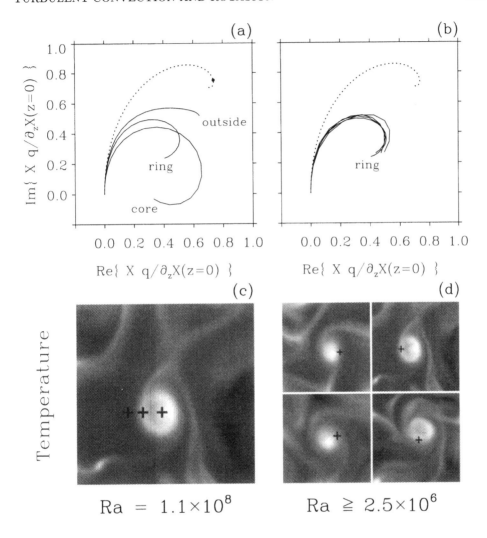

Figure 11.7: (a,b) Velocity hodographs $(v(x_i, y_i, z)$ versus $u(x_i, y_i, z)$ at a single horizontal location $(x_i, y_i))$ associated with plumes over no-slip boundaries. (a) Hodographs within the 'core' of a plume, in the 'ring' region at the edge of a plume, and just 'outside' a plume for $Ra = 1.1 \times 10^8$. (b) Hodographs in the ring region of plumes for high Ra ($Ra = 2.5 \times 10^6$, 8.4×10^6, 2.8×10^7 and 1.1×10^8). (c) Temperature on a plane just above the hot thermal boundary layer for $Ra = 1.1 \times 10^8$. Hotter fluid is depicted with lighter shades. Crosses indicate the locations of the hodographs in (a). (d) Temperature above the boundary layer for the four values of Ra depicted in (b). Ra increases from left to right with the highest Ra shown in the upper right corner and the lowest in the lower left.

indicated in Fig. 11.7c on a plane positioned just above the mean thermal boundary layer. These locations are chosen to include the 'core' of a plume, the 'ring' region at the edge of a plume, and the region just 'outside' a plume for $Ra = 1.1 \times 10^8$. The solid dot on the classic solution marks the height to which the simulation results are plotted, roughly twice the thermal boundary-layer thickness. Hodographs are normalized by $q/\partial_z X_0$ and rotated by $45°$ so they may be compared easily.

While the core and outside regions exhibit substantial variability from plume to plume (due to different plume 'ages' and different instantaneous flow conditions external to the plume), the ring region appears better defined with similar Ekman spirals for different plumes. In fact, for sufficiently high Ra ($Ra \geq 2.5 \times 10^6$), the tightness of the Ekman spiral (i.e., the nonlinearity) in the ring region appears to saturate, producing spirals of nearly identical shape for plumes at different Ra; see Figs. 11.7b and 11.7d. Solutions with $Ra < 2.5 \times 10^6$ exhibit weaker Ekman spirals, tending toward the classic linear solution as Ra (and hence, the nonlinearity) is decreased.

11.5.6   Plume-plume interactions

Having discussed the manner in which the dynamics of isolated plumes are changed by external rotation, it is useful now to characterize how rotation molds the interactions between plumes. As is stated above, all of the plumes in our simulations rotate cyclonically. Therefore, mutual advection of neighboring plumes induces co-rotation. When sufficiently close, co-rotating plumes eventually merge, as Zhong et al. observe. Furthermore, if the dynamical time-scale for advection is short compared to diffusive processes, mergers must be accompanied by filamentation and expulsion of some of the vorticity from the plumes (Melander et al., 1988). Figure 11.8 illustrates the merger and associated filamentation of three neighboring plumes in simulations with no-slip boundaries.

The mutual advection, filamentation, and merger dynamics of a multi-vortex system provides an efficient means to stir the flow field laterally, mixing fluid properties in horizontal planes. Because plumes and vortices are more numerous and more closely packed at higher Ta, vortical interactions and, hence, lateral mixing will be enhanced. The end result is dilution of plume temperature (and an accompanying enhancement of the background temperature) as plumes are mixed with ambient fluid. As a result, only very few plumes survive the tortuous journey across the layer, explaining the absence of boundary bombardments when rotation is sizeable (Fig. 11.5b). Furthermore, at each incremental level away from the boundaries, fewer plumes survive, the spent plumes having deposited all of their thermal content into the background nearer the boundary from which they originated. We should expect, therefore, a reduction in the background intensity of temperature as we move away from the boundaries, and hence, a mean unstable stratification when lateral mixing is intense.

Figure 11.9 shows profiles for temperature with depth at increasing Ra and Ta (holding Ra/Ta fixed). Higher Ra curves possess sharper boundary layers at $z = \pm 1$. The expected mean stratification in the interior is observed and persists (and in fact saturates) at high Ra. Mean thermal gradients have also been observed in laboratory rotating convection (Fernando et al., 1991; Boubnov & Golitsyn, 1990; Hart & Ohlsen, 1996; Ecke, 1996). In contrast, in nonrotating Rayleigh-Bénard convection, temperature is adiabatically stratified with a negligible mean gradient observed outside the thermal boundary layers (e.g., Belmonte et al., 1993).

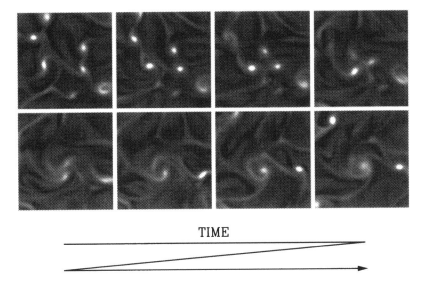

TIME

Figure 11.8: Merger of three vortex filaments over a no-slip boundary. $Ra = 8.4 \times 10^6$. Temperature on a plane just above the hot boundary layer is shown with lighter shades depicting hotter fluid. Since plumes intensify cyclonic vorticity, the light spots also indicate centers of strong vorticity.

11.5.7   Rotating hard turbulence

Despite the many changes rotation brings to turbulent convection, the heat-transport law is virtually unchanged. Figure 11.10 shows the computed convective heat transfer, $Nu - 1$, versus Ra for both no-slip and stress-free solutions. Remarkably, the no-slip solutions exhibit exactly the same scaling behavior (2/7 law) as nonrotating convection experiments. The scaling behavior of the rms vertical velocity and temperature at mid-layer, $i.e.$, u_c and Δ_c, is also identical to the nonrotating case (Julien et $al.$, 1996b).[5]

Stress-free results are also included in the figure. These numerical solutions do not have analogous laboratory counterparts because the slippery surfaces simulated are unphysical. Interestingly however, at the onset of scaling behavior, Priestley's 1/3 law emerges. By $Ra = 4 \times 10^7$ the heat transport becomes slightly less efficient, apparently when the Ekman and thermal boundary layers cross (Julien et $al.$, 1996b). Of course with such a limited range of possible scaling with Ra (10^6–10^8), it is difficult to distinguish two power laws from a different functional form; however, it is clear that the initial behavior of Nu for $Ra > 10^6$ differs between the no-slip and stress-free cases. For more details on the rotating hard-turbulent state, see Julien et $al.$, 1996a,b and Hart & Ohlsen, 1996.

---

[5] After we reported the 2/7 law for rotating convection at fixed Ra/Ta (Julien et $al.$, 1996a,b), Ecke et $al.$ were the first to observe the result in laboratory experiments (Ecke, 1996). Hart & Ohlsen also study rotating hard turbulence, but they do not directly measure Nu (Hart & Ohlsen, 1996).

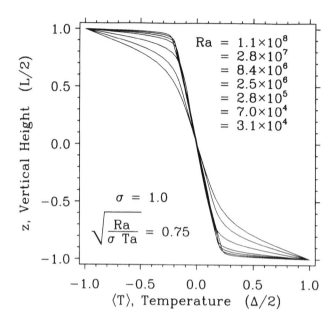

Figure 11.9: Profiles of mean temperature with height z. All solutions shown possess no-slip boundaries.

11.6 Conclusions

In brief summary, the many changes rotation brings to turbulent convection include vortical plumes of ringed shape, Ekman pumping, reduced horizontal length scales, vortex-vortex interactions, mutual advection, enhanced lateral mixing, and mean thermal gradients in the bulk. Despite these myriad changes, rotating solutions with no-slip boundaries exhibit the same 2/7 law for heat transport observed for nonrotating laboratory convection. This is a remarkable result. With plume dynamics dramatically different for rotating and nonrotating flows, identical scaling suggests the relative unimportance of the detailed dynamics of plumes to a general theory for transport in turbulent convection. This is all the more confounding in light of the fact that the appearance of plumes heralds a change in the global transport (*e.g.*, the emergence of the 2/7 law) in nonrotating laboratory experiments.

Sense can be made of these facts when we consider that stress-free solutions exhibit a different heat-transport law. Apparently, boundary details are crucial. Though coherent structures in the form of plumes are required to organize motions necessary for the 2/7 law, reduced heat transport does not occur unless the organized shearing motions act on no-slip boundaries. Of the four turbulent transport theories currently competing to derive the 2/7 law, only Shraiman & Siggia's predicts different behavior for stress-free and no-slip boundaries.

In the opening remarks of NCAR's 1995 summer meeting inspiring these collected works, Kerry Emanuel mentioned laboratory convection and the 2/7 exponent. He speculated that

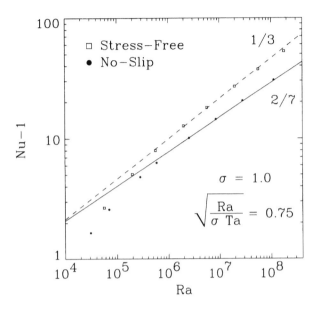

Figure 11.10: Convective heat transport $(Nu - 1)$ versus Ra for both no-slip and stress-free boundaries. The uncertainty in Nu results from the finite numerical integration times used and is less than or equal to the size of the plotting symbols.

differences in laboratory and real-world boundary conditions could prove crucial in determining the applicability (or inapplicability) of the 2/7 law to atmospheric convection. As we see from the work reported here, Emanuel's assessment is accurate. Even so, the correct transport laws for turbulent atmospheric, oceanic, and solar convection remain unknown. To advance we must evaluate the nature of coherent motions on small scales in natural flows and address how the dynamics couple with real-world boundaries to set flux-transport laws. Work addressing topography relevant to the Earth's planetary boundary layer has already begun with convection experiments over 'rough' surfaces (Shen *et al.*, 1996). Nevertheless, much fundamental fluid-dynamic research remains to be done. In future work addressing geophysical and astrophysical fluids (for which boundaries are neither perfectly flat nor no-slip), we may find utility in individual aspects of several of the turbulent-transport theories discussed here.

Acknowledgements

I have learned much about rotating convection from my enthusiastic experimental friends Dan Ohlsen and Bob Ecke, and my collaborators Keith Julien, Sonya Legg and James McWilliams. This work was partially supported by NCAR's Advanced Study Program, NCAR's Visiting Scientist Program, and NSF HPCC grant ECS9217394. Computations were carried out at the Pittsburgh Supercomputing Center under grant MCA935010P. NCAR is sponsored by the National Science Foundation.

Bibliography

Belmonte, A., Tilgner, A. & Libchaber, A. "Boundary layer length scales in thermal turbulence." Phys. Rev. Lett. 70, 4067–4070 (1993).

Boubnov, B. M. & Golitsyn, G. S. "Experimental study of convective structures in rotating fluids." J. Fluid Mech 167, 503–531 (1986).

Boubnov, B. M. & Golitsyn, G. S. "Temperature and velocity field regimes of convective motions in a rotating plane fluid layer." J. Fluid Mech. 219, 215–239 (1990).

Carrier, G. F. "Swirling flow boundary layers." J. Fluid Mech. 49, 133–144 (1971).

Castaing, B., Gunaratne, G., Heslot, F., Kadanoff, L., Libchaber, A., Thomae, S., Wu, X.-Z., Zaleski, S. & Zanetti, G. "Scaling of hard thermal turbulence in Rayleigh-Bénard convection." J. Fluid Mech. 204, 1–39 (1989).

Chandrasekhar, S. "Hydrodynamic and Hydromagnetic Stability." Oxford University Press (1961).

Ecke, R. private communication (1996).

Ekman, V. W. "On the influence of the earth's rotation on ocean-currents." Arkiv. Matem. Astr. Fysik, Stockholm, 2–11, 1–52 (1905).

Fernando, H. J. S., Chen, R-R. & Boyer, D. L. "Effects of rotation on convective turbulence." J. Fluid Mech. 228, 513–547 (1991).

Gill, A. E. "Atmosphere-Ocean Dynamics," Academic Press, New York (1982).

Hart, J. E. & Ohlsen, D. R. "Laboratory experiments on rotating turbulent convection." preprint (1996).

Heslot, F., Castaing, B. & Libchaber, A. "Transitions to turbulence in helium gas." Phys. Rev. A 36, 5870–5873 (1987).

Hide, R. "The viscous boundary layer at the free surface of a rotating baroclinic fluid." Tellus 16, 523–529 (1964).

Howard, L. N. "Convection at high Rayleigh number." In Proc. of the 11th Int. Congr. of Appl. Mech. Munich (Germany) (ed. H. Gortler). pp. 1109-1115. Springer, Berlin (1966).

Julien, K., Legg, S., McWilliams, J. & Werne, J. "Hard turbulence in rotating Rayleigh-Bénard convection." Phys. Rev. E 53, 5557R–5560R (1996a).

Julien, K., Legg, S., McWilliams, J. & Werne, J. "Rapidly rotating turbulent Rayleigh-Bénard convection." J. Fluid Mech. 322, 243-273 (1996b).

Kolmogorov, A. N. Akad. Nauk SSSR 32, 19 (1941).

Krishnamurti, R. "Low Frequency Oscillations in Turbulent Rayleigh-Bénard Convection" preprint (1995).

Malkus, W. V. R. "Outline of a theory of turbulent convection." In Theory and Fundamental Research in Heat Transfer, (ed. J. A. Clark). Pergamon Press, Tarrytown (1963).

Maxworthy, T. & Narimousa, S. "Unsteady deep convection in a homogeneous rotating fluid." J. Phys. Oceanogr. 24, 865–887 (1994).

McWilliams, J. C. "The boundary layer dynamics of symmetric vortices." Ph.D. thesis, Harvard University, unpublished (1971).

Melander, M. V., Zabusky, N. J. & McWilliams, J. C. "Symmetric vortex merger in two dimensions: causes and conditions." J. Fluid Mech. 195, 303–340 (1988).

Moeng, C.-H. & Rotunno, R. "Vertical Velocity Skewness in the Buoyancy-Driven Boundary Layer." J. Atmos. Science 47, 1149-1162 (1990).

Nakagawa Y. & Frenzen, P. "A theoretical and experimental study of cellular convection in rotating fluids." Tellus 7, 1–21 (1955).

Prandtl, L. "Meteorologische Anwendungen der Strömungslehre." Beitr. Physik fr. Atmos. 19, 188–202 (1932).

Priestley, C. H. B. "Convection from a large horizontal surface." Austral. J. Phys. 7, 176–201 (1954).

Priestley, C. H. B. "Turbulent transfer in the lower atmosphere." The University of Chicago Press. Chicago & London (1959).

Sano, M., Wu, X.-Z. & Libchaber, A. "Turbulence in helium-gas free-convection." Phys. Rev. A 40, 6421–6430 (1989).

She, Z.-S. "On the scaling laws of thermal turbulent convection." Phys. Fluids A 1, 911–913 (1989).

Shen, Y., Xia, K.-Q. & Tong, P. "Measured Velocity Boundary Layers in Turbulent Convection" Phys. Rev. Lett. 75, 437 (1995).

Shen, Y., Xia, K.-Q. & Tong, P. "Turbulent Convection over Rough Surfaces" Phys. Rev. Lett. 76, 908–911 (1996).

Shraiman, B. & Siggia, E. "Heat transport in high-Rayleigh-number convection." Phys. Rev. A 42, 3650–3653 (1990).

Spiegel, E. A. "On the Malkus theory of turbulence." Mécanique de la turbulence, Colloque Internationaux de CNRS a Marseille, pp. 181-201. Éditions CNRS (1962).

Tennekes, H. & Lumley, J. L. "A First Course in Turbulence" MIT, Cambridge, MA (1972).

Werne, J. "The structure of hard-turbulent convection in two dimensions: numerical evidence." Phys. Rev. E 48, 1020–1035 (1993).

Werne, J. "2D Convection in Tall Narrow Containers: Implications for Theories of Heat Transport in Hard Turbulence." in preparation (1996).

Wu, X.-Z. & Libchaber, A. "Scaling relations in thermal turbulence—the aspect-ratio dependence." Phys. Rev. A 45, 842–845 (1992).

Yakhot, V. "4/5 Kolmogorov law for statistically stationary turbulence: application to high-Rayleigh-number Bénard convection." Phys. Rev. Lett. 69, 769–771 (1992).

Zhong, F., Ecke, R. & Steinberg, V. "Rotating-Bénard convection: asymmetric modes and vortex states." J. Fluid Mech. 249, 135–159 (1993).

Chapter 12

HELICAL BUOYANT CONVECTION

DOUGLAS LILLY

School of Meteorology
University of Oklahoma
Norman, OK 73019

Many of the severe and tornado-producing thunderstorms in the U. S. incorporate strong rotation about a vertical axis and helicity. The formation processes are moderately well understood, as a result of numerical simulations compared with Doppler radar observations. Moderately strong vertical shear of the mean horizontal wind is required, which may be unidirectional or with strong directional changes with height, but the helicity maintenance mechanisms are apparently more efficient as hodograph curvature increases. The storms are strongly three dimensional, and show little or no tendence to be oriented downshear, as does convection between differentially moving parallel plates or in a sheared boundary layer. It is speculated that this difference is due to effects of water phase change, particularly conditional instability, even though these effects are not obviously required in the dynamic descriptions.

12.1 Rotating thunderstorms and tornadoes

The early explorers and emigrants crossing the North American plains experienced a wide variety of dramatic and hazardous weather phenomena, with severe thunderstorms, hail, lightning, and tornadoes among the greatest risks. One of the documented expressions of the covered wagon emigrants was the "elephant," which came to represent all their overwhelming and potentially deadly experiences while crossing the plains. More specifically, it is sometimes believed that this expression originated in observations of tornadoes pendant from giant cumulonimbus clouds. In any event the central U. S. is the world's bullseye for tornadoes. Again it is partly speculation, but I believe that the extremely prompt and efficient rescue response to the Oklahoma City bombing of April 19, 1995 was largely due to preparations made for tornado disasters, which occur most frequently during the spring and on a similar time and space scale.

The existence of strong rotation in some intense thunderstorms, and its direct association with tornadic vortices, has mainly been recognized since the advent of meteorological radar. The first well-documented storm of this type was observed, oddly, over southern England by Browning and Ludlam (1962), but the mid-U. S. plains soon became the center for observations of them. An established phenomenology was developed after Doppler radar became available for atmospheric research. The recent acquisition by the U. S. Weather Service of operational Doppler radars (NEXRAD or WSR88D) was largely justified for the purpose of tornado warning. For more than a decade after their observational discovery, the meteorological dynamics of thunderstorms with rotation were poorly understood. The majority of the storms rotate cyclonically (counterclockwise in the northern hemisphere) and propagate in a direction to the right of the mean wind vector (right-movers), while a few left-movers with anticyclonic circulations have been described. The rotation is not as a solid body, however, nor does it arise from concentration of earth's angular momentum, as does that of hurricanes and middle latitude cyclonic storms. The center of rotation is, however, observed to be nearly coincident with the center of vertical motion, as is true for hurricanes and middle latitude cyclones near the surface. Cyclonic rotation in updrafts is usually accompanied by anticyclonic rotation in downdrafts. Initially non-rotating storms are sometimes observed to split into two oppositely rotating progeny, which then drift apart as right- and left-movers, although most left movers do not survive long. The splitting process was for a time regarded as similar to the Karman vortex development behind an obstacle, and the lateral motion of rotating storms was thought to be produced by the Magnus effect (Fujita and Grandoso, 1968). Those interpretations are no longer widely accepted, but their schematic (Figure 12.1) remains at least historically interesting, and nearly convincing. The main instantly visible difference between these sketches and real evolution is that the splitting storms usually develop beside the initial updraft, not in its wake.

The most intense and largest storms are usually somewhat isolated, and were dubbed by Browning (1964) "supercells," to distinguish them from the smaller single cell or multicell structures of less intense storms. Supercell and "organized multicell" storms typically have an orderly and durable structure, sometimes remaining quasi-steady or cyclically evolving for hours. When tornadoes form within them, the spatial location and sequencing is fairly predictable, which allows tornado chasers to operate with moderate safety. In the Great Plains the convective available potential energy (CAPE), a thermodynamic estimate of the maximum energy available for conversion to kinetic form, is often extremely large, several thousand $m^2 s^{-2}$, and observed updrafts are correspondingly intense, 50 ms^{-1} or higher. Tornado bearing storms are not always that energetic, however, e.g. in landfallen hurricanes (McCaul, 1991). The common denominator in nearly all well-documented tornadic events is fairly strong vertical shear of the ambient wind, usually with a total vector change of 20 ms^{-1} or more over the lowest 1-4 km, sometimes almost unidirectional and sometimes with considerable direction change. A breakthrough in understanding arose from results of a series of successful numerical simulations by Klemp and Wilhelmson (1978; also Wilhelmson and Klemp, 1978), and by Schlesinger (1980), accompanied by definitive dual Doppler radar studies (Klemp, et al, 1981). This is one of the more dramatic examples of numerical simulation results leading to new concepts. The dynamical processes now believed important are presented in the next section and are illustrated with a small selection of numerical and observational examples. A fairly recent review was prepared by Klemp (1987).

In about the last decade it has been recognized that storms with strong rotation are also quite strongly helical, that is the vorticity and velocity vectors are close to parallel over much of the storm (Lilly, 1986; Brandes, et al, 1988). The dynamic significance of this fact is still somewhat uncertain. Helicity in fluid motion was recently reviewed by Moffatt and Tsinober (1992). Figure 12.2 shows, in a cartoon form, two vortices linked together. The closed contours represent vortex tubes and the partial circles around them (K1 and K2)

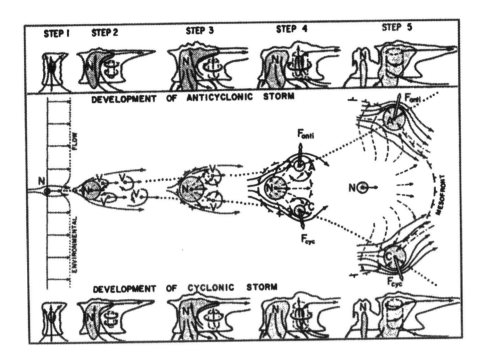

Figure 12.1: A proposed schematic of splitting storms, based on an assumption of wake vortex formation and subsequent lateral motion by the Magnus effect. From Fujita and Grandoso, 1968.

indicate the fluid motion vectors. It can be seen that the flow field of each vortex tends to be parallel to the vortex lines on the other in the vicinity of their intersection. Although Moffatt and Tsinober did not obviously have geophysical applications in mind, the divergent motions of a convective storm are conceptually similar to the circulation around vortex #2, while the rotating updraft in its middle levels and return flow outside could be indicated by vortex #1.

Andre and Lesieur (1977) showed that helical flows are resistant to formation of an inertial range, so that the dissipation rate for the same total energy and outer turbulent scale is reduced. Lilly (1986) suggested that this might contribute to the high energy and durability of these storms. It is not an easy hypothesis to prove, however, as it seems impossible to create, or even imagine, two storms that are alike except for their helicity. Results of a preliminary study aimed at testing the hypothesis was presented by Droegemeier, Bassett, and Lilly (1996). The increased energy associated with helical blocking of the inertial range was found, but results regarding durability were inconclusive.

The convective storm environment of the U.S. plains spawn a large variety of storm types, including nearly two-dimensional squall lines, arc-shaped "bow echo" storms, and many other kinds of organized or chaotic structures. It is sometimes unclear whether these are deter-

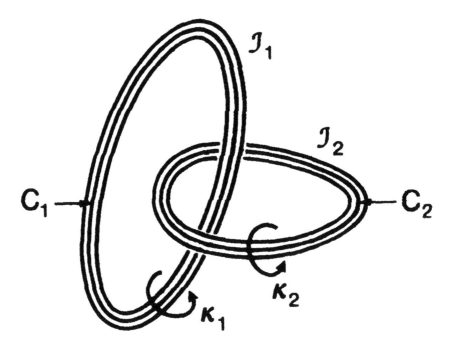

Figure 12.2: Linked vortex tubes that generate helicity. From Moffatt and Tsinober (1992).

mined mostly by the environmental profiles or by the varied and complex storm initiating mechanisms. The emphasis here is on the types which develop rotation around vertical axes, which also are those most likely to produce tornadoes. The mechanisms by which tornadoes are formed from the larger and less intense elevated circulations are somewhat complex and still uncertain, and will not be considered here except for a brief descriptive paragraph at the end. Since presenting this paper orally, I have become aware of a considerable degree of controversy regarding the significance of helicity in convective storms. Although the description of the mechanisms for generating rotating storms is not in much dispute, my emphasis on helicity should be considered a personal view, shared in many respects by many colleagues but at least partially disputed by others.

12.2   Analysis and Illustrations

The mechanism for generating mesocyclones, the strong rotary circulations in supercell and organized multicell storms, is generally believed to be upward tilting of the horizontal vortex tubes associated with vertical shear of the mean horizontal wind. This is not unrelated to the prototypical mechanism for generation of boundary layer shear turbulence, the horseshoe or hairpin vortex, discussed in a review by Robinson (1991). In the shear driven case the vortex tubes are lifted by a self-interaction process. If an initially straight horizontal tube is lifted a

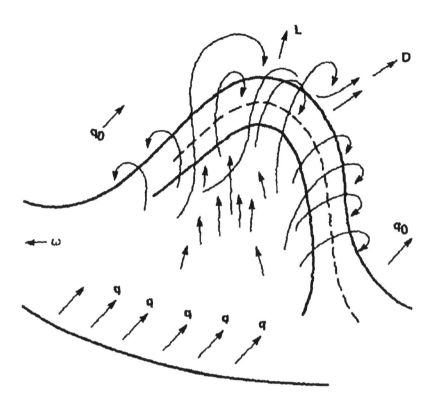

Figure 12.3: Schematic of a horseshoe vortex, formed in boundary layer turbulence. The vortex is tilted at about a 45° angle downshear. From Robinson (1991) and originally Theodorsen (1952).

little, it extends downstream a little, following the mean shear. The circulation generated at the edges of the lifted region then act to lift the central part further. The result is illustrated in Figure 12.3. Lifting by buoyant convection greatly amplifies or replaces this process in mesocyclone evolution, and allows the vortex tubes to become upright and persist without being shredded apart or tilted severely by the mean shear. Figure 12.4 shows a conceptual description. The process is complicated by the density loading and evaporative cooling effects of precipitation as it forms and falls through the cloud. Even when those effects are numerically suppressed in simulations, however, the same general evolution occurs. It may be seen that if the white arrows are closed by a descent branch to symbolize the center of another vortex tube, that tube expands rapidle to the right and left. Although it passes through the updraft tubes the interaction helicity is weak. Later the updraft splits and the two halves become coincident with the vortex pair, with strong helicity. A more detailed description of the storm dynamics follows.

The dynamical outline is based on the Boussinesq equations of motion and continuity, written in inviscid form as

$$\frac{du_i}{dt} = -\frac{\partial \pi}{\partial x_i} + \delta_{i3} b \tag{12.1}$$

$$\frac{\partial u_i}{\partial x_i} = 0 \tag{12.2}$$

where $\pi = (p - \bar{p})\,\bar{\rho}$ and $b = -g\,(\rho - \bar{\rho})\,\bar{\rho}$, with $\bar{\rho}$ a reference density and $\bar{p}$ a pressure profile hydrostatically consistent with $\bar{\rho}$. This system neglects atmospheric compressibility, but still often provides fairly realistic dynamics. Closure of the system requires an evolution equation for buoyancy, which for the atmosphere demands careful consideration of the effects of moisture phase changes. Detailed treatment of those processes is not within the present scope. It is important to recognize, however, that a convectively active environment in the atmosphere above the boundary layer is usually classified as conditionally unstable (see the contributions by Kerr and Lemone in this volume). This requires, among other things, that the rate of temperature decrease with height is between those for dry and for moist adiabatic processes, where the latter is smaller than the former due to release of the latent heat of condensation. For a small upward displacement, an initially unsaturated air parcel will cool with height from adiabatic expansion faster than does its environment, and so will tend to return to its original level. However, if a parcel is displaced upward far enough to start condensing its water vapor, the released latent heat of condensation allows the parcel to become warmer than its environment and accelerate upward. Downward moving parcels are generally stable, unless cooled by evaporating cloud or rain water and/or melting snow. Intense and sometimes dangerous thunderstorm downdrafts are produced by these processes, but they are usually confined to levels fairly close to the surface and are not as strong contributors to kinetic energy generation as are the moist updrafts.

The equation for vector vorticity, ω_i, may be written as the curl of (12.1) in the form

$$\frac{d\omega_i}{dt} = \omega_j \frac{\partial u_i}{\partial x_j} + \varepsilon_{ij3} \frac{\partial b}{\partial x_j} \tag{12.3}$$

The first term on the rhs includes the so-called "tilting" and "stretching" terms, with the stretching identified as that with j=i. The second term shows that a horizontal buoyancy gradient leads to generation of horizontal vorticity normal to that gradient. Three quadratic quantities, kinetic energy, enstrophy, and helicity, arise from the flow variables considered here, and evolution equations for them can be obtained from (12.1) and (12.3). Here I single out the helicity, $h = u_i \omega_i$ (strictly speaking, helicity density) for attention, since it is a special feature of the kinds of storms under consideration. Within the Boussinesq approximation, the helicity evolution equation is obtained by scalar multiplication of (12.1) by ω_i and of (12.3) by u_i and addition to yield

$$\frac{dh}{dt} = \omega_j \frac{\partial}{\partial x_j}\left(\frac{u_i^2}{2} - \pi\right) + b\zeta + \varepsilon_{ij3} u_i \frac{\partial b}{\partial x_j} \tag{12.4}$$

The first term on the rhs can move helicity around but not generate it except at boundaries, since ω_i is a non-divergent vector. The negative sign inside the parenthesis is interesting. If the flow tends to satisfy a Bernoulli equation, then the two quantities add. The two buoyancy terms appear qualitatively distinct, with the first generating helicity in a buoyant vortex while the second generates it if a horizontal flow is parallel to isolines of buoyancy. The volume integrals of the two terms are, however, the same. Moffatt and Tsinober (1992) present equations for vorticity and helicity which do not include the buoyancy terms, apparently using an unstated assumption of constant density.

A simple explanation of development of storm rotation, as illustrated in Figure 12.4, is based on the vertical part of (12.3), linearized about mean horizontal wind profiles, $\bar{u}(z)$ and $\bar{v}(z)$, as presented by Lilly (1979).

$$\frac{\partial \zeta}{\partial t} + \bar{u}_i \frac{\partial \zeta}{\partial x_i} = \bar{\omega}_i \frac{\partial w}{\partial x_i} = \varepsilon_{ij3} \frac{\partial \bar{u}_i}{\partial z} \frac{\partial w}{\partial x_j} \tag{12.5}$$

It is seen that positive vertical vorticity is generated on the right side of an updraft and negative on the left (oriented with respect to the mean shear vector), with reversed signs in a downdraft. If the mean shear is unidirectional, the horizontal vorticity is normal to that shear direction. Then as the mean flow passes through an updraft and is lifted, the vortex tubes are tilted upward on both sides of it. If the lifting is great enough to reach saturation, the upward tilting will continue and be enhanced by non-linear stretching. Along the outer edges of the updraft, where lifting is slight, the dry stability will allow that part of the vortex tube to settle back to a horizontal axis. At some point the rising and sinking parts of the vortex tubes presumably must fracture, with the undisplaced tubes reconnecting to the rear of the storm and the lifted tubes on both sides connecting at the top, although that process is not illustrated in Figure 12.4.

Rotunno and Klemp (1985) have shown that the linear analysis can be extended to non-linear flow, leading to the expression

$$\zeta = \omega_i \frac{\partial h}{\partial x_i} \tag{12.6}$$

where h is the height or height displacement of a surface of equivalent potential temperature, θ_e. The gradients are confined to the horizontal but ω_i is not linearized. This is derived by assuming conservation of θ_e, using it to form an equation for equivalent potential vorticity, and assuming that gradients of θ_e are proportional to buoyancy gradients, which is strictly true only after condensation. The equivalent potential vorticity is assumed zero in the initially shearing mean flow and supposedly remains that way as the vortex tubes are lifted to the vertical.

I have here implicitly assumed, following linear analysis and simulations of early storm formation, that a convectively driven updraft will move approximately with the mean flow at some level which it crosses. For reasons to be described, however, the updraft tends to develop a motion component normal to the mean shear. In this circumstance the vorticity generated at the upstream edge (with respect to this normal velocity component) is advected into the center and attains its maximum value there before being removed as it passes across the downstream edge. The horizontal vorticity has a component in the same direction as (or opposite to) the flow velocity, and is described as "streamwise vorticity" (Davies-Jones, 1984), as distinct from the "crosswise vorticity" of Figure 12.4. Since the vertical vorticity also becomes correlated with the vertical velocity, the total flow tends to become helical, as measured by the relative helicity, $h|u||\omega|$, or its arccos, the angle between the velocity and vorticity vectors. From Doppler radar analysis and simulation results it has been found that the vector vorticities and velocities are well correlated in many storms (Brandes, et al, 1988, Wu, 1990). Helicity is, however, not independent of the reference frame, which leads to conceptual ambiguities. It is usually considered that the appropriate frame is that of the moving storm, although that is unknown in advance.

Meteorologists working with convective storms often present mean wind data in the form of a hodograph, in which the horizontal wind vector is plotted at consecutive levels and a line is drawn between them. A merit of such plots is that the shape of the hodograph depends only on the shear profile, not the frame of velocity reference. Since the amplitude and directional variability of the wind shear is found to be more important than the ground-relative wind speed and direction, the hodograph plot is considered the most definitive. Figure 12.5 shows a composite mean hodograph from observations taken close to tornadic storms, mostly in the Great Plains. The mean flow shear direction is strongly curved in the boundary layer, but becomes more unidirectional at higher levels. Most hodographs are curved in the lower levels due to the Ekman effect, but that shown in Figure 12.5 has more curvature and considerably greater shear than would be present in a seasonally averaged sounding in the same area.

Within the constraints of linearized dynamics, a source of streamwise vorticity and helicity occurs when the mean flow hodograph is strongly curved. With hodograph curvature, a convective updraft will usually move in a vector direction on the concave side of the hodograph. In the extreme case of a nearly circular hodograph, a buoyant updraft is expected, from linear analysis, to move with a velocity nearly into the center of that hodograph, and rather similar results are found from numerical simulations. The vorticity is then streamwise at every level, and the upward tilting of vortex tubes produces vertical vorticity uniformly correlated with vertical velocity. Nearly circular hodographs are uncommon, but their occurrence seems to be well correlated with tornadic storms. One of the early storms simulated by Klemp, et al (1981), the "Del City tornado" event, was of this type. More recently McCaul (1991) has shown that landfallen hurricanes which produce tornadoes typically have a nearly circular hodograph of rather large velocity amplitude. Figure 12.6 shows a comparison of McCaul's "top ten" composite, along with a supercell composite similar, though not identical, to that in Figure 12.5.

Several other examples of unexpected tornadic storms have been correlated with highly curved hodographs. It should be noted that the flow in a circular hodograph is a Beltrami (purely helical) flow in the frame of the hodograph center, and its vertically integrated helicity is independent of the reference frame. Kanak and Lilly (1996) investigated the linear stability properties of buoyant convection in a circular hodograph, finding that the most unstable modes typically exhibit large disturbance helicity. In this case the disturbance helicity is obtained from the mean helicity through a Reynolds stress-type term.

Non-linear dynamical principles offer a number of important results, including a probably more important cause for storm motion across the hodograph, at least for unidirectional shear. Lilly (1986) investigated the properties of a postulated Beltrami disturbance flow that has superficial resemblances to supercell storms, with inflow in the low levels, a rotating updraft in the middle, and outflow aloft. Although such a flow is highly ideaized, and does not even allow buoyancy effects, its full non-linearity is attractive and has encouraged further exploration. Further analysis of the helicity equation, (12.4), shows that an unambiguous exchange between mean and disturbance helicity can occur, analogous to the exchange of kinetic energy between sheared mean flows and disturbances. The transformation term is given by $-\overline{u_i w} \partial \bar{\omega}_i / \partial z$, and is analogous to the exchange between mean and disturbance kinetic energies, with the mean flow vorticity replacing velocity in the vertical gradient. If the mean flow relative helicity is large (positive or negative), the energy and helicity exchanges are likely to be of the same (or opposite) signs, generally leading to transfer from means to disturbances. This occurs when the mean flow direction changes with height. For unidirectional shear, the vertical gradients of mean flow velocity and vorticity are not correlated.

The process which apparently gives rise to storm motion lateral to a unidirectionally sheared flow can be described from pressure forcing or azimuthal vorticity generation. The former was shown by Rotunno and Klemp (1982), and will be followed approximately here. In an anelastic or Boussinesq flow, the pressure is found by taking the divergence of (12.1) and solving it as a Poisson equation for π. The time derivatives vanish because of (12.2), and the result may be written

$$\frac{\partial^2 \pi}{\partial x_i^2} = \frac{\partial b}{\partial z} - \frac{\partial}{\partial x_i}\left(u_j \frac{\partial u_i}{\partial x_j}\right) \tag{12.7}$$

The part of the solution arising from the first term on the rhs is a pressure field whose vertical gradient tends to reduce the effects of buoyancy in the vertical equation of motion. Physically it corresponds to the transfer of kinetic energy from vertical motion to horizontal inflow and outflow. The second term may be rewritten as

$$\frac{\partial}{\partial x_i}\left(u_j \frac{\partial u_i}{\partial x_j}\right) = \frac{\partial u_j}{\partial x_i}\frac{\partial u_i}{\partial x_j} = \frac{1}{4}\left(\frac{\partial u_j}{\partial x_i} + \frac{\partial u_i}{\partial x_j}\right)^2 - \frac{1}{4}\left(\frac{\partial u_j}{\partial x_i} - \frac{\partial u_i}{\partial x_j}\right)^2 \tag{12.8}$$

The first term on the rhs of the second equality is one-half the square of the deformation tensor and the second one-half the square of the vorticity vector. Under the assumption that the solution of a Poisson equation has maxima and minima of opposite signs to those of the Laplacian (which is typically fairly valid in the mid-levels of a buoyant thermal), (12.7) and (12.8) indicate that vorticity maxima of either sign correspond to pressure minima, while deformation maxima of either sign correspond to pressure maxima. These features are colloquially termed "spin" and "splat." The analysis by Rotunno and Klemp (1982) differed slightly from the above in the parameter combinations used, but arrived at the same result.

Thus the vorticity maxima found near or within updraft centers can be expected (and in numerical simulations are found) to have low pressure near their centers, which, from (12.4), are usually near the levels of maximum updraft. The consequence is an upward vertical acceleration beneath them or a downward acceleration above them. For the case of a circular hodograph, with a vorticity maximum at or near the updraft maximum, this effect simply amplifies the updraft. For the unidirectional shear case, with the vorticity at the edge of the updraft, the consequences are a little less clear, since either a new updraft or downdraft could ensue or the pressure forcing could tend to displace the original updraft. The possibility of a downdraft is largely removed by conditional instability, as discussed later. The options of either creating a new updraft or encouraging migration of an existing one into it remain, however.

An estimate of the amplitude of a new updraft can be made through use of the azimuthal equation of vorticity, i.e.

$$\frac{d\omega_\phi}{dt} = \omega_r \frac{\partial v_\phi}{\partial r} + \omega_\phi \frac{\partial v_\phi}{\partial \phi} + \zeta \frac{\partial v_\phi}{\partial z} - \frac{\partial b}{\partial r} \tag{12.9}$$

Assuming circular symmetry, $\omega_\phi = -\partial w/\partial r$, $\omega_r = -\partial v_\phi/\partial z$, and $\zeta = r^{-1}\partial(rv_\phi)/\partial r$. Neglecting the buoyancy term and assuming that $d/dt = w\partial/\partial z$, this may be written

$$w\frac{\partial}{\partial z}\left(\frac{\partial w}{\partial r}\right) = \frac{\partial v_\phi}{\partial z}\frac{\partial v_\phi}{\partial r} - \frac{1}{r}\frac{\partial(rv_\phi)}{\partial r}\frac{\partial v_\phi}{\partial z} = -\frac{v_\phi}{r}\frac{\partial v_\phi}{\partial z} \tag{12.10}$$

As a simple and not too unrealistic example I choose v_ϕ to be sinusoidal in the vertical and the derivative of a gaussian function in the horizontal, i.e.

$$v_\phi = V \frac{r}{R} \exp\left(-\frac{r^2}{2R^2}\right) \sin\frac{\pi z}{H} \tag{12.11}$$

where V is the maximum azimuthal velocity amplitude, which occurs at radius R, and H is the column height. Then a solution for vertical velocity vanishing at the top and bottom and with maximum amplitude at r=0 is

$$w = \pm V \exp\left(-\frac{r^2}{2R^2}\right) \sin\frac{\pi z}{H} \tag{12.12}$$

This indicates that the maximum vertical velocity induced by the vortex low pressure is about equal to the maximum vortex circulation velocity, but the existence of the buoyancy gradient term in 12.8) will strongly amplify it. The above approach was used by Lilly (1986).

Apparently both the new updraft and migrating updraft options are possible. In the original Wilhelmson-Klemp (1978) simulations, an initial cloudy updraft in a unidirectional shear forms vortical circulations on either side of it. After penetration of a rain-driven downdraft, the original updraft appears to split, with the two halves moving laterally into the position of the vortices, after which both of the rotating convective storms move apart and establish separate quasi-steady identities, with motion components lateral to the hodograph. An alternative evolution was shown in a radar-based observational analysis by Brown (1992), in which a sequence of storm updrafts develop (see Figure 12.7). Although this case has not been successfully simulated (perhaps due to resolution inadequacies), I postulate a dynamic evolution consistent with the observations (though admittedly different from that proposed by Brown). Each updraft center generates a cyclonic vortex to its right, which induces a new updraft center and then another vortex in a cyclic process. Each updraft scion also generates anticyclonic vorticity to its left (looking downshear), which tends to remove or reverse the spin of its ancestor. In this way, the sequence of updrafts moves to the right. A similar sequence was described observationally by Browning and Ludlam (1962) for organized multicell storms (see Fig 12.8). At the time this schematic model was proposed, neither Doppler radar nor 3-dimensional numerical simulations were available and the rotational aspects of the cells were difficult or impossible to discern. The mechanism for new cell formation was assumed to be outflow from downdrafts.

12.3   Further discussion

As above outlined, the dynamics of helical thunderstorms are not particularly bizarre, being largely explained from the standard Boussinesq equations and fluid processes. Yet the process seems unfamiliar to other areas of fluid dynamics. In particular, laboratory and numerical simulations of buoyant convection in shear almost inevitably generate downshear rolls, rather than 3-dimensional elements with strong rotation, as do atmospheric boundary layer flows. In my opinion, two special circumstances associated with conditional instability of a moist atmosphere are largely responsible. First is the maintenance of strong shear in a convectively active environment. Vortex motion will not appear in a well-mixed fluid with strong boundary gradients, because the generation of vertical vorticity requires existence of a mean flow shear near the level of maximum updraft. The mixing out process typical of turbulent flows does not happen in strongly forced regions of the atmosphere, at least not so fast, because of the suppressing effects of conditional instability and the related tendency for a stable lid to develop above the boundary layer turbulence and just below cloud base. The regions of active convection, although intense, typically cover only a few percent of the domain, and only slowly relax an initially strong shear.

The importance of this factor is indicated from results of a set of numerical experiments by Wu et al. (1992), which simulate numerically the evolution of convection between parallel plates with a curved mean flow hodograph. The mean flow is generated by a combination of relative plate motion (Couette flow) and channel (Poisseuille) flow at right angles to the shear direction between the plates. This combination produces a parabolic hodograph in an undisturbed viscous fluid. For small supercritical Rayleigh number, Ra=15000, the mean state remains fairly similar to the original viscous state, the convective cells showed mid-level rotation and the helicity is strong, with an rms angle of $35°$ between the velocity and vorticity vectors. Although there is no real turbulence, the helicity is found to reduce the transfer of energy from the mean flow to the disturbances from that in either a Couette or Poisseuille flow (or their average) in the same conditions. At $Ra \approx 10^6$, however (reported in Wu, 1990), the active convection soon brings most of the flow into a well-mixed state, with the shear confined to thin boundary layers.

The other effect of conditional instability is the bias toward upward motion into and through a dynamically generated low pressure center in the middle or lower troposphere. Without this bias one would expect either ascent from below or descent from above, or possibly both adjacent to each other, or one above the other. With it, downdrafts are discouraged by static stability, and the effect of the central low pressure is to generate updraft motion through it to the tropopause approximately. The updraft is amplified by latent heat release and the vortex generation process is maintained.

Bibliography

Andre, J. C. and M. Lesieur, 1977: Influence of helicity on the evolution of isotropic turbulence at high Reynolds number. J. Fluid Mech., 81, 1987-2007.

Brandes, E. A., R. P. Davies-Jones and B. C. Johnson, 1988: Streamwise vorticity effects on supercell morphology and persistence. J. Atmos. Sci., 45, 947-963.

Brown, R. A.,1992: Initiation and evolution of updraft rotation within an incipient supercell thunderstorm. J. Atmos. Sci., 49, 1997-2014.

Browning, K. A. and F. H. Ludlam, 1962: Airflow in convective storms. Quart. J. Roy. Meteor. Soc., 88, 117-135.

Browning, K. A., 1964: Airflow and precipitation trajectories within severe local storms which travel to the right of the mean wind. J. Atmos. Sci., 21, 634-639.

Davies-Jones, R. B., Streamwise vorticity: the origin of updraft rotation in supercell storms. J. Atmos. Sci., 41, 2991-3006.

Droegemeier, K. K., G. M. Bassett and D. K. Lilly, 1996: Does helicity really play a role in supercell longevity? Preprint volume, Americal Meteorological Society 18th Conference on Severe Local Storms, Feb. 19-23, 1996, pp. 205-209.

Fujita, T. T. and H. Grandoso, 1968: Split of a thunderstorm into cyclonic and anticyclonic storms and their motion as determined from numerical model experiments. J. Atmos. Sci., 25, 416-439.

Kanak, K. M. and D. K. Lilly, 1996: The linear stability and structure of convection in a circular mean shear. J. Atmos. Sci., 53, 2578-2593.

Klemp, J. B., 1987: Dynamics of tornadic thunderstorms. Ann. Rev. Fluid Mech., 19, 369-402.

Klemp, J. B., and R. B. Wilhelmson, 1978: The simulation of three-dimensional convective storm dynamics. J. Atmos. Sci., 35, 1070-1096.

Klemp, J. B., R. B. Wilhelmson and P.S. Ray, 1981: Observed and numerically simulated structure of a mature supercell thunderstorm. J. Atmos. Sci., 38, 1558-1580.

Lilly, D. K., 1979: The dynamical structure and evolution of thunderstorms and squall lines. Annual Review of Earth and Planetary Sciences, 7, 117-161, Annual Review, Inc., Palo Alto, CA.

Lilly, D. K., 1986: The Structure, Energetics and Propagation of Rotating Convective Storms. Part II: Helicity and Storm Stabilization. J. Atmos. Sci., 43, 126-140.

Maddox, R. A., 1976: An evaluation of tornado proximity wind and stability data. Mon. Wea. Rev., 104, 133-142.

McCaul, E. W., Jr., 1991: Buoyancy and shear characteristics of hurricane tornado environments. Mon. Wea. Rev., 119, 1954-1978.

Moffatt, H. K. and A. Tsinobar, 1992: Helicity in laminar and turbulent flow. Ann. Rev. Fluid Mech., 24, 281-312.

Robinson, S. K., Coherent motions in the turbulent boundary layer. Ann. Rev. Fluid Mech., 23, 601-639.

Rotunno, R., and J. B. Klemp, 1982: The influence of the shear-induced pressure gradient on thunderstorm motion. Mon. Wea. Rev., 110, 136-151.

Rotunno, R., and J. B. Klemp, 1985: On the rotation and propagation of simulated supercell thunderstorms. J. Atmos. Sci., 42, 271-292.

Schlesinger, R. E., 1980: A three-dimensional numerical model of an isolated thunderstorm. Part II: Dynamics of updraft splitting and mesovortex couplet evolution. J. Atmos. Sci., 37, 396-420.

Wilhelmson, R. B. and J. B. Klemp, 1978: A three-dimensional numerical simulation of splitting that leads to long-lived storms. J. Atmos. Sci., 35, 1037-1063.

Wu, W.-S., 1990: Helical buoyant convection. Ph. D. dissertation, University of Oklahoma. 161 P.

Wu, W.-S., D. K. Lilly and R. M. Kerr, 1992: Helicity and thermal convection with shear. J. Atmos. Sci., 1800-1809.

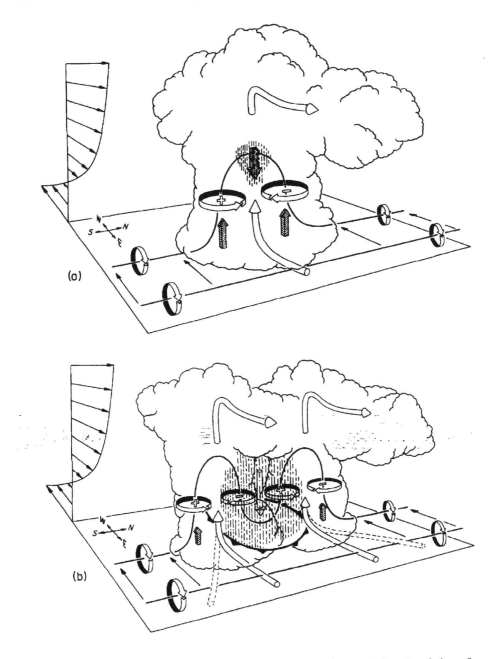

Figure 12.4: Schematic of a convective updraft (white arrows) in a unidirectional shear flow, with wind speed profile shown on the left. The vortex tubes are lifted and generate vertical vorticity along the edges of the updraft. Rain is forming in the center which will soon generate a central downdraft. From Klemp (1987).

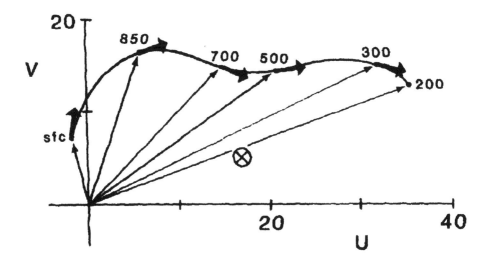

Figure 12.5: Mean wind hodograph for data taken in 62 cases of nearby tornadic storms. The axes are labelled in ms^{-1}. The mean storm motion is shown by the circled X. The heights are shown in Hectopascal pressure units, with 700 corresponding approximately to 3 km and 300 to 10 km. Surface pressure is about 1000. From Maddox (1976).

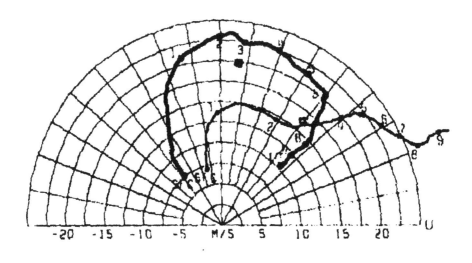

Figure 12.6: Composite hodographs for a set of hurricane-generated tornadoes (the large nearly circular line) and an Oklahoma supercell composite (the lower line veering off to the right). The radial coordinate is velocity in ms^{-1}.

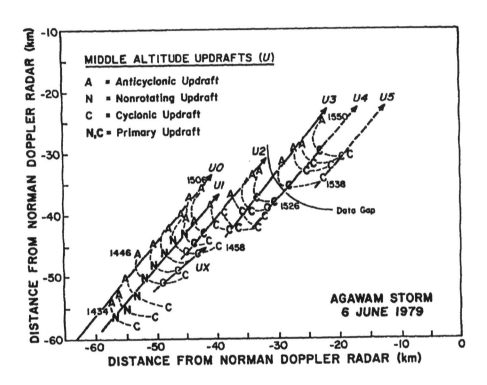

Figure 12.7: Cell tracks from a convective storm near Agawam, OK. Each track runs from southwest to northeast, with times given along the tracks. The notations "A" (anticyclonic), "N"(none), and "C" (cyclonic) indicate rotation of the cell. Note that some cells start with cyclonic rotation to the right of an older cell, then lose the cyclonic rotation and develop anticyclonic rotation as new cells develop to their right. From Brown (1992).

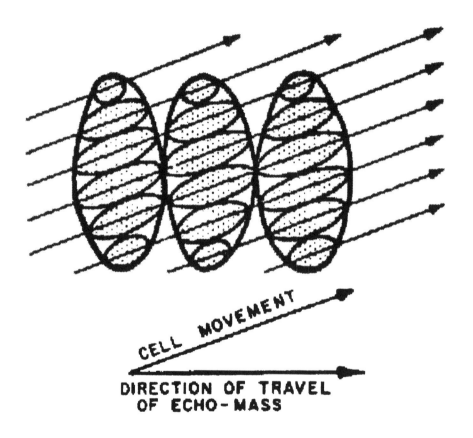

Figure 12.8: Schematic evolution of multicell storms, with similar sequencing to those in Figure 12.7. From Browning and Ludlam (1962).

Chapter 13

MODELING MANTLE CONVECTION: A SIGNIFICANT CHALLENGE IN GEOPHYSICAL FLUID DYNAMICS

D. A. YUEN

Minnesota Supercomputer Institute
Department of Geology and Geophysics
University of Minnesota, Minneapolis, MN, 55415-1227

S. BALACHANDAR

Department of Theoretical and Applied Mechanics
University of Illinois, Urbana, IL 61801

U. HANSEN[1]

Department of Theoretical Geophysics
Earth Sciences Institute
University of Utrecht
TA 3508 Utrecht, Holland

Mantle convection describes the creeping flow that occurs on geological timescale in the mantle extending from the surface of the Earth to the core-mantle boundary. The equation for momentum conservation governing this motion is elliptic in character because of the infinite Prandtl number of the mantle material. The principal time-dependence in mantle convection comes from the energy equation. Outstanding features of mantle convection are: (1) the strongly temperature and strain-rate dependent rheology (2) phase transitions that occur in a narrow zone (3) the presence of both thermal and compositional buoyancy forces (4) various sources of heating, such as internal heating from radioactive decay and local heating through viscous dissipation and adiabatic compression. (5) thermal-coupling with the convecting core and (6) the thermal-mechanical coupling to the moving lithosphere. Here we outline some of the significant advancements achieved in recent years in mantle

---

[1]Present address: Institut für Geophysik, Univ. Münster, D.48149 Münster, Germany.

convection along with the present-day computational requirements. Several interesting phenomena in mantle convection are also presented. They involve: (1) variable viscosity convection (2) localized heating by viscous dissipation and (3) bistable transitions in thermo-chemical mantle convection with a high surface temperature. Also discussed are the prospects for future computational work in mantle convection.

13.1   Introduction

Mantle convection has a long history going back to the original proposal by Holmes (1931) in order to explain the phenomenon of continental drift postulated by Alfred Wegener in the early part of this century. This subject matter remained dormant for the next twenty years, except for some occasional investigation (e.g. Vening-Meinesz 1956), until the advent of the plate-tectonic revolution in the mid-sixties (Wilson 1965 and Morgan 1968). The landmark papers on boundary-layer theory of mantle convection by Turcotte & Oxburgh (1967) and the thermal history model by Tozer (1967) have helped to launch the new field of mantle convection. The seventies saw the beginning of numerical modeling of mantle convection in the finite-amplitude regime (Torrance & Turcotte 1971, McKenzie *et al.* 1974 and Parmentier *et al.* 1976) and the boundary-layer regime (Yuen & Schubert 1976 and Schubert *et al.* 1976). A new school in geophysical fluid dynamics was started at that time with a series of doctoral theses (Takeuchi & Sakata 1970, Peltier 1973 and Richter 1973).

From that time on, the field of mantle convection has taken off, especially since the mid-eighties with the establishment of the mantle convection workshop at Los Alamos and the bi-annual European mantle convection workshop, which have produced two notable benchmark papers (Blakenbach *et al.* 1989 and Busse *et al.* 1993). The pace of mantle convection research has been extremely rapid and a torrent of recent work on various aspects of mantle convection has been conducted since the last review paper (Schubert 1992). For the last quarter of a century, one major outstanding issue in mantle convection, provoked by geo-chemical and seismic data, has been whether the mantle has or would convect in one or two layers (Richter & Johnson 1974, O'Connell 1977 and Richter & McKenzie 1981).

In this decade, one of the most spectacular development in mantle convection has been the rediscovery of a very potent instability (Christensen & Yuen, 1985) caused by an endothermic phase transition, coined the "mantle flush" instability (Machetel & Weber 1991, Tackley *et al.* 1993, Honda *et al.* 1993a, 1993b, Steinbach *et al.* 1993, Weinstein 1993 and Solheim & Peltier 1994). The flush instability has put the question of layered versus whole mantle convection under a new light, as the mantle could switch from one state to the other in a quasi-periodic fashion (e.g. Machetel & Weber 1991, Honda *et al.* 1993a and Tackley *et al.* 1993).

Through the eyes of seismic tomography (e.g. Su *et al.* 1994, Grand 1994, van der Hilst 1995 and Zhou 1996), mantle convection modelers can now obtain a glimpse of the interior heterogeneities inside the mantle to the extent that certain tantalizing features, such as slabs interacting with the transition zone and megaplumes under the central Pacific and Africa are beginning to be discernible. Global tomographic models (Zhou 1996) can now approach the resolution of a few degrees characteristic of the regional models (van der Hilst *et al.* 1991 and van der Hilst 1995). Seismic studies have also revealed so prominently the horizontally lying slabs above and under the transition zone at 660 km depth (Fukao *et al.* 1992, van der Hilst *et al.* 1991, Kawakatsu & Niu 1994 and Niu & Kawakatsu 1997), thus providing valuable constraints on the present degree of layering in mantle convection. The issues concerning the existence of megaplumes in the lower mantle, first hinted at by the early models (e.g. Dziewonski 1984), have been receiving greater attention again by the more recent tomographic models (Su *et al.* 1994, Li & Romanowicz 1996 and Zhou, 1996), since there are many mechanisms capable of explaining the existence of long-wavelength circulation and large plumes in the lower mantle, such as depth-dependent properties (Davies & Richards 1992, Balachandar *et al.* 1992, Hansen *et al.* 1993, Zhang & Yuen 1995, 1996 and Bunge *et al.* 1996, 1997) and phase transitions (Tackley 1996a). With the deployment of more local arrays, seismic tomography will be probing the interior structure and dynamics of the Earth's mantle with ever-increasing accuracy, providing a calibrated laboratory in which geodynamical processes can be studied under difficult physical conditions.

Mantle convection involves creeping flow with many peculiarities, such as the strongly nonlinear nature of mantle rheology, phase transitions, the thermomechanical coupling to the core, the possibilities for both thermal and chemical buoyancy, and strong viscous heating with attendant coupling to temperature-dependent viscosity. Its closest kin may be polymer processing in chemical engineering. In fact, one of the strengths in the development of mantle convection has been the constant flux of outstanding researchers from other areas, such as aerospace engineering (Turcotte & Oxburgh 1967, Zebib *et al.* 1980 and Trompert & Hansen 1996), applied mathematics (Fowler 1985), applied mechanics (Balachandar *et al.* 1992), astrophysics (Glatzmaier 1988) and computer science (Malevsky & Yuen, 1991 and van Keken *et al.* 1995). Many of the challenging aspects of mantle convection, such as boundary layer flows involving variable viscosity and multiple timescale phenomena, are certainly of interest to other areas of fluid mechanics. Our purpose in writing this article is to communicate to researchers in the other areas of geophysical fluid dynamics community that there are definitely present some interesting problems in mantle convection, which offer significant scientific and computational challenges.

It must be pointed out that the parameter space covered by mantle convection is quite wide and must be studied systematically in order to develop an understanding of the nonlinear interaction between the different physical mechanisms. One cannot simply add and subtract out different physical processes and hope to isolate an individual mechanism. The superposition of the different effects, for instance internal heating along with depth-dependent viscosity, cannot be so straightforwardly understood in terms of individual (single-effect) models. The great diversity of possibilities in the transport and thermodynamic properties distinguishes mantle convection from the usual thermal convection in environmental sciences and engineering, which are much better constrained. One cannot simply fit parameters from a simple model to the observations without a proper appreciation of the complexity in the physics, because of the highly non-unique nature of the problem.

Our plan is to discuss first the approximations made in the mathematical formulation of the mantle convection problem and present the resulting governing equations. Next we will go over the achievements attained up to now in the mantle convection community and the computational requirements and challenges. Section 3 will offer a venue to some of our own recent research efforts, in particular, convection with variable viscosity and viscous heating. The final section will be our perspective on the near future of mantle convection.

13.2   Model and Numerical Techniques

Owing to the complex nature of mantle convection varying levels of simplifying assumptions are often made in its mathematical modeling. At the simplest level we have Boussinesq convection with isothermal boundary conditions at the hot bottom core-mantle boundary and at the cold top surface. In the limit of infinite Prandtl number appropriate for mantle convection, the only time dependence arises from the energy equation. The Boussinesq approximation ignores all compressibility effects and assumes constant transport properties (i.e., viscosity, η, and thermal diffusivity, κ are assumed constant). Density variations are considered to be only due to temperature perturbations and important only in the buoyancy term. In all other terms, density is considered to be a constant equal to the reference density. From theoretical and laboratory investigations, there is mounting evidence to suggest that the thermodynamic, such as the thermal expansivity, and transport properties of the mantle material, especially viscosity, strongly vary with depth, owing to the orders of magnitude change in temperature and pressure across the mantle (Chopelas & Boehler 1992 and Weertman 1970). A common extension to the above formalism which includes the dominant depth-dependent property variations is the extended Boussinesq approximation (Turcotte *et al.* 1973 and Christensen & Yuen 1985). In the extended Boussinesq approximation compressibility effects are accounted for in the energy equation in terms of the viscous and adiabatic compressional heating terms, but without including the compressibility effects in the continuity equation.

13.2.1   Anelastic Liquid Model

A rational approximation to the fully compressible equations which includes the compressibility effects on convective time scale but filters out the compressibility effect on the acoustic time scale is given by the anelastic liquid approximation (Jarvis & McKenzie 1980 and Glatzmaier 1988). Effectively the speed of sound is considered to be infinite and consequently the sound waves are filtered out. Thus the stringent time-step restriction arising from the sound waves is avoided in this formulation. In the anelastic liquid model, the reference state for the thermodynamic variables can be chosen to be time independent and only a function of depth given by the adiabatic and hydrostatic condition. The adiabatic density variation, ρ_a, for the mantle is generally taken to be given by the Adams-Williamson equation (Birch 1952)

$$\frac{1}{\rho_a}\frac{d\rho_a}{dz} = -\frac{\alpha(\rho)gh}{\gamma(\rho)C_p} \tag{13.1}$$

where C_p is the specific heat at constant pressure, h is the depth of the layer, $\gamma(\rho)$ is the Grüneisen parameter, which measures the degree of compressibility with $\gamma = \infty$ representing the incompressibility limit, and $\alpha(\rho)$ is the thermal expansivity (Chopelas & Boehler 1992). The corresponding adiabatic temperature and pressure distributions are governed by

$$\frac{dT_a}{dz} = -\frac{\alpha(\rho)T_a g}{C_p} \quad \text{and} \quad \frac{dp_a}{dz} = -\rho_a g. \tag{13.2}$$

Given a particular form for the implicit depth dependences of α and γ the adiabatic density variation can be obtained from Eq (13.1). The adiabatic temperature variation can then be obtained by integrating the above equation (Eq (13.2)) and the constant of integration is set by the parameter, T_0, the dimensional surface temperature. The density, temperature and pressure perturbations about the adiabatic state are defined as:

$$\rho' = \rho - \rho_a = \rho_a\left(-\alpha\theta' + \chi p'\right), \quad \theta' = T - T_a, \quad p' = P - p_a \tag{13.3}$$

Here it is assumed that the density perturbations about the adiabatic state scale linearly with the temperature and pressure perturbations with the coefficient of thermal expansion and isothermal compressibility given by α and χ, respectively. More complex forms of mantle equation of state for the perturbation quantities can be considered (see Ita &King 1994). Comparisons between the extended-Boussinesq and the anelastic formulation have been carried out by Steinbach *et al.* (1989) and Ita & King (1994).

The above expressions for the thermodynamic variables can be substituted in the governing mass, momentum and energy equations and nondimensionalized with the convective layer depth, h, as the length scale; κ_s/h, as the velocity scale, $\kappa_s\eta_s/h^2$, as the pressure scale and ΔT as the temperature scale. Where ΔT is the temperature difference across the convective layer, κ is the thermal diffusivity, η is the dynamic viscosity and subscript s denotes properties evaluated at the surface. The thermodynamic and transport properties are nondimensionalized by their respective values at the surface and the nondimensional values are represented by an overbar (for example $\bar{\alpha} = \alpha/\alpha_s$). The anelastic liquid approximation is obtained in the limit of the nondimensional parameter, $\varepsilon = \alpha_s\Delta T$, being very small. The resulting mass and momentum conservation equations for the nondimensional velocity, temperature and pressure perturbations, $(\mathbf{u},\ \theta,\ p)$, away from the adiabatic hydrostatic state are

$$\nabla \cdot (\bar{\rho}_a \mathbf{u}) = 0 \qquad (13.4)$$

$$-\frac{\partial p}{\partial x_i} + \frac{\partial \tau_{ki}}{\partial x_k} + \delta_{i3} Ra \bar{\rho}_a \bar{\alpha} \theta = 0 \qquad (13.5)$$

$$\bar{\rho}_a \frac{D\theta}{Dt} + Di\bar{\alpha}\bar{\rho}_a w\theta - \Phi - \frac{\partial}{\partial x_i}\left(\bar{k}\frac{\partial \theta}{\partial x_i}\right) - \frac{\partial}{\partial z}\left(\bar{k}\frac{\partial \bar{T}_a}{\partial z}\right) = R \qquad (13.6)$$

$$\tau_{ik} = \bar{\eta}\left(\frac{\partial u_i}{\partial x_k} + \frac{\partial u_k}{\partial x_i} - \frac{2}{3}\delta_{ik}\nabla\cdot\mathbf{u}\right), \qquad (13.7)$$

where D/Dt is the total derivative in time, $Ra = \rho_s\alpha_s g\Delta T h^3/\eta_s\kappa_s$ and $Di = \alpha_s gh/C_p$ are the Rayleigh number and dissipation number evaluated at the top surface. Here R is the nondimensional rate of internal volumetric heat generation due to radioactive decay (see Leitch & Yuen 1989 for an estimate of magnitude). In the momentum equation, the inertia term is absent because of the limit of infinite Prandtl number appropriate for mantle convection. As noted above, the only time derivative is then for the temperature and therefore only the energy equation is prognostic, while the equations of mass and momentum conservation are diagnostic. Also, in the energy equation, F is viscous heating due to dissipation given by (Batchelor 1967)

$$\Phi = \frac{Di}{Ra}2\eta\left[\frac{1}{2}\left(\frac{\partial u_i}{\partial x_j} + \frac{\partial u_j}{\partial x_i}\right)\frac{1}{2}\left(\frac{\partial u_i}{\partial x_j} + \frac{\partial u_j}{\partial x_i}\right) - \frac{1}{3}\left(\frac{w d\bar{\rho}_a}{\bar{\rho}_a dz}\right)^2\right]. \qquad (13.8)$$

This term is of importance in mantle convection because of its potential nonlinear coupling to temperature-dependent viscosity (e.g., Balachandar *et al.* 1995) and non-Newtonian rheology (Larsen *et al.* 1996b). Many different approximations of mantle rheology can be accommodated within the above formalism. Different efforts in the past have made assumptions from a constant viscosity mantle (for example, Jarvis & McKenzie 1980) to one in which viscosity is a function of temperature, pressure and strain-rate (for example, Christensen 1984).

13.2.2   Internal Solid-State Phase Transitions

Phase transitions are known to exert strong influences on mantle convection (Schubert *et al.* 1975). These solid to solid phase transitions depend both on temperature and pressure and are governed by the Clapeyron slope. A simple way in which the internal phase transitions of the mantle from olivine to spinel and from spinel to perovskite can be incorporated is through an effective depth-dependent thermal expansivity to account for the effects of local changes in buoyancy and of latent-heat release across the phase transitions (Christensen & Yuen 1985; Steinbach & Yuen 1992). This approximation neglects the effect of phase boundary distortion due to the dynamics (see Christensen & Yuen 1984, Machetel & Weber 1991). With this approximation the thermal expansivity, α, consists of two parts, one describing the background variation, which is mainly the dominant pressure-dependent variation with depth as prescribed from the laboratory experiments (Chopelas & Boehler 1992). The second prescribing the rapid variation in α across the phase transition zones. A simple form of thermal expansivity is then given by

$$\alpha = \alpha_b + \alpha_1(z) + \alpha_2(z) \tag{13.9}$$

where α_b is given by the background pressure dependence (Honda et $al.$ 1993a, Reuteler et $al.$ 1993), and $\alpha_1(z)$ and $\alpha_2(z)$ are appropriate localized functions representing respectively the olivine to spinel and spinel to perovskite phase transitions. This formalism assumes that the two phase transitions to occur at a constant depth of approximately 400 km and 670 km respectively, owing to the dominant effect of depth-dependent hydrostatic pressure on phase transition. Extensions to more complex models involving both pressure and temperature-dependent phase transitions are quite feasible. Ita & King (1994) have compared this formalism with the phase boundary distortion approach.

13.2.3   Thermal-Chemical Convection

A number of enhancements to the above basic mantle convection model have been attempted in the past in order to incorporate additional complex physics, in particular those which take place near the surface and near the core-mantle boundary. One such complex process that will be discussed in greater detail below is double-diffusive convection (DDC) which has been investigated in the past (Hansen & Yuen 1988, 1989, 1994, 1995 Lenardic & Kaula 1995, Kellogg & King 1993) to study the thermal-chemical instabilities at the core-mantle boundary and near the surface. With the addition of compositional inhomogeneity the linearized equation of state for the perturbation density (Eq (13.3)) can be enhanced as

$$\rho' = \rho - \rho_a = \rho_a \left(-\alpha\theta' + \chi p' + \beta C'\right), \tag{13.10}$$

where β is the compositional counterpart to the coefficient of thermal expansion and C is the compositional perturbation from a suitable reference state. In the momentum equation in addition to the thermal buoyancy term (third term on the right hand side of equation 13.5) a compositional buoyancy term, $-\delta_{i3}Ra_c\bar{\rho}_a\bar{\beta}C$, also needs to be included, where $\bar{\beta}$ and C are the nondimensional β and the nondimensional compositional perturbation field. The compositional Rayleigh number is given by, $Ra_c = \rho_s\beta_s g\Delta C h^3/\eta_s\kappa_s$, where ΔC is the compositional difference across the depth of the convection layer. Furthermore an additional concentration equation for the composition, given below, must also be solved

$$\frac{DC}{Dt} = \frac{1}{Le}\nabla^2 C. \tag{13.11}$$

Here for convenience it is assumed that the Lewis number, $Le = \kappa/\mathcal{D}$, is a constant, where $\mathcal{D}$ is the concentration diffusivity and note that there is no source term for C in eqn (13.7).

This will give rise to strong heat production from the liberation of compositional buoyancy (Hansen & Yuen 1996). For mantle and magma chambers Le is an extremely small number of $O(10^{-8})$ thus Eqn. (13.11) is basically hyperbolic. Chemical boundary layers are thin compared to thermal boundary layers. This aspect is a major computational challenge in the numerical modeling of mantle convection.

13.2.4 Mantle Rheology

The rheological behavior of mantle material is much more complex than of water or air. It depends on many factors, most prominently temperature, strain-rate and also on depth or the overburden pressure. Mantle rheology has been studied extensively both experimentally and theoretically. A good introduction can be found in the books by Poirier (1991) and Ranalli (1995). A synopsis of the current issues in mantle rheology can be found in a recent review by Karato (1997). Because of the formation of dislocations and defects in solids, mantle material deforms under applied stress by a strongly temperature-dependent thermally activated process under the high temperatures expected in the mantle. The strain-rate dependence on temperature and pressure is of the Arrehnius form and the dependence on applied stress is nonlinear. The deviatoric stress tensor, σ_{ij}, is related to the strain-rate tensor, e_{ij}, by

$$e_{ij} = AS^{n-1}\exp\left[-\left(E^* + pV^*\right)/RT\right]\sigma_{ij} \tag{13.12}$$

where A is a material constant, T is the temperature in degrees Kelvin, n is the power-law index, usually 3 for upper-mantle material, E^* is the activation energy, p is the overburden pressure, R is the gas constant, V^* is the activation volume. The parameter n can also vary as a function of depth depending on the creep mechanism (van den Berg et al. 1993 and Karato & Li 1992). Both E^* and V^* can also be functions of depth (Sammis et al. 1977). The second invariant of the deviatoric stress tensor is given by S. Computationally the effective viscosity is better cast in terms of the strain-rate. The effective viscosity is defined by

$$\eta = \frac{\sigma_{ij}}{e_{ij}} = \left(\frac{1}{AS^{n-1}}\right)\exp\left(\frac{E^* + pV^*}{RT}\right). \tag{13.13}$$

Mantle viscosity is thus dependent on stress, temperature and pressure. Viscosity in terms of the second invariant of the strain-rate tensor, e, is obtained by replacing S^{n-1} in the above equation and solving for η. Viscosity then takes the form

$$\eta = \frac{1}{A^{1/n}e^{(n-1)/n}}\exp\left(\frac{E^* + pV^*}{nRT}\right). \tag{13.14}$$

We note that in non-Newtonian rheology the explicit dependence of the viscosity on temperature and pressure is attenuated by a factor of n in the argument of the exponential and is hence weaker. The dominance of non-Newtonian rheology over temperature-dependent rheology has been demonstrated in numerical simulations by Larsen et al. (1993) and van den Berg & Yuen (1995). In these flows it was found that much larger variations in the viscosity can be generated dynamically from fluctuations in the strain-rate than from the thermal perturbations. The dominant role played by non-Newtonian rheology in different dynamical situations has recently been demonstrated by Houseman & Molnar (1997) and Houseman & Gubbins (1997). Rheology may also depend on grain size, which must be solved self consistently (Kameyama et al. 1997). In sum, mantle rheology depends on temperature, pressure, strain rate, grain-size (Karato 1997, Kameyama et al. 1997) and volatile content.

13.2.5 Numerical Methodologies

The initial numerical techniques in solving mantle convection problems in the nonlinear regime were based on second-order finite-difference methods (e.g., Torrance & Turcotte 1971, Turcotte *et al.* 1973, McKenzie *et al.* 1974 and Parmentier *et al.* 1975). Splines with 4th order accuracy were used in the early eighties to solve 2-D problems with variable viscosity (Christensen, 1984) and finite-element methodology with first to second-order accuracy were developed for the same purpose (Hansen & Ebel 1988 and King *et al.* 1990).

The first 3-D code in spherical shell was based on finite-difference and spherical-harmonic expansion (Machetel *et al.* 1986) and this was followed by Glatzmaier (1988) who adopted his 3-D astrophysical convection code, based on Chebyshev polynomials and spherical-harmonic expansion to mantle problems. A multigrid finite-element method for 3-D spherical convection has been developed by Bunge & Baumgardner (1995). Zhang & Yuen (1995, 1996) have devised a 3-D code for spherical geometry based on higher-order finite-difference in the radial direction and spectral expansion for the field variables in spherical harmonics along the circumferential directions.

The first Cartesian 3-D codes (Cserepes *et al.* 1988 and Travis *et al.* 1990) were based on second- to fourth-order accurate finite-difference in the vertical direction and Fourier spectral expansion along the horizontal directions. Balachandar & Yuen (1994) developed a spectral-transform 3-D Cartesian code based on Chebyshev expansion in the vertical and Fourier expansion along the horizontal directions. This method has also been extended to variable viscosity problems (Balachandar *et al.* 1995a,b). Recently, Malevsky (1996) has developed a 3-D convection code based on cubic splines and Larsen *et al.* (1997) have shown the importance of using higher-order (eighth-order) methods in high Rayleigh number three-dimensional convection. It has been popular among geophysicists to use the second-order finite-volume algorithm of Patankar (1980) in solving 3-D convection problems with strongly variable viscosity (Ogawa *et al.* 1991, Iwase, 1996 and Moresi & Solomatov 1995). A multigrid method with finite-volume discretization has also been used in the 3-D variable viscosity case (Tackley 1993 and Trompert & Hansen 1996).

With temperature-dependent viscosity, the momentum equation becomes a linear elliptic equation with a strongly varying coefficient due to the lateral variations of the viscosity. This combined with the diagnostic nature of the momentum equation requires iterative methodologies for the computation of the velocity field from the thermal field. With the presence of non-Newtonian rheology, the momentum equation becomes a strongly nonlinear elliptic equation and many iterations are needed at each timestep for the solution of the momentum equations. From a computational standpoint, non-Newtonian rheology is more challenging than the purely temperature-dependent viscosity (Malevsky & Yuen 1992) and hence it is important that suitable preconditioners be developed to accelerate the convergence of the iterative scheme. Shared memory architectures would bring the direct method back into vogue, as memory of 100 Gbytes can be used for matrices associated with 3-D geometries.

13.3 Past Achievements and Computational Challenges

In the last decade, with the increasing availability of high-performance computers of vector and parallel architecture, mantle convection modeling is now approaching nearly the same level of sophistication as other neighboring disciplines, such as astrophysics and the atmospheric sciences. Steady-state models, which were quite popular until the early nineties, are giving way to time-dependent solutions. The idea of hard-turbulence in thermal convection

(Castaing *et al.* 1989) has also spread to the mantle convection community (Hansen *et al.* 1990). Yuen *et al.* (1993) gave a summary of the role played by hard-turbulent convection on the thermal evolution of the Earth. Rayleigh numbers up to 10^{10} have been attained in 2D simulations with constant viscosity (Yuen & Malevsky 1992). In the meantime, three-dimensional work on mantle convection has become commonplace, since the initial work begun in the eighties (Machetel *et al.* 1986, Glatzmaier, 1988, Bercovici *et al.* 1989 and Travis *et al.* 1990). Three-dimensional calculations for constant or depth-dependent viscosity have reached the range of realistic Rayleigh number of the present day mantle, between 10^7 and 10^8, both in Cartesian (Malevsky & Yuen 1993, Balachandar *et al.* 1993, Paramentier *et al.* 1994 and Yuen *et al.* 1994) and spherical (Zhang & Yuen 1995, 1996 and Bunge *et al.* 1996, 1997) geometries. We note that it was not until 1994 that higher Rayleigh numbers, on the order of $O(10^8)$ with more complicated physics, such as phase transitions, were attainable in Cartesian 3D geometry (Yuen *et al.* 1994).

13.3.1   Sample Past Results

Figure 13.1 shows the temperature fields associated with 3D Cartesian convection (Yuen *et al.* 1994) with multiple phase transitions for averaged Rayleigh numbers spanning two orders in magnitude from $O(10^6)$ to $O(10^8)$. The sharp interface is caused by the endothermic phase transition, which is located about 1/3 of the layer depth from the top surface in this model. The effects of mantle phase transitions in deflecting vertical flows are due to changes in the density of about 10% across the phase transition, and the latent heat release due to phase change. The role of phase transition in mantle convection is analogous to that of hydrogen ionization in astrophysical convection, which causes drastic changes in the equation of state (e.g. Rast & Toomre 1993 and Rast's contribution in this volume). In figure 13.1, the tendency of the mantle to be layered with increasing Rayleigh number can be clearly seen. Also to be noted are the presence of large, thick plumes at lower Rayleigh numbers, in the range of $O(10^6)$. Imaging of the interior of the mantle (Su *et al.* 1994) has revealed two major plumes in the lower mantle suggesting an Ra in the range of 10^6. One can also observe that the plumes become extremely thin at high Ra and cannot reach the interface associated with the phase transitions in the upper mantle. Flush events have also been captured in a 3-D spherical shell geometry (Tackley *et al.* 1993) where the instabilities are more regional than global in character.

There is definitely energy exchange between the core and the mantle. This, along with the time-dependent internal heating due to radioactive decay, contributes to the non-equilibrium nature of mantle convection. The non-equilibrium issue becomes especially acute in the face of phase transitions, as first demonstrated by Steinbach *et al.* (1993), who showed that thermal-coupling between the core and the mantle could be responsible for producing multiple timescales in the thermal evolution of the mantle. Under non-equilibrium conditions the Nusselt-Rayleigh number relationship may be non-unique and can be significantly different from the power-laws obtained under equilibrium conditions (Yuen *et al.* 1995). The computation of mantle convection with non-equilibrium boundary conditions takes much more CPU time than regular convection with a steady-state boundary condition at the core-mantle boundary, because of the increased stiffness of the equations and the much thinner boundary layers at the core-mantle boundary.

Variable viscosity in mantle convection is another area in which there has been a strong focus of activities, because of the effects on inducing fast and thin upwellings and the development of a strong surface boundary layer. Most of the work done up to now has focused

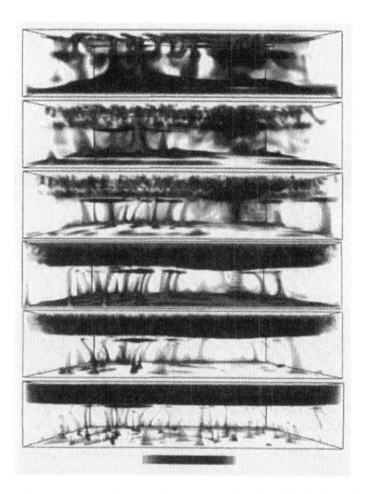

Figure 13.1: Temperature fields spanning a range of surface Rayleigh numbers between 2×10^6 and 4×10^8 in a box with $5\times5\times1$ aspect-ratio. Rayleigh numbers are 2×10^6, 10^7, 4×10^7, 6×10^7, 10^8 and 4×10^8 from the top to the bottom. To convert the surface Rayleigh number to the averaged Rayleigh number one multiplies the surface Rayleigh number by 0.2263. All frames are taken at the last stages well beyond the transient regime. Further details can be found in Yuen *et al.* (1994).

on the Boussinesq approximation in which there is no coupling between viscous heating and temperature-dependent viscosity, a point to be taken up below in section 4. A great deal of understanding has been developed recently by the numerical and asymptotic work on steady-state 2-D convection with large viscosity contrasts and high Rayleigh numbers (Hansen & Yuen 1993, Solomatov 1995 and Moresi & Solomatov 1995). Recent 3-D work (Ratcliff *et al.* 1997) has verified the 2-D findings. There exist several regimes in flows with variable viscosity: (1) the stagnant regime for large enough viscosity contrast, (2) the intermediate regime; and (3) the small viscosity regime in which the heat transport is controlled by the viscosity of the isothermal core (Hansen & Yuen 1993). From these asymptotic results, a new scheme for Nusselt-Rayleigh number relationship for temperature-dependent viscosity has been proposed, using, the local Rayleigh and Nusselt numbers at each boundary layer (Honda 1996). The combined effects of temperature and pressure dependent rheology tend to flatten out the $Nu(Ra)$ curves (van den Berg and Yuen 1998).

Figure 13.2 shows 3D Cartesian convection in an aspect-ratio 8 box for variable viscosity in the Boussinesq formulation for an intermediate Rayleigh number of 10^5 (Tackley 1993). Here the temperature-dependence of the viscosity is taken to be of the Arrhenius, $\exp(A/T)$, form. The constant viscosity case is illustrated at the upper-left panel and is to be compared with the temperature-dependent viscosity cases with rigid (upper right) and stress-free (bottom left) top and bottom boundary conditions. A temperature and stress-dependent rheology with $n = 3$ is shown at the bottom right panel. With constant viscosity there is greater connectivity between the downwellings than the upwellings and there is some tendency for spoke-pattern convection. The effects of rigid boundary conditions are to reduce the wavelengths of the convecting cells. With stress-free boundary conditions we see a completely different pattern with much longer wavelengths and large quasi-cylindrical downwellings. There is a stability to the overall long wavelength pattern and yet at the same time there are highly time-dependent features associated with the local small scale up and downwelling. Finally with the stress-dependent rheology the upper boundary layer is softened, which makes an easier reentry of the cold material into the interior. This results in smaller cells. Upwelling sheets and descending plumes appear in this case of non-Newtonian and temperature-dependent rheology.

The time-dependence of non-Newtonian flows has been much more extensively studied in two-dimensional configuration because of computational reasons. Numerical results (Christensen & Yuen 1989, Malevsky & Yuen 1992, Yuen & Malevsky 1992, Larsen *et al.* 1996a) have demonstrated that there is far more richness in the spatio-temporal dynamics of non-Newtonian flows. There is a multiplicity of timescales present associated with very fast and slow flows involving a wide variety of cell sizes. Figure 13.3 shows the temporal development of the velocity magnitude in Boussinesq, non-Newtonian and temperature-dependent convection in a wide two-dimensional box (Larsen *et al.* 1996a). Both basal heating ($R = 0$) and internal-heating ($R = 7$) configurations have been considered. The abrupt changes in the flow configuration and their magnitudes are evident. Their surface and bottom heat fluxes for the internal heating case as a function of time are shown in figure 13.4. Here one can discern the sharp changes in the time-dependence of the flow. Timescales are much smaller at the bottom boundary layer because of the higher local Rayleigh number resulting from lower viscosity due to the temperature-dependent part of the rheology. Because of the nature of this abrupt time-dependence and the associated finer spatial features, the calculation of non-Newtonian convection takes much more computer time than for constant or even purely temperature-dependent viscosity convection.

Thermal-chemical convection studies have been carried out to hopefully explain many facets of mantle dynamics, such as the thermal-chemical instabilities at the core-mantle boundary (Hansen & Yuen 1988, 1989 and Kellogg & King 1993), at the 660 km discontinuity (Kellogg 1991 and Hansen & Yuen 1994) and in the crust (Lenardic & Kaula 1995). A recent review of the role played by double-diffusive convection in the geological sciences can be found in Hansen and Yuen (1995). Thermal-chemical convection with partial melting (Ogawa 1993, 1994 and Kameyama *et al.* 1996) is an extremely computationally intensive problem because of the wide range of timescales present in the problem due to the different thermochemical and thermodynamical processes. For instance, a typical two-dimensional run requires about 10^5 to 10^6 timesteps to reach a statistically stationary state (Kameyama *et al.* 1996). Recently 10^7 timesteps were found to be necessary for cases with a phase transition (Ogawa & Nakamura 1998). Thus even two-dimensional computations, involving non-Newtonian rheology or thermal-chemical convection with melting included, can easily consume more CPU time than 3D convection with simple rheology.

The problem of mantle convection is further complicated by the presence of surface plates and the associated plate tectonics. Aside from the Jovian moons, the Earth is the only planet known to have plate tectonics. Subduction of plates is one of the important features of plate tectonics and is also one of the more challenging aspects to model. Very early on, a zone of weakening at the plate margin was used to simulate subduction (Schmeling & Jacoby 1981). A more sophisticated manner to handle faults in the lithosphere within the framework of global mantle convection has been devised by Zhong & Gurnis (1994, 1995). Figure 13.5 shows the effects of incorporating a fault in a descending slab on mantle convection with a single phase transition at 660 km depth. The presence of a fault separating the motion of the descending slab and the overlying lithosphere significantly influences the dynamical interaction between the slab and the underlying mantle. This type of model has recently been extended to 3-D (Zhong & Gurnis 1996). Implicitly important in these models is the geological history of pre-existing faults, which, in a way, represent a memory effect of the rheology of plates. This can be caused by pre-existing fluids. Viscous heating may also contribute to the lubrication of slabs at plate margins (Schmeling 1989, Larsen *et al.* 1996b and Bercovici 1996).

13.3.2   Computational Requirements

The computational requirements of any problem depend on the questions posed, the numerical method chosen and the demands imposed by visualization of time-dependent processes. In the case of mantle convection, one can divide the issues surrounding computational needs into two types of problem: global dynamics and local flow processes. For global dynamics, one would like to have for 3-D convection in Cartesian geometry a resolution of around $400 \times 400 \times 400$ grid points and a similar resolution in 3-D spherical geometry as well. Such kinds of resolution are today only available on state-of-the-art massively parallel computers, such as the IBM-SP2, the CRAY-T3E and SGI clusters. Such 3-D problems, involving phase transitions, melting and variable viscosity, most certainly belong to the category of grand-challenge problems. Two-dimensional mixing problems involving over 1000×1000 grid points have already been solved (Ten *et al.* 1996, Ten *et al.* 1997) and the challenge here lies in extending the mixing statistics further to $10^4 \times 10^4$ grid points. In this way there will be an improvement in the statistics necessary for fractal analysis of mixing dynamics (Sreenivasan *et al.* 1989). The kind of local problems of geophysical interest are those concerned with thermal-chemical convection and viscous heating, as both processes would encompass a multitude of spatial scales, spanning at least 3 to 4 orders in magnitude. It seems that for mantle convective processes the number of grid points or number of modes possible along each direction with 20th century computers and data storage devices is about 10^4 in 2-D and 10^3 in 3-D. These limitations will put definite bounds on the possible range of spatial resolution which we hope to gain in the quest for including more physics and chemistry in the problem.

13.4 Results

13.4.1 Viscous heating in mantle convection

The effects of viscous heating and adiabatic cooling/heating due to expansion/compression are included only in the compressible models of mantle convection. In these models, the strength of viscous heating and adiabatic effects are controlled by the dissipation number, Di. It has long been recognized that viscous dissipation may play a potentially important role in mantle dynamics, especially for local flow situations such as the corner flow induced by the subducting slab (Toksoz et $al.$ 1971), shear flows in the asthenosphere below the lithosphere (Melosh, 1976 and Yuen & Schubert 1977) and conduit flows (Fujii & Uyeda 1974). Because of the localized nature of viscous heating, it was not until recently with the usage of high-resolution grids that the importance of viscous heating has been revealed in a self-consistent manner in 3-D (Balachandar et $al.$ 1993, Balachandar et $al.$ 1995a, 1995b, 1996 and Zhang & Yuen 1996) and 2-D (Steinbach & Yuen 1994, Larsen et $al.$ 1995, 1996b, and Kincaid & Silver 1996) numerical simulations. The potency of shear heating has also been demonstrated in the boundary-layer (Schubert et $al.$ 1976), mean-field (Yuen et $al.$ 1987) and shallow-water (Bercovici 1996) approximations to the mantle convection problem.

First we will present results from a three-dimensional Cartesian model of mantle convection performed in a box of 5×5 square planform and unit height. The surface Rayleigh number is 10^5 and the dissipation number is 0.3. In this model, viscosity varies with both temperature and depth as $\exp(c_1 z - c_2 T)$, where the constants c_1 and c_2 are chosen such that viscosity increases by factor 10 with depth, z, from the surface to the core-mantle boundary and decreases by a factor of 250 as nondimensional temperature varies from 0 to 1. Thus viscosity is a factor of 25 lower at the core-mantle boundary than at the surface and correspondingly the local Rayleigh number in the lower mantle is significantly higher than the surface Rayleigh number. Other thermodynamic and transport properties are assumed to dependent only on depth (see Balachandar et $al.$ 1993). Both viscous and adiabatic heatings have been included in this compressible formulation. Phase transitions and internal heat generation by present-day radioactive decay were not included in this model.

Figure 13.6 shows temperature contours in greyscale and superimposed velocity vector plot on a vertical $(x - z)$ plane passing through the computational box. This plane has been chosen to cut through an intense hot plume which can be seen near $x=3.5$ and the vertical velocity associated with the plume can be seen to be significantly larger than elsewhere on this plane. Other weaker cold downwellings and hot upwellings can also be observed on this plane. In mantle convection the reflectional symmetry about the midplane is broken due to the depth and temperature-dependent properties and in the present simulation the upwellings are generally cylindrical and plume-like, while the descending cold fluid is sheet-like. Although viscosity associated with the strong upwelling is lower than the surrounding due to the higher temperature, it is observed that viscous heating can be strong in the ascending plumes as well, since the intense velocity gradient associated with the thermal plume more than compensates for the decrease in viscosity.

Figure 13.7 shows shaded contours of viscous heating along with the velocity vector plot. Only the region close to the strong plume seen in figure 13.6 is shown. Intense viscous heating can be observed all along the plume forming a tube from the base of the plume to near the top where the plume impinges the surface.

In the middle of the mantle, the intense heating associated with the ascending plume is around the plume periphery, not at the center of the plume, owing to increased local shear and viscosity. This effect can be clearly seen in the three-dimensional surface plot of viscous heating shown in figure 13.8. Also plotted in this figure are two different isosurfaces of temperature corresponding to hot and cold fluid. The lighter shaded surface is the region of intense viscous heating. It can be seen around the sides of a hot upwelling plume and near where the plume impinges the surface. As pointed out above, significant viscous heating is present even at the base of the plume, but in this picture this region is hidden under the isosurface of hot temperature. On the right, a surface of a descending cold flow can be seen. Significant shear heating can be seen both at the base of the descending cold flow and near where it impinges the core-mantle boundary.

Figure 13.9 shows contours of viscous heating plotted on a horizontal plane at the surface ($z=1$). Overlaid on top are the velocity vectors. It can be seen that the flow convergence occurs in regions of interconnected thin sheet. These are the base of cold downwelling sheets and are associated with significant viscous heating. Viscous heating intensifies in magnitude at the intersection of these downwelling sheets. But the most intense viscous heating occurs at the centers of flow divergence. These are the regions where the hot plumes impinge the top surface and spread horizontally. It is somewhat puzzling that viscous heating in the hot plumes be larger than in the cold downwelling sheets. In fact, in the limit of constant α, dominant viscous heating is observed at the cold downwelling sheets and viscous heating in the hot plumes is significantly weaker (Balachandar et al. 1996). Here, with the introduction of realistic depth-dependent thermal expansivity (α decreases with depth) the hot plumes increase in buoyancy on their way up and therefore accelerate upward (Balachandar et al. 1992, 1995a). This effect focuses the hot plumes and the resulting increased vigor and higher stresses are responsible for such intense viscous heating. Furthermore, the magnitude of viscous heating at the surface in figure 13.9 is nearly an order of magnitude higher than that shown in figure 13.7. It is also observed that this intense heating occurs in a very thin layer at the surface where large stresses are locally generated due to the stagnation point nature of the horizontally spreading hot plume and viscosity is high due to the rapid cooling of the hot plume.

The peak value of nondimensional viscous heating observed at this time instance is as high as 700 and it occurs where the hot ascending plume impinges the surface. This corresponds to approximately a dimensional local heating rate of about 3600 K/Byr. This shows, that when compared to the estimated internal heating rate of 120 K/Byr due to radioactive decay in the Earth's mantle, local heating due to friction can be more than an order of magnitude greater than chondritic heating. Even using horizontal averages, the averaged viscous heating is comparable in magnitude to the present-day radioactive heating.

It must be noted that, although viscous heating is significant, unlike internal heating by radioactive decay, viscous heating itself can not initiate and drive convection on its own, since its origin lies in the convective motion itself but it can cause local instabilities. In terms of global energy balance, the net heat input from the positive definite viscous heating is precisely counter balanced by the net adiabatic cooling arising from expansion against background adiabatic pressure. Just like viscous heating, locally adiabatic cooling can be much larger than internal radioactive heating. Furthermore, viscous heating is dominant near the top and bottom of the convecting layer, while the adiabatic cooling is stronger in the interior. The stress field is focused wherever there is strong viscous heating. This preferential heating and cooling tends to destabilize the flow in the lower part of the mantle but tends to stabilize the flow near the top. Thus viscous heating and adiabatic cooling can have a strong effect on not only the local dynamics but also on the overall convective motion and global heat transfer.

A similar scenario emerges in mantle convection studies in a spherical shell as well. In figure 13.10 we show the viscous heating distribution superimposed on the temperature structures of plumes in 3-D spherical-shell convection with an endothermic phase transition at 660 km depth. The focus of viscous heating inside the plumes arises from the stratification of the lower mantle viscosity in this model (Zhang & Yuen 1996), which has a local viscosity maximum about a quarter of the way up from the core-mantle boundary. Large stresses are developed, as the plumes plough their way through a stiff lower mantle, thus producing strong viscous heating with a magnitude about 10 times greater than that produced by normal radioactivity arising from chondritic mantle rocks. Large amounts of viscous heating and stress can be produced in the transition zone from plumes impinging on the endothermic phase transition (Steinbach & Yuen 1994, 1995) and by plumes impinging on the lithosphere (Larsen et al. 1996b, Balachandar et al. 1995b, Tackley 1996b).

13.4.2   High Rayleigh Number Thermal-chemical Convection

There are some interesting physics associated with the high Rayleigh number regime in the extended Boussinesq approximation. They deal explicitly with the influences of the surface temperature T_0 (see discussion under Eqn 13.2) on adiabatic heating. Turcotte et al. (1974) have pointed out that viscous dissipation and adiabatic heating would have a stabilizing influence on the flow. Steinbach (1991) demonstrated from linear stability analysis and also from finite-amplitude results the increased stability from viscous and adiabatic heating and the differences on the stability from depth-dependent thermal expansivity. Here we will show interesting scenarios in high Rayleigh number convection with significant contributions from viscous dissipation and adiabatic heat transfer (see Eqns. 13.5 and 13.5).

The importance of the adiabatic heating term increases with greater magnitude of the surface temperature T_0 and larger values of the dissipation number Di. The strength of viscous dissipation is determined by Di/Ra times the viscosity and the square of the strain-rate. When the surface temperature is high compared to the temperature drop across the convecting layer, one can expect significant effects for adiabatic heating. For the Earth's mantle, T_0 is around 0.3 to 0.4 for the oceanic mantle and this value can be twice as large under the continental tectosphere. For the planet Mars, because of the thick lithosphere T_0 can be around 1.2. For the Jovian satellite Io, T_0 can be as large as 3.5. For the Earth's core T_0 can be as large as 4, which is to say that the ratio of the surface temperature at the top of the core to the temperature drop across the Earth's outer core is 4!

In figure 13.11 we show the snapshots of the streamfunction, temperature field and the vertical temperature profiles for $T_0=5$ $Di=0.5$ at three different Rayleigh numbers. The buoyancy source is assumed to decrease with depth according to the geophysically relevant decrease in ag across the outer core. Here it is taken to be $1/10$. In spite of the high Rayleigh numbers employed here (Ra ranging from 10^7 to 5×10^{12}), convection appears to be weak in the sense that the temperature fields are hardly distorted by the flow. Nevertheless, the amplitude of the velocities is quite significant. Because of the strong contribution of the adiabatic heating, the temperature field does not serve as a scalar field to map out the fluid motion. High rising material cools adiabatically and the amplification of the small temperature differences by the high Rayleigh number leads to further fluid motion. In all cases, the temperature profiles resemble those of a static internally heated system, except at the bottom boundary layer, where convection is quite vigorous and can homogenize the fluid. This similarity to an internally heated system can be understood from the viewpoint that the adiabatic term functions here as a source term.

It may be surprising that convection is strongest at the bottom, even though the thermal expansivity decreases with depth. But this unusual effect comes about because of the destabilizing influences induced by the adiabatic heating term with depth (Balachandar *et al.* 1996). This decrease in the local stability more than makes up for the decrease in the thermal expansivity. The time histories of the surface Nusselt numbers for Ra from 10^7 to 5×10^{12} are shown in figure 13.12. In general, one can see a transition in the flow complexity from a stationary pattern at $Ra = 10^7$ to a chaotic configuration at higher Ra. As a gauge for comparison, values of Nu in Boussinesq convection are around 45 for $Ra = 10^7$ and 95 for $Ra = 10^8$ for aspect-ratio one flow (Hansen *et al.* 1992). At higher Ra, from around 10^{11}, the flow develops into a bistable mode that fluctuates between two states with superimposed high frequency components. This phenomenon is better illustrated in figure 13.13 in the time trace of Nu for $Ra = 10^{12}$ and 10^{14}. The extrema of Nu correlates well with two different types of flow structures. As shown in figure 13.13 maxima in Nu are attained when the flow has developed a structure of two cells. These phases are separated by periods during which a disordered circulation prevails. These periods are associated with the troughs in Nu. Other simulations at larger aspect-ratios of three also show this bistable type of behavior. Thus we believe that bistability is a characteristic feature of time-dependent convection in this high T_0 regime.

A range of possible dissipation numbers Di can be present in geophysical systems. Large-scale phenomena like whole-mantle convection or convection in the outer core may take place with Di ranging from 0.1 to 0.5, while more local instabilities such as the ascent of thermal-chemical diapirs in the lithosphere can be characterized by a Di of O(0.01). We will now look at the influences of Di on the flow for $T_0 = 5.0$. Three snapshots, taken after the transients have died away, are shown in figure 13.14, where the influences of increasing Di are clearly shown. The temperature fields become smoother with larger values of Di. Hansen & Yuen (1996) have pointed out that the role of viscous heating may be particularly strong when buoyancy is compositional in nature instead of thermal in origin. This is due to the greater efficiency with which chemical gravitational energy can be converted into mechanical heat by viscous dissipation. In order to demonstrate the differences in the efficacy of viscous dissipation, we have carried out two experiments in which one is driven dominantly by thermal buoyancy, while in the second compositionally induced forces dominate. In figure 13.15 are shown the evolution of the thermal fields (top row), the compositional fields (middle row) and the streamfunction (bottom row). In both cases we have assumed an initial condition which consists of an initially cold fluid, depleted in the heavy component and at the same time being heated from below and being supplied by an influx of heavy material from above. For both situations the compositional diffusivity is ten times smaller than the thermal diffusivity. The thermally driven case is shown in figure 13.15a. Initially a thermal boundary layer develops at the bottom. The whole system then becomes unstable, giving rise to a break-off in the plume-like development. At the same time, a boundary layer in the compositional field also develops and the instability at the top is then driven by the thermally induced flow at the right. Additional instabilities also develop in the concentration field. These instabilities are swept away by the thermally-driven flow. Over a long timescale the temperature fields look very similar to the one shown in figure 11a. This shows that the thermal field is nearly not affected by the presence of the C-field.

In the case where the C-field plays the dominant role (Figure 13.15b) a totally different picture emerges. Here we see that the compositionally driven instabilities from the descending flow is strongly heated by both downward adiabatic compression and shear heating. Both of these mechanisms help to heat up the whole system. This heating effect is manifested in the $Nu(t)$ plots for both scenarios, shown in figure 13.16. Here we show the values of the Nusselt numbers at the top and bottom surfaces and also their difference. For the thermally driven case Nu approaches a nearly constant value of around 2. In contrast, in the compositionally driven case, so much heat is produced by mechanical heating that the system can lose heat through both the top and bottom. This phenomenon comes from the asymmetry in the conservation equations for T and C, arising from the presence of the heating terms in the T equation and lack of source terms in the C-equation (compare eqns. 13.5 and 13.8). In short, compositional heterogeneities can become dynamically working heat-sources for the T equation.

13.5   Perspectives and Future Directions

Over the past 14 years, research employing high-performance computing has come of age under the High-Performance Computing and Communications (HPCC) initiative. Mantle convection, like other fields, has certainly taken advantage of this major focus and substantial progress has been made since 1984. We have already cited the great improvements made in 3-D convection, which have gone from the weakly nonlinear regime (Machetel *et al.* 1986) to Rayleigh numbers close to realistic values (Balachandar *et al.* 1996, Zhang & Yuen 1996 and Bunge *et al.* 1997). But in the face of the current budget pressure and changes in the computer industry, it is reasonable for the geophysical modeling community to be concerned about the future momentum in high-performance computing and the coming generation of hardware. It seems unlikely in the near future that specialized high-performance computers, devoted to scientific computing, will be developed. Instead, the industrial trend seems to be the usage of the same chip for computers of all sizes. This strategy would definitely put an upper bound on the expectant increase in the maximum speed of the large computers even in national centers. By the year 2001 we may not see an increase of speed by a factor of 50 over the fastest machine today in 1998. With greater computational speed, we still face a severe bottleneck in the speed of the storage device. Unless this imbalance between speed of the computer and storage device is properly resolved, this will severely limit the effectiveness of large future machines to solve very large problems. On the other hand, great improvement is expected in the area of scientific visualization, where the large fallout from the developing commercial market can be expected. The usage of interactive visualization (Haimes, 1994, Jordan *et al.* 1996) among collaborators is expected to gain a foothold. The present paradigm of postprocessing large amounts of data may need to be revised in view of the growing development of interactive visualization.

In the coming years, we can anticipate a different class of computations, significantly different from those of the past, because of greater participation by geoscientists of all ranks in numerical modeling. The range of timescales in the problems and the spatial resolution involved will become wider than before. The dynamical range in timescales would go up by at least an order in magnitude, as well as the range in spatial scales. We can expect to resolve features of O(km) and timescales in the range of 10^4 years in local mantle and lithospheric dynamics. There is a need to encompass local fine-structure grids within a coarser mesh in problems such as the thermal-kinetic coupling in subducting slabs (Daessler & Yuen 1993). Since the general purpose hardware is not expected to improve so dramatically, it is thus essential to devote greater efforts on getting "more bang for the buck" in computing accurate solutions. The usage of higher methods, such as higher order finite-difference (Fornberg 1996, Larsen *et al.* 1997) cubic-spline (Malevsky 1996) and spectral-transform (Balachandar & Yuen 1994) should be tested more against lower-order methods, such as the finite-volume technique (Patankar 1980). For spherical geometry recent advances in pseudo-spectral techniques (Fornberg and Merrill 1997) holds much promise. We should also consider wavelet methods, which has the potential for handling problems of multiple scales accurately and economically. Here adaptivity and the multiple-scale nature are the key. Recent work on viscoelastic flow using adaptive wavelets has shown that this method can capture local variations of the viscosity spanning over 12 orders in magnitude (Vasilyev *et al.* 1997).

By the year 2001, some calculations, such as back-arc spreading and thermal fields in subducting region, will become routine and will allow more people outside the immediate modeling community to participate in modeling. We see the growing popularity of easy-to-use PDE packages, such as PDE2D, which had been used in mantle convection research (Leitch *et al.* 1991) or the finite element package ABAQUS (Regenauer-Lieb & Yuen 1998). By then, we will undoubtedly obtain a self-consistent plate-mantle convection model by coupling the rheology of the lithosphere to the mantle flow (Tackley 1998). The excitation of toroidal flows by non-Newtonian, temperature-dependent rheology under high Rayleigh number situations is a worthwhile problem in computational geophysics. Some of the lower-mantle processes, such as the cause of megaplumes and possible melting inside plumes with composite rheology (Yuen *et al.* 1996), would be a challenging problem. The coupling of melting and melt-segregation to mantle convection will require a better theoretical formulation, before further progress can be made on a faster computer. What is clear now is that the days of using simple models to study mantle convection are over. These models have been educational and now it is time to march on to more sophisticated models, which incorporate more nonlinear physics.

Acknowledgements

We thank Shije Zhong, Paul Tackley, Shuxia Zhang and Tine Larsen for providing some of the figures. This research has been supported by Geophysics program and Division of Mathematical Sciences (NSF DMS 93-01042 and DMS 96-22889) of National Science Foundation, Geoscience program of Department of Energy.

Bibliography

Balachandar, S., Yuen, D.A. and D.M. Reuteler, Time-dependent three-dimensional compressible convection with depth-dependent properties, *Geophys. Res. Lett.*, **19**, 2247-2250, 1992.

Balachandar, S., Yuen, D.A. and D.M. Reuteler, Viscous and adiabatic heating effects in three-dimensional compressible convection at infinite Prandtl number, *Phys. Fluids*, **A 5**, 2938-2945, 1993.

Balachandar, S. and D.A. Yuen, Three-dimensional fully spectral numerical method for mantle convection with depth-dependent properties, *J. Comput. Phys.*, **113**, 62-74, 1994.

Balachandar, S., Yuen, D.A. and D.M. Reuteler, Localization of Toroidal motion and shear heating in 3D high Rayleigh number convection with temperature-dependent viscosity, *Geophys. Res. Lett.*, **22**, 477-480, 1995a.

Balachandar, S., Yuen, D.A., Reuteler, D.M. and G. Lauer, Viscous dissipation in three dimensional convection with temperature- dependent viscosity, *Science*, **267**, 1150-1153, 1995b.

Balachandar, S., Yuen, D.A. and D.M. Reuteler, High Rayleigh number convection at infinite Prandtl number with weakly temperature-dependent viscosity, *Geophys. Astrophys. Fluid Dyn.*, **83**, 79-117, 1996.

Batchelor, G.K., *An Introduction to Fluid Dynamics*, Chapter 3, Cambridge Univ. Press, 1967.

Bercovici, D., Schubert, G. and G.A. Glatzmaier, Three- dimensional spherical models of convection in the earth's mantle, *Science*, **244**, 950-955, 1989.

Bercovici, D., Plate generation in a simple model of lithosphere- mantle flow with dynamic self-lubrication, *Earth Planet. Sci. Lett.*, **144**, 41-51, 1996.

Birch, F., Elasticity and constitution of the Earth's interior, *J. Geophys. Res.*, **57**, 227-286, 1952.

Blankenbach, B. *et al.*, A benchmark comparison of mantle convection codes, *Geophys. J. Int.*, **98**, 23-38, 1989.

Bunge, H.-P. and J.R. Baumgardner, Mantle convection modelling on parallel virtual machines, *Computers in Physics*, **9**, No2, 207-215, 1995.

Bunge, H.-P., Richards, M.A. and J.R. Baumgardner, Effect of depth-dependent viscosity on the planform of mantle convection, *Nature*, **379**, 436-438, 1996.

Bunge, H.-P., Richards, M.A. and J.R. Baumgardner, A sensitivity study of 3-D spherical mantle convection at 10^8 Rayleigh number: effects of depth-dependent viscosity, heating mode and an endothermic phase change, *J. Geophys. Res.*, **102**, 11991-12008, 1997.

Busse, F.H., Christensen, U.R., Clever, R. Cserepes, L., Gable, C., Giannandrea, E., Guillou, L., Houseman, G., Nataf, H.C., Ogawa, M., Parmentier, E.M., Sotin, C. and B. Travis, 3-D convection at infinite Prandtl number in Cartesian geometry: A benchmark comparison, *Geophys. Astrophys. Fluid Dyn.*, **75**, 39-59, 1993.

Castaing, B., Gunaratne, G., Heslot, F., Kadanoff, L., Libchaber, A., Thomae, S., Wu, X.Z., Zaleski, S.,and B. Zanetti, Scaling of hard thermal turbulence in Rayleigh-Benard convection, *J. Fluid Mech.*, **204**, 1-30, 1989.

Chopelas, A. and R. Boehler, Thermal expansivity of the lower mantle, *Geophys. Res. Lett.*, **19**, 1983-1986, 1992.

Christensen, U.R., Convection with pressure- and temperature- dependent non-Newtonian rheology, *Geophys. J.R. Astr. Soc.*, **77**, 343-384, 1984

Christensen, U.R. and D.A. Yuen, Layered convection induced by phase transitions, *J. Geophys. Res.*, **90**, 10 291-10 300, 1985.

Christensen, U.R. and D.A. Yuen, Time-dependent convection with non-Newtonian viscosity, *J. Geophys. Res.*, **94**, 814-820, 1989.

Cserepes, L., Rabinowicz, M. and C. Rosemberg-Borot, Three- dimensional infinite Prandtl number convection in one and two layers with implication for the Earth's gravity field, *J. Geophys. Res.*, **93**, 12009-25, 1988.

Daessler, R. and D.A. Yuen, The effects on phase transition kinetics on subducting slabs, *Geophys. Res. Lett.*, **20**, 2603-2606, 1993.

Davies, G.F. and M.A. Richards, Mantle convection, *J. Geology*, **100**, 151-206, 1992.

Dziewonski, A.M., Mapping the lower mantle: Determination of lateral heterogeneities in P velocity up to degree and order 6, *J. Geophys. Res.*, **89**, 5929-5952, 1984.

Fornberg, B. and D. Merrill, Comparison of finite difference and pseudo spectral methods for convective flow over a sphere, *Geophys. Res. Lett.*, **24**, 3245-3248, 1997.

Fowler, A.C., Fast thermoviscous convection, *Stud. Appl. Math.*, **72**, 189-219, 1985.

Fujii, N. and S. Uyeda, Thermal instabilities during flow of magma in volcanic conduits, *J. Geophys. Res.* **79**, 3367 -3369, 1974.

Fukao, Y., Obayashi, M., Inoue, H. and M. Nenbai, Subducting slabs stagnant in the mantle transition zone, *J. Geophys. Res.*, **97**, 4809-4822, 1992.

Glatzmaier, G.A., Numerical simulations of mantle convection: time-dependent, three-dimensional, compressible, spherical shell, *Geophys. Astrophys. Fluid Dyn.*, **43**, 223-264, 1988.

Grand, S.P., Mantle shear structure beneath the Americas and surrounding oceans, *J. Geophys. Res.*, **99**, 11, 591-11, 621, 1994.

Haimes, R. pV3: A distributed system for large-scale unsteady CFD visualization, *AIAA*, 94-0321, 1994.

Hansen, U. and A. Ebel, Time-dependent thermal convection - a possible explanation for a multi-scale flow in the Earth's mantle, *Geophys. J.*, **94**, 181-191, 1988.

Hansen, U. and D.A. Yuen, Numerical simulations of thermal- chemical instabilities and lateral heterogeneities at the core- mantle boundary, *Nature*, **334**, 237-240, 1988.

Hansen, U. and D.A. Yuen, Dynamical influences from thermal- chemical instabilities at the core-mantle boundary, *Geophys. Res. Lett.*, **16**, 629-632, 1989.

Hansen, U., Yuen, D.A. and S.E. Kroening, Transition to hard turbulence in thermal convection at infinite Prandtl number, *Phys. Fluids*, **A 2**, 2157-2163, 1990.

Hansen, U., Yuen, D.A., Kroening S.E. and T.B. Larsen, Dynamical consequences of depth-dependent thermal expansivity and viscosity on mantle circulations and thermal structure, *Phys. Earth. Planet. Inter.*, **77**, 205-223, 1993.

Hansen, U. and D.A. Yuen, High Rayleigh number regime of temperature-dependent viscosity convection and the Earth's early thermal history, *Geophys. Res. Lett.*, **20**, 2191-2194, 1993.

Hansen, U. and D.A. Yuen, Effects of depth-dependent thermal expansivity on the interaction of thermal-chemical plumes with a compositional boundary, *Phys. Earth Planet. Inter.*, **86**, 205-221, 1994.

Hansen, U. and D.A. Yuen, Formation of layered structures in double-diffusive convection as applied to the Geosciences, in *Double-Diffusive Convection*, Geophysical Monograph 94, pp 135 -149, Amer. Geophys. Union, Washington, D.C., 1995.

Hansen, U. and D.A. Yuen, Potential role played by viscous heating in thermal-chemical convection in the outer core, *Geochim et Cosomochim. Acta*, **60**, 1113-1124, 1996.

Holmes, A., Radioactivity and Earth movements, *Trans. Geol. Soc.*, Glasgow, **18**, 559-606, 1931.

Honda, S., Yuen, D.A., Balachandar, S. and D.M. Reuteler, Three- dimensional instabilities of mantle convection with multiple phase transitions, *Science*, **259**, 1308-1311, 1993a.

Honda, S., Balachandar, S., Yuen, D.A. and D.M. Reuteler, Three-dimensional mantle dynamics with an endothermic phase transition, *Geophys. Res. Lett.*, **20**, 221-224, 1993b.

Honda, S., Local Rayleigh and Nusselt numbers for Cartesian convection with temperature-dependent viscosity, *Geophys. Res. Lett.*, **23**, 2445-2448, 1996.

Houseman, G. and P.H. Molnar, Gravitational (Rayleigh-Taylor) instability of a layer with non-linear viscosity and convective thinning of continental lithosphere, *Geophys. J. Int.*, **128**, 125-150, 1997.

Houseman, G. and D. Gubbins, Deformation of subducted oceanic lithosphere, *Geophys. J. Int.*, in press, 1997.

Ita, J. and S.D. King, Sensitivity of convection with an endothermic phase change to the form of the governing equations, initial conditions, boundary conditions, and equation of state, *J. Geophys. Res.*, **99**, 15,919-15,938, 1994.

Iwase, Y.A., Three-dimensional infinite Prandtl number convection in a spherical shell with temperature-dependent viscosity, *J. Geomag. Geoelectr.*, **48**, 1499-1514, 1996.

Jarvis, G.T. and D.P. McKenzie, Convection in a compressible fluid with infinite Prandtl number, *J. Fluid Mech.*, **96**, 515-583, 1980.

Jordan, K.E., Yuen, D.A., Reuteler, D.M., Zhang, S. and R. Haimes, Parallel interactive visualization of 3-D mantle convection, *IEEE Computational Science & Engineering*, **3** (**4**), 29-37, 1996.

Kameyama, M., Fujimoto, H. and M. Ogawa, A thermo-chemical regime in the upper mantle in the early Earth inferred from a numerical model of magma migration in the convecting upper mantle, *Phys. Earth Planet. Inter.*, **94**, 187-216, 1996.

Kameyama, M., Yuen, D.A. and H. Fujimoto, The interaction of viscous heating with grain-size dependent rheology in the formation of localized slip zones, *Geophys. Res. Lett.*, **24**, 2523-2526, 1997.

Karato, S. and P. Li, Diffusion creep in perovskite: implications for the rheology of the lower mantle, *Science*, **255**, 1238- 1240, 1992.

Karato, S., Phase transformations and rheological properties of mantle minerals, Chapter 8, in *Earth's Deep Interior*, edited by D. Crossley and A.M. Soward, Gordon-Breach, pp. 223-272, 1997.

Kawakatsu, H. and F. Niu, Seismic evidence of a 920-km discontinuity in the mantle, *Nature*, **371**, 301 -305, 1994.

Kellogg, L.H., Interaction of plumes with a compositional boundary at 670 km, *Geophys. Res. Lett.*, **18**, 865-868, 1991.

Kellogg, L.H. and S.D. King, Effects of mantle plumes on the growth of D'' by reaction between the core and mantle, *Geophys. Res. Lett.*, **20**, 379-382, 1993.

Kincaid, C. and P. Silver, The role of viscous dissipation in the orogenic process, *Earth Planet. Sci. Lett.*, **142**, 271-288, 1996.

King, S.D., Raefsky, A. and B.H. Hager, ConMan: Vectorizing a finite-element code for incompressible two-dimensional convection in the Earth's mantle, *Phys. Earth Planet Interior*, **59**, 195-207, 1990.

Larsen, T.B., Malevsky, A.V., Yuen, D.A. and J.J. Smedsmo, Temperature-dependent Newtonian and non-Newtonian convection: implications for lithospheric processes, *Geophys. Res. Lett.*, **20**, 2595-2598, 1993.

Larsen, T.B., Yuen, D.A. and A.V. Malevsky, Dynamical consequences on fast subducting slabs from a self-regulating mechanism due to viscous heating in variable viscosity convection, *Geophys. Res. Lett.*, **22**, 1277-1280, 1995.

Larsen, T.B., Yuen, D.A., Malevsky, A.V. and J.J. Smedsmo, Dynamics of strongly time-dependent convection with non-Newtonian temperature-dependent viscosity, *Phys. Earth Planet. Inter.*, **94**, 75-103, 1996a.

Larsen, T.B., Yuen, D.A., Smedsmo, J.J. and A.V. Malevsky, Thermomechanical modeling of pulsation tectonics and consequences on lithospheric dynamics, *Geophys. Res. Lett.*, **23**, 217-220, 1996b.

Larsen, T.B., Moser, J., Yuen, D.A. and B. Fornberg, A higher-order finite-difference method applied to large Rayleigh number mantle convection, *Geophys. Astrophys. Fluid Dyn.*, **84**, 53-83, 1997.

Leitch, A.M. and D.A. Yuen, Internal heating and thermal constraints on the mantle, *Geophys. Res. Lett.*, **16**, 1407-1410, 1989.

Leitch, A.M., Yuen, D.A. and G. Sewell, Mantle convection with internal heating and pressure-dependent thermal expansivity, *Earth Planet. Sci. Lett.*, **102**, 213-232, 1991.

Lenardic, A. and W.M. Kaula, Mantle dynamics and the heat flow into the Earth's continents, *Nature*, **378**, 709-711, 1995.

Li, X.-D. and B. Romanowicz, Global mantle shear-velocity model developed using nonlinear asymptotic coupling theory, *J. Geophys. Res.*, **101**, 22245-22272, 1996.

Machetel, P., Rabinowicz, M. and P. Bernadet, Three-dimensional convection in spherical shells, *Geophys. Astrophys. Fluid Dyn.*, **37**, 57-84, 1986.

Machetel, P. and P. Weber, Intermittent layered convection in a model mantle with an endothermic phase change at 670 km, *Nature*, **350**, 55-57, 1991.

Malevsky, A.V. and D.A. Yuen, Characteristics-based methods applied to infinite Prandtl number thermal convection in the hard turbulent regime, *Phys. Fluids*, A **3**, 2105-2115, 1991.

Malevsky, A.V. and D.A. Yuen, Strongly chaotic non-Newtonian mantle convection, *Geophys. Astro. Fluid Dyn.*, **65**, 149-171, 1992.

Malevsky, A.V. and D.A. Yuen, Plume structures in the hard turbulent regime of three-dimensional infinite Prandtl number convection, *Geophys. Res. Lett.*, **20**, 383-386, 1993.

Malevsky, A.V., Spline-characteristic method for simulation of convective turbulence, *J. Comp. Phys.* **123**, 466-475, 1996.

Melosh, H.J., Plate motion and thermal instability in the asthenosphere, *Tectonophysics*, **35**, 363-390, 1976.

McKenzie, D.P., Roberts, J.M. and N.O. Weiss, Convection in the earth's mantle: toward a numerical simulation, *J. Fluid Mech.*, **62**, 465-538, 1974.

Moresi, L.-N. and V.S. Solomatov, Numerical investigation of 2D convection with extremely large viscosity variations, *Phys. Fluids*, **7**, 2154 -2162, 1995.

Morgan, W.J., Rises, trenches, great faults and crustal blocks, *J. Geophys. Res.*, **73**, 1959 -1982, 1968.

Niu, F. and H. Kawakatsu, Depth variation of the mid-mantle seismic discontinuity, *Geophys. Res. Lett.*, **24**, 429-432, 1997.

O'Connell, R.J., On the scale of mantle convection, *Tectonophys.* **38**, 119-136, 1977.

Ogawa, M., Schubert, G. and A. Zebib, Numerical simulations of three-dimensional thermal convection in a fluid with strongly temperature-dependent viscosity, *J. Fluid Mech.*, **233**, 299-328, 1991.

Ogawa, M., A numerical model of a coupled magmatism-mantle convection system in Venus and the Earth's mantle beneath Archaean continental crusts. *Icarus*, **102**, 40-61, 1993.

Ogawa, M., Effects of chemical fractionation of heat-producing elements on mantle evolution inferred from a numerical model of coupled magmatism-mantle convection system, *Phys. Earth Planet Inter.* **83**, 101-127, 1994.

Ogawa, M. and H. Nakamura, The thermo-chemical regime of early earth inferred from numerical models of coupled magnetism-mantle convection system with the solid-solid phase transitions at depths around 660km, *J. Geophys. Res.*, in press, 1998.

Parmentier, E.M., D.L. Turcotte and K.E. Torrance, Numerical experiments on the structure of mantle plumes, *J. Geophys. Res.*, **80**, 4417-4425, 1975.

Parmentier, E.M., Turcotte, D.L. and K.E. Torrance, Studies of finite amplitude non-Newtonian convection with applications to convection in the Earth's mantle, *J. Geophys. Res.*, **81**, 1839- 1846, 1976.

Parmentier, E.M., Sotin, C. and B.J. Travis, Turbulent 3-D thermal convection in an infinite Prandtl number, volumetrically heated fluid: implications for mantle dynamics, *Geophys. J. Int.*, **116**, 241 -251, 1994.

Patankar, S.V., *Numerical Heat Transfer and Fluid Flow*, Hemisphere, New York, 1980.

Peltier, W.R., Penetrative convection in the planetary mantle, *Geophys. Astrophys. Fluid Dyn.*, **5**, 47 -88, 1973.

Poirier, J.-P., *Introduction to the Earth's Interior*, Cambridge Univ. Press, 1991.

Ranalli, G., *Rheology of the Earth*, second edition, Chapman and Hall, 1995.

Rast, M.P. and J. Toomre, Compressible convection with ionization. II. Thermal boundary-layer instability, *The Astrophysical J.* **419**, 240-254, 1993.

Ratcliff, J.T., Tackley, P.J., Schubert, G. and A. Zebib, Transitions in thermal convection with strongly variable viscosity, *Phys. Earth Planet. Inter.*, **102**, 201-212, 1997.

Regenauer-Lieb, K. and D.A. Yuen, Rapid conversion of elastic energy into shear heating and localized deformation of the continental lithosphere, submitted to *Geophys. Res. Lett.*, 1997.

Reuteler, D.M., Yuen, D.A., Balachandar, S. and S. Honda, Three-dimensional mantle convection: effects of depth-dependent properties and multiple phase transitions, *Int. Video J. Eng. Research*, **3**, 47-63, 1993.

Richter, F.M., Dynamical models for sea floor spreading, *Rev. Geophys. & Space Phys.*, **11**, 223-287, 1973.

Richter, F.M. and C.E. Johnson, Stability of a chemically layered mantle, *J. Geophys. Res.*, **79**, 1635-1639, 1974.

Richter, F.M. and D.P. McKenzie, On some consequences and possible causes of layered mantle convection, *J. Geophys. Res.*, **86**, 6133- 6142, 1981.

Sammis, C.G., Smith, J.C., Schubert, G. and D.A.Yuen, Viscosity- depth profile in the Earth's mantle: effects of polymorphic phase transitions, *J. Geophys. Res.*, **85**, 3747-3761, 1977.

Schmeling, H. and W.R. Jacoby, On modelling the lithosphere in mantle convection with nonlinear rheology, *J. Geophys.*, **50**, 89- 100, 1981.

Schmeling, H. Compressible convection with constant and variable viscosity: the effect on slab formation, geoid and topography, *J. Geophys. Res.*, **94**, 12,463-12,481, 1989.

Schubert, G., Yuen, D.A. and D.L. Turcotte, Role of phase transitions in a dynamic mantle, *Geophys. J. R. Astron. Soc.*, **42**, 705-735, 1975.

Schubert, G., Froidevaux, C. and D.A. Yuen, Oceanic lithosphere and asthenosphere: thermal and mechanical structure, *J. Geophys. Res.*, **81**, 3525-3541, 1976.

Schubert, G., Numerical models of mantle convection, *Ann. Rev. Fluid Mech.*, **24**, 359 -294, 1992.

Solheim, L.P. and W.R. Peltier, Avalanche effects in phase transition modulated thermal convection: A model of Earth's mantle, *J. Geophys. Res.*, **99**, 6997-7018, 1994.

Solomatov, V.S., Scaling of temperature- and stress-dependent viscosity convection, *Phys. Fluids*, **7**, 266-274, 1995.

Sreenivasan, K.R., Ramshankar, R. and C. Meneveau, Mixing, entrainment and fractal dimensions of surfaces in turbulent flows, *Proc. R. Soc. Lond.* **A 421**, 79-108, 1989.

Steinbach, V., Hansen, U. and A. Ebel, Compressible convection in the Earth's mantle: a comparison of different approaches, *Geophys. Res. Lett.*, **16**, 633-636, 1989.

Steinbach, V. *Numerische Experimente zur Konvektion in kompressiblen Medien*, Doktorsarbeit, Univ. zu Koeln, 1991.

Steinbach, V. and D.A. Yuen, The effects of multiple phase transitions on Venusian mantle convection, *Geophys. Res. Lett.*, **19**, 2243-2246, 1992.

Steinbach, V., Yuen, D.A. and Zhao, W., Instabilities from phase transitions and the timescales of mantle evolution, *Geophys. Res. Lett.*, **20**, 1119-1122 1993.

Steinbach,V. and D.A. Yuen, Melting instabilities in the transition zone, *Earth Planet. Sci. Lett.*, **127**, 67-75, 1994

Steinbach, V.C. and Yuen, D.A., The effects of temperature- dependent viscosity on mantle convection with the two major phase transitions, *Phys. Earth Planet. Inter.*, **90**, 13-36, 1995.

Su, W-J., Woodward, R.L. and A.M. Dziewonski, Degree 12 model of shear velocity heterogeneity in the mantle, *J. Geophys. Res.* **99**, 6945- 6980, 1994.

Tackley, P.J., Stevenson, D.J., Glatzmaier, G.A. and G. Schubert, Effects of an endothermic phase transition at 670 km. depth on spherical mantle convection, *Nature*, **361**, 699-704, 1993.

Tackley, P.J., Effects of strongly temperature-dependent viscosity on time-dependent, three-dimensional models of mantle convection, *Geophys. Res. Lett.*, **20**, 2187-2190, 1993.

Tackley, P.J., On the ability of phase transitions and viscosity layering to induce long wavelength heterogeneity in the mantle, *Geophys. Res. Lett.*, **23**, 1985-1988, 1996a.

Tackley, P.J., Effects of strongly variable viscosity on three- dimensional compressible convection in planetary mantles, *J. Geophys. Res.*, **101**, 3311-3332, 1996b.

Tackley, P.J., Self-consistent generation of tectonic plates in three dimensional convection, *Earth Planet. Sci. Lett.*, in press, 1998.

Takeuchi, H. and S. Sakata, Convection in a mantle with variable viscosity, *J. Geophys. Res.*, **75**, 921-927, 1970.

Ten, A., Yuen, D.A., Larsen, T.B. and A.V. Malevsky, The evolution of material surfaces in convection with variable viscosity as monitored by a characteristics-based method, *Geophys. Res. Lett.*, **23**, 2001-2004, 1996.

Ten, A., Yuen, D.A., Podlachikov, Yu.Yu., Larsen, T.B., Pachepsky, E. and A.V. Malevsky, Fractal features in mixing of non-Newtonian and Newtonian mantle convection, *Earth Planet. Sci. Lett.*, **146**, 401-414, 1997.

Torrance, K.E. and D.L. Turcotte, Thermal convection with large viscosity variations, *J. Fluid Mech.*, **47**, 113-125, 1971.

Toksoz, M.N., Minear, J.W. and B.R. Julian, Temperature field and geophysical effects of a downgoing slab, *J. Geophys. Res.*, **76**, 1113-1138, 1971.

Tozer, D.C., Towards a theory of thermal convection in the mantle, in *The Earth's Mantle*, T.F. Gaskell, ed., pp. 325-353, Academic Press, London, 1967.

Travis, B., Olson, P. and Schubert, G., The transition from two- dimensional to three-dimensional planforms in infinite Prandtl- number thermal convection, *J. Fluid Mech.* **216**, 71-91, 1990.

Trompert, R.A. and U. Hansen, The application of a finite volume multigrid method to 3-D flow problems in a highly viscous fluid with a variable viscosity, *Geopys. Astrophys. Fluid Dyn.*, **83**, 263-293, 1996.

Turcotte, D.L. and E.R. Oxburgh, Finite amplitude convective cells and continental drift, *J. Fluid Mech.*, **28**, 29-42, 1967.

Turcotte, D.L., K.E. Torrance and A.T. Hsui, Convection in the Earth's mantle, in *Methods in Computational Physics*, 13, 431-451, Academic Press, 1973.

Turcotte, D.L., Hsui, A.T., Torrance, K.E. and G. Schubert, Influence of viscous dissipation on Benard convection, *J. Fluid Mech.*, **64**, 369-374, 1974.

van den Berg, A.P., van Keken, P.E. and D.A. Yuen, The effects of a composite non-Newtonian and Newtonian rheology in mantle convection, *Geophys. J. Int.*, **115**, 62-78, 1993.

van den Berg, A.P. and D.A. Yuen, Convectively induced transition in mantle rheological behavior, *Geophys. Res. Lett.*, **22**, 1549-1552, 1995.

van den Berg, A.P. and D.A. Yuen, Temperature at the Core-mantle boundary (CMB) as a control variable in modelling planetary dynamics, *Phys. Earth Plant. Inter.*, in press, 1998.

van der Hilst, R., Engdahl, R., Spakman, W. and G. Nolet, Tomographic imaging of sub-ducted lithosphere below northwest Pacific island arcs, *Nature*, **353**, 37-43, 1991.

van der Hilst, R., Complex morphology of subducted lithosphere in the mantle beneath the Tonga trench, *Nature*, **374**, 154-157, 1995.

Van Keken, P.E., D.A. Yuen and L.R. Petzold, DASPK: a new high order and adaptive time-integration technique with applications to mantle convection with strongly temperature- and pressure- dependent rheology. *Geophys. Astrophys. Fluid Dyn.*, **80**, 57-74, 1995.

Vasilyev, O.V., Yuen, D.A. and Yu.Yu. Podlachikov, Applicability of wavelet algorithm for geophysical viscoelastic flow, *Geophys. Res. Lett.*, **24**, 3097-3100, 1997.

Vening-Meinesz, F.A., Convective instability induced by phase transitions in the mantle, *Proc. Kon. Ned. Akad.*, **B59**, 1-22, 1956.

Weertman, J., The creep strength of the Earth's mantle, *Rev. Geo- phys. Space Phys.*, **8**, 145-168, 1970.

Weinstein, S.A., Catastrophic overturn in the Earth's mantle driven by multiple phase changes and internal heat generation, *Geophys. Res. Lett.*, **20**, 101-104 1993.

Wilson, J.T., A new class of faults and their bearing on continental drift, *Nature*, **207**, 343 -347, 1965.

Yuen, D.A. and G. Schubert, Mantle plumes: a boundary layer approach for Newtonian and non-Newtonian temperature-dependent rheologies, *J. Geophys. Res.*, **81**, 2499-2510, 1976.

Yuen, D.A. and G. Schubert, Asthenospheric shear flow : thermally stable or unstable?, *Geophys. Res. Lett.*, **4**, 503-506, 1977.

Yuen, D.A., Quareni, F. and H.J. Hong, Effects from equation of state and rheology in dissipative heating in compressible mantle convection, *Nature*, **326**, 67-69, 1987.

Yuen, D.A. and A.V. Malevsky, Strongly chaotic Newtonian and non-Newtonian mantle convection, in *Chaotic Processes in the Geological Sciences*, ed. by D.A. Yuen, pp 71-88, Springer Verlag, New York, 1992.

Yuen, D.A., Hansen, U., Zhao, W., Vincent, A.P. and A.V. Malevsky, Hard turbulent thermal convection and thermal evolution of the mantle, *J. Geophys. Res.*, **98**, E3, 5355 -5373, 1993.

Yuen, D.A., D.M. Reuteler, S. Balachandar, V. Steinbach, A.V. Malevsky and J.J. Smedsmo, Various influences on three-dimensional mantle convection with phase transitions, *Phys. Earth Planet Inter.*, **86**, 185-203, 1994.

Yuen, D.A., Balachandar, S., Steinbach, V.C., Honda, S., Reuteler, D.M., Smedsmo, J.J. and Lauer, G.S., Non-equilibrium effects of core cooling and time-dependent internal heating on mantle flush events, *Nonlinear Processes in Geophysics*, **2**, 206-221, 1995.

Yuen, D.A., Cadek, O., Van Keken, P.E., Reuteler, D.M., Kyvalova, H. and B.S. Schroeder, Combined results from mineral physics, tomography and mantle convection and their implications on global geodynamics, in *Seismic Modelling of Earth Structure*, ed. by E. Boschi, G. Ekstroem and A. Morelli, pp. 463-505, Editrice Compositori, Bologna, Italy, 1996.

Zebib, A.F., Schubert, G. and J.M. Straus, Infinite Prandtl number thermal convection in a spherical shell, *J. Fluid Mech.*, **97**, 257-277, 1980.

Zhang, S. and D.A. Yuen, The influences of lower-mantle viscosity stratification on 3-D spherical-shell mantle convection, *Earth Planet. Sci. Lett.*, **132**, 157-166, 1995.

Zhang, S. and D.A. Yuen, Various influences on plumes and dynamics in time-dependent compressible mantle convection in 3-D spherical shell, *Phys. Earth Planet. Inter.*, **94**, 241 -267, 1996.

Zhao, W., Yuen, D.A. and S. Honda, Multiple phase transitions and the style of mantle convection, *Phys. Earth Planet. Inter.*, **72**, 185 -210, 1992.

Zhong, S. and M. Gurnis, Controls on trench topography from dynamic models of subducted slabs, *J. Geophys. Res.*, **99**, 15683-15695, 1994.

Zhong, S. and M. Gurnis, Mantle convection with plates and mobile, faulted plate margins, *Science*, **267**, 838-843, 1995.

Zhong, S. and M. Gurnis, Weak faults and non-Newtonian rheology produce plate tectonics in 3D models of mantle flow, *Nature*, **383**, 245-247, 1996.

Zhou, H.W., A high-resolution P-wave model for the top 1200 km of the mantle, *J. Geophys. Res.*, **101**, 27791-27810, 1996.

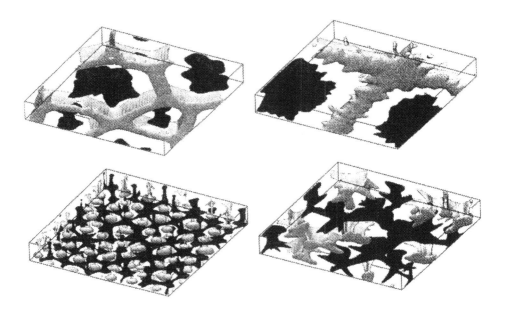

Figure 13.2: Isocontours of the lateral thermal anomalies, showing where the temperature is higher (light) or lower (dark) than the horizontally averaged value by 10 to 15%. Top left is the constant viscosity case with stress-free boundary conditions; top right is the Newtonian temperature-dependent viscosity case with a total contrast of 1000 and rigid boundary conditions; bottom left has the same rheology as for the top right case except the boundary conditions are stress-free, and the bottom right is the non-Newtonian ($n = 3$) case with a viscosity contrast of 1000 for the temperature and stress-free boundary conditions. Further details can be found in Tackley (1993).

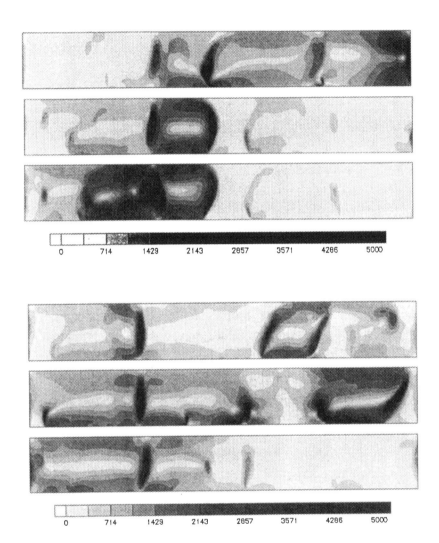

Figure 13.3: (a) Three snapshots of the magnitude of the velocity for non-Newtonian ($n = 3$) convection with a temperature-dependent viscosity contrast of 100 and an averaged Rayleigh number of 2.2×105, no internal heating ($R = 0$). Top panel, t=0.0039; middle panel, t=0.0044; bottom panel, t=0.0046. (b) Three snapshots of the magnitude of the velocity for non-Newtonian ($n = 3$) convection with a temperature-dependent viscosity contrast of 100 , internal heating with $R = 7$ and an averaged Rayleigh number of 8.4×10^5. Top panel, t=0.0084; middle panel, t=0.0087; bottom panel, t=0.010. Further details can be found in Larsen *et al.* (1996a).

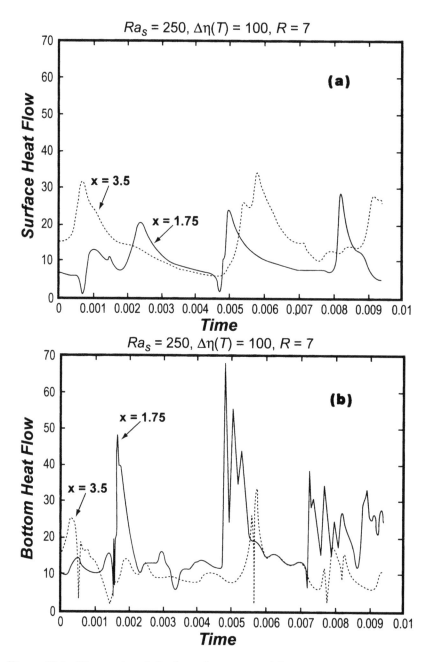

Figure 13.4: Time-series of the heat flow at two different points $x=1.75$ and $x=3.5$ in a aspect-ratio 7 box. The case corresponds to that given in figure 13.2b. Top and bottom panels represent respectively the surface and bottom heart flow.

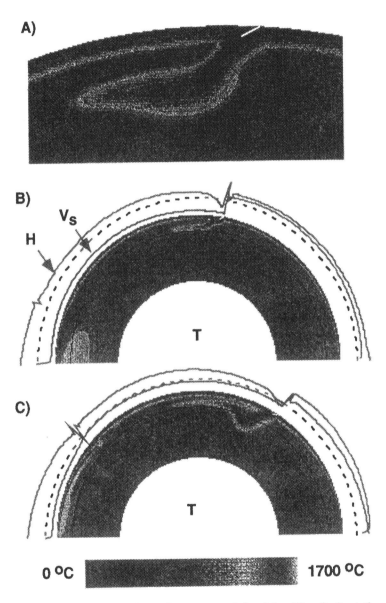

Figure 13.5: (a) the zoomed-in thermal structure of the slab with a fault at the converging margin. The temperature , dynamic topography (H) and surface velocity for before (b) and (c) after the start of slab penetration. Further details can be found in Zhong & Gurnis (1995).

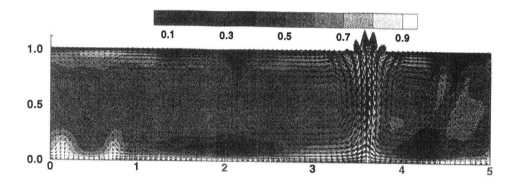

Figure 13.6: Temperature contours in greyscale and superimposed velocity vector plot on a vertical $(x - z)$ plane passing through the computational box. This plane has been chosen to cut through an intense hot plume which can be seen near x=3.5. The surface Rayleigh number for this case is 10^5 and Di=0.3. The simulation employed a high resolution of $180 \times 180 \times 80$ points.

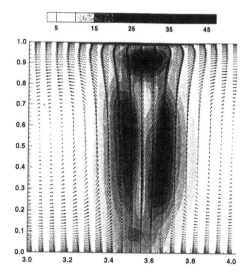

Figure 13.7: Plot showing shaded contours of viscous heating along with the velocity vectors on a vertical $(x - z)$ plane same as the one in figure 13.6. Only the region close to the hot upwelling plume is shown.

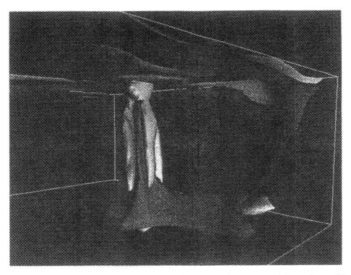

Figure 13.8: Three-dimensional surface plot of viscous heating. Also plotted in this figure are two different isosurfaces of temperature corresponding to hot and cold fluid. The lighter shaded surface is the region of intense viscous heating. It can be seen around the sides of a hot upwelling plume and near where the plume impinges the surface. On the right, surface of a descending cold flow can be seen.

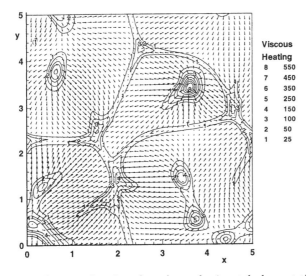

Figure 13.9: Contours of viscous heating plotted on a horizontal plane at the surface. Overlaid on top are the velocity vectors. It can be seen that the flow convergence occurs in regions of interconnected thin sheets. These are the base of cold downwelling sheets and are associated with significant viscous heating. The most intense heating occurs at the centers of flow divergence. These are the regions where the hot plumes impinge the top surface.

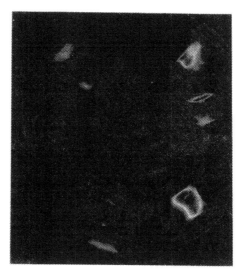

Figure 13.10: The viscous heating distribution superimposed on the temperature structures of plumes in 3-D spherical-shell convection with an endothermic phase transition at 660 km depth.

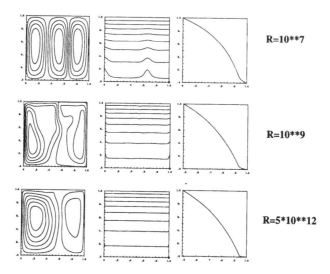

Figure 13.11: Streamfunction, temperature and horizontally averaged temperature profile. Parameters are $T_0=5.0$, $Di=0.5$. A grid of 55×55 elements has been used for $Ra = 10^7$ and a grid of 75×75 elements has been used for the calculation at Ra from 10^9 and higher. Thermal expansivity decreases by a factor of 10 across the layer.

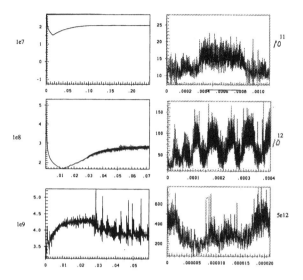

Figure 13.12: Time series of Nusselt numbers for different Rayleigh numbers. Other parameters are as specified above.

Thermally driven convection

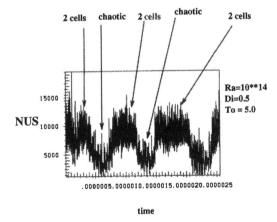

Figure 13.13: Time trace of the surface Nusselt number

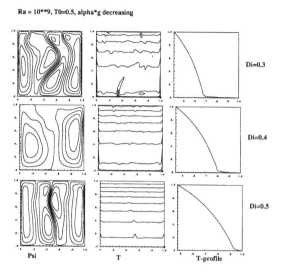

Figure 13.14: Snapshots of streamfunction, temperature and vertical profile of the temperature for $Ra = 10^9$, T_0=5.0 and Di varying from 0.3 to 0.5. The buoyancy term ag decreases by a factor of 10 across the layer.

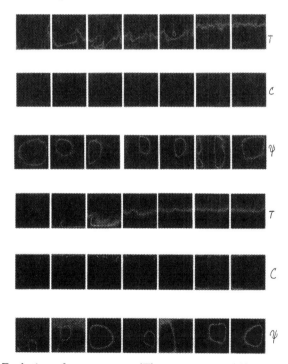

Figure 13.15: (a) Evolution of temperature (T), composition (C) and streamfunction (y) for thermochemical convection. The motion of the flow is dominated by thermally induced density differences. Parameters are $Ra = 10^7$, $Ra_c = 10^6$, $Le = 1$. T_0 is 5.0 and Di=0.5. A grid consisting of 55×55 elements has been employed. (b) Same as for a) but for $Ra = 10^6$ and $Ra_c = 10^7$ (compositionally dominated case).

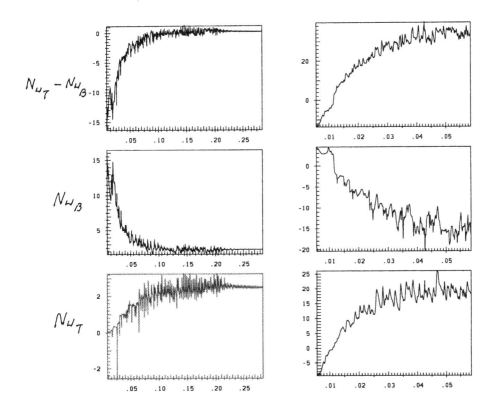

Figure 13.16: Time history of the surface Nusselt number (bottom panel) and Nu at the bottom (middle panel) of the system for thermally driven convection (left column) and compositionally driven flow (right column). The top panels show the effective internal heating rate (Nu-top - Nu-bottom) Parameters for thermally and compositionally dominated flow same as the ones specified in figure 13.15.

Chapter 14

TURBULENT TRANSPORT IN ROTATING COMPRESSIBLE CONVECTION

NICHOLAS H. BRUMMELL

Joint Institute for Laboratory Astrophysics,
Department of Astrophysical and Planetary Sciences
University of Colorado, Boulder, CO 80309-0391

The outer envelope of our nearest star, the sun, is a very complicated fluid dynamics laboratory. Observations reveal highly turbulent flows and fields, and yet a remarkable degree of large-scale coherence emerges from this chaos. High-performance computing is allowing simulations of the compressible gas dynamics to enter the turbulent regime at last, and the results are yielding transport properties which are at considerable variance with earlier laminar models. Such turbulent modelling is providing insight into the generation of the largest scales from small-scale disorder and thus also into the processes underlying the observed solar phenomena.

14.1 Introduction

Stars like our sun exhibit observable phenomena which vary tremendously in scale through time and space. Granules 1Mm in size overturn on the solar surface in 5 minutes and yet there exists a differential rotation of the star as a whole (radius 700Mm), where the poles rotate with a period of 33 days compared to 25 days at the equator. Magnetic fields live as tiny bundles in the inter-granular lanes but also exhibit a global polarity law which reverse on a distinct 11 year cycle. The source for much of these wide-ranging phenomena is believed to be the motions of thermal convection, which is the mechanism of choice for carrying the core-created heat flux to the surface in the outer 30% of the star. Such fluid motions interact with the rotation of the star and its magnetic field, and are further complicated by an intricate dependence of the compressible gaseous medium on ionization state, opacity and radiative transfer. However, a fundamental difficulty is that the convection itself is spread over a huge range of scales and at every scale it is intensely turbulent, due to the small diffusivities, large physical scale and high flow velocities.

Such an intricate web of physics denies access to theoretical interpretation in general, although simple models (linear and mixing length theory, see e.g. Busse, 1970; Baker and

Spiegel 1975; Gough and Weiss 1976; Rosner and Weiss, 1985; Hathaway, 1984; Durney, 1987; Rüdiger, 1989) have provided remarkable insight into global (mean) properties. Stellar fluids are compressible and operate at such incredibly high forcing that laboratory experiments are extremely difficult. Observations initially only allowed a peek at the surface regime until the subject of helioseismology (see review by Gough and Toomre, 1991) recently blossomed and began attempting to quantize the interior via the acoustic emissions. Consequently, many of the major thrusts toward an understanding the dynamics of the solar convection zone come through numerical models, where parameters can be varied, physics can be added or removed, observables can be compared and the interior can be seen in detail. Despite the flexibility of numerical simulations however, most solar phenomena still defy a complete understanding. An outstanding example of this is the solar differential rotation problem, wherein the observational data and numerical results for the profile of azimuthally-averaged zonal angular velocity in the convection zone distinctly disagree. The helioseismic inversions of p-mode frequency-splitting observations suggest a profile that is roughly constant on radii (Libbrecht, 1989) whereas ground-breaking and more recent spherical shell simulations (see e.g. Gilman and Miller, 1986; Glatzmaier, 1987; Glatzmaier and Toomre, 1995) all hold that velocity more constant on cylinders.

This dichotomy may reflect the overwhelming difficulties of solar numerical modelling, namely the resolution of the vast the range of scales and the highly turbulent nature of the motions. Current numerical models operating on forefront hardware technology can utilise at most 10^3 degrees of freedom per spatial direction giving a dynamic range of three orders of magnitude, whereas the observed phenomena cover at least six in each physical dimension. The Reynolds numbers for the individual phenomena are also so high (probably greater than 10^{12}) that present simulations fall short of these solar values by six or so orders of magnitude. Attempting to resolve the solar differential rotation puzzle really requires simulations of convection in the correct full spherical geometry. However, in such a global model the largest scale is pinned, and so scales down to only three orders of magnitude below that global size can be resolved. Much of the small-scale turbulent dynamics, equivalent to the supergranulation and granulation in the sun for example, are by necessity then largely omitted. One alternative is to reduce the fixed maximum scale by studying smaller portions of the full problem and utilizing the three orders of magnitude to encompass the dynamical range of turbulent scales. There are clear tradeoffs: the global models operate in the correct geometry but cannot encompass enough of a dynamical range to admit fully turbulent solutions, whereas the local models are able to study fully turbulent convection in localized domains only.

Described here is one such local approach. Within this framework, the physics of the solar problem has been prioritised, retaining the minimum set to achieve some understanding of the underlying dynamics. For the stellar problem, the density contrast across the convection zone is large and therefore the compressibilty of the gas must be taken into account. If the solar differential rotation is of interest, then some form of local rotation must also be included. Presented here then are the novel insights that such simple, local, compressible, rotating convection simulations have revealed in relation to the nature of transport properties of the turbulent rather than laminar regimes.

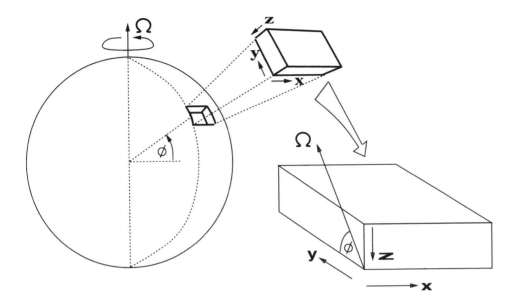

Figure 14.1: Local f-plane model positioned at latitude ϕ used to study flows in a portion of a spherical shell rotating with angular velocity Ω_o. The rectilinear coordinate system has the horizontal variables x increasing eastward and y poleward, the depth z increasing downward; the tilted rotation vector $\mathbf{\Omega}$ thus lies in the $y - z$ plane.

14.2   Local modelling of rotating compressible convection

Fundamentally, this paper reviews the results found in papers Cattaneo *et al.* (1991) and Brummell *et al.* (1996, 1998), hereinafter referred to as Papers 0, I and II. These models are all built upon a similar framework, described briefly here. For more details of this formulation and all of the following results, please consult those references.

A rectilinear domain containing a fully compressible but ideal gas is confined between two horizontal, impenetrable, stress-free walls a distance d apart. The upper surface is held at a fixed temperature T_0 whereas a constant temperature gradient Δ is maintained through the lower boundary. The flow is assumed to be periodic in the two horizontal directions. Fluid properties, such as the specific heats c_p and c_v, shear viscosity μ and thermal conductivity K are assumed constant as is the gravitational acceleration g. In hydrostatic balance, the temperature, T_p, density ρ_p and pressure p_p can exist in a polytropic state

$$T_p/T_o = (1 + \theta\tilde{z}/d), \tag{14.1}$$

$$\rho_p/\rho_o = (1 + \theta\tilde{z}/d)^m, \tag{14.2}$$

$$p_p/p_o = (1 + \theta\tilde{z}/d)^{m+1}, \tag{14.3}$$

for $0 \le \tilde{z} \le d$, where ρ_o is the density at the upper boundary, $p_o = (c_p - c_v)T_o\rho_o$, $m = -1 + g/\Delta(c_p - c_v)$ is the polytropic index and $\theta = d\,\Delta/T_o$. The equations of conservation of mass, momentum and energy and the perfect gas equation can be non-dimensionalized using d as the unit of length, the isothermal sound crossing time $[d^2/(c_p - c_v)T_o]^{1/2}$ as the unit of time, and T_o and ρ_o as the units of temperature and density, to produce

$$\partial_t \rho + \nabla \cdot \rho \mathbf{u} = 0, \tag{14.4}$$

$$\partial_t \rho \mathbf{u} + \nabla \cdot \rho \mathbf{u} \mathbf{u} + C_k P_r T_{a_0}^{1/2} \left(\hat{\mathbf{\Omega}} \times \rho \mathbf{u} \right) =$$
$$-\nabla p + P_r C_k \left(\nabla^2 \mathbf{u} + \tfrac{1}{3} \nabla (\nabla \cdot \mathbf{u}) \right) + \theta (m+1) \rho \hat{\mathbf{z}}, \tag{14.5}$$

$$\partial_t T + \mathbf{u} \cdot \nabla T + (\gamma - 1) T \nabla \cdot \mathbf{u} = \frac{\gamma C_k}{\rho} \nabla^2 T + V_\mu, \tag{14.6}$$

$$p = \rho T. \tag{14.7}$$

Here $\mathbf{u} = (u, v, w)$ is the velocity, T, ρ and p are the temperature, density and pressure respectively, and $V_\mu = \frac{(\gamma-1)C_k}{\rho} \sigma \partial_i u_j (\partial_i u_j + \partial_j u_i - \tfrac{2}{3} \nabla \cdot \mathbf{u} \delta_{ij})$ is the viscous heating. Rotation enters the momentum equation in a modified f-plane formulation in this local model via the rotation vector

$$\mathbf{\Omega} = \Omega_0 \hat{\mathbf{\Omega}} = (\Omega_x, \Omega_y, \Omega_z) = (0, \Omega_o \cos\phi, -\Omega_o \sin\phi) \tag{14.8}$$

where ϕ is the latitudinal positioning of the domain on the true sphere, as exhibited in figure 14.1. In the presentation of this paper, the rotation sense has been changed for ease of comparison with the solar case. In the z-downward coordinate system, positive rotation is clockwise when viewed from above the North pole. This paper adopts the more familiar anti-clockwise rotation sense by setting $\Omega_0 \to -\Omega_0$ (equivalently $u \to -u, x \to -x$) when exhibiting the results.

The dimensionless numbers parameterizing the problem are the Rayleigh number,

$$R_a(z) = \frac{\theta^2 (m+1)}{P_r C_k^2} \left(1 - \frac{(m+1)(\gamma-1)}{\gamma} \right) (1 + \theta z)^{2m-1}, \tag{14.9}$$

involving the thermal dissipation parameter $C_k = K/\{d\rho_o c_p [(c_p - c_v)T_o]^{1/2}\}$ (the ratio of the sound crossing time to the thermal relaxation time) and $\gamma = c_p/c_v$ (the ratio of the specific heats), together with the Prandtl number $\sigma = \mu c_p / K$ (the ratio of the diffusivities of heat and momentum) and the Taylor number

$$T_{a_0} = \frac{4\Omega_0^2 d^4}{(\mu/\rho_0)^2} = \left(\frac{\rho}{\rho_0} \right)^2 T_a. \tag{14.10}$$

This paper quotes R_a and T_a (the more usual Taylor number) as evaluated at mid-layer in the initial polytrope. A measure of the influence of the rotation on global motions derived in terms of these parameters is the *convective Rossby number*

$$R_o = \left(\frac{R_a}{T_a P_r} \right)^{1/2}. \tag{14.11}$$

This parameter gives an indication of the influence of rotation from the parameters of the problem in advance. A value of R_o less than unity implies a significant influence of the rotation since then in the time a fluid element is driven across the box by buoyancy it can execute more than one inertial rotation. A true Rossby number R_{o_t} may be determined from the results of a calculation in terms of the ratio of the rms vorticity generated to the background vorticity i.e.

$$R_{o_t} = \omega_{\mathrm{rms}}/2\Omega. \tag{14.12}$$

It is found that in these simulations that the convective and true Rossby numbers are similar in magnitude.

At the upper and lower boundaries the following conditions are applied

$$\rho w = \partial_z u = \partial_z v = 0 \text{ at } z = 0, 1, \tag{14.13}$$

$$T = 1 \text{ at } z = 0, \ \partial_z T = \theta \text{ at } z = 1, \tag{14.14}$$

which ensure that the mass flux and mechanical energy flux vanish on the boundaries so that the total mass in the computational domain is conserved and the heat flux is the only flux of energy in and out of the system.

Averaging the momentum equations (14.5) over the two periodic horizontal directions x and y produces the equations for the mean zonal and meridional flows

$$\partial_t \overline{\rho u} = -\partial_z \overline{\rho u w} - f_z \overline{\rho v} + \sigma C_k \partial_z^2 \overline{u}, \tag{14.15}$$

$$\partial_t \overline{\rho v} = -\partial_z \overline{\rho v w} + f_z \overline{\rho u} + \sigma C_k \partial_z^2 \overline{v}. \tag{14.16}$$

Here an overbar denotes the horizontal average of a quantity and $f_z = \sigma C_k T_{a_0}^{1/2} \sin \phi$ is the inertial frequency for horizontal momentum about the vertical. These mean flow components represent the f-plane model equivalents of the zonal and meridional differential rotations with depth of that local latitude in a spherical fluid shell. In particular, when averaged in time, $\overline{u}(z)$ becomes $\langle u(z) \rangle$, which, within the limitations of the model, represents the local version of what for the sun is commonly referred to as *the solar differential rotation*, i.e. the time-averaged zonal angular velocity.

Equations 14.4 – 14.7 are solved numerically as an initial value problem by a semi-implicit, hybrid finite-difference/pseudospectral scheme. A sparse covering of the four-dimensional parameter space (Rayleigh number R_a, Taylor number T_a, Prandtl number σ and latitude ϕ) has been investigated. The emphasis has been on the role of turbulent as opposed to laminar transport, both with and without rotational influence, for a fixed reference state ($\theta = 10, m = 1, \gamma = 5/3$, aspect ratio 1 : 4 : 4). The parameter survey covers $T_a \leq 2 \times 10^7, R_a \leq 10^7, P_r \geq 0.1, \phi_o = (15°, 45°, 90°)$, with grid resolutions up to $256 \times 256 \times 130$. The major results stemming from these local simulations are now described.

14.2.1 Turbulent transport of convective energy

Increasing the degree of turbulence by reducing P_r or raising R_a in such local simulations results in a dramatic transition from laminar cellular overturning flow throughout the domain to a different state. The new topology, as in the perspective view of vertical velocities in figure 14.2 a, is such that the near-surface portion of the domain remains as a connected network of downflows surrounding broad upflows. However, this nearly laminar upper surface layer is purely a consequence of the expansion of fluid elements rising through a rapidly decreasing density stratification near the top, and serves to disguise a fully turbulent interior. Away from the upper boundary, the domain contains fast small-scale, isotropic motions. Powerful downflows occur at the interstices of the upper network and pierce the interior turbulence, spanning the multiple scale heights from the top to the bottom of the domain. Despite being mobile, especially in the rotating case, these strong downflowing structures are spatially and temporally coherent, coexisting with the interior turbulence for many turnover times.

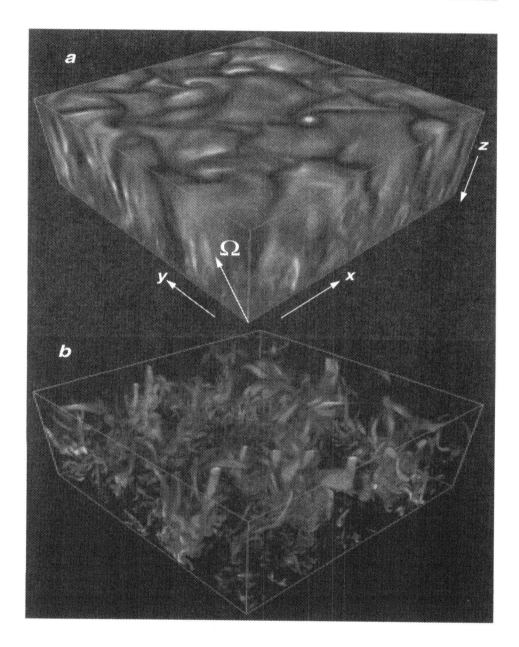

Figure 14.2: *a.* A typical overview of the computational domain shown in vertical velocity, with darker tones representing downflowing fluid and lighter tones upflowing fluid. *b.* The same domain shown in enstrophy, where the value is assigned a color and an opacity such that strong enstrophy is light and opaque and weak enstrophy is dark and translucent. ($R_a = 10^7, T_a = 2 \times 10^7, P_r = 0.1, R_o = 2.2, \phi = 45°$)

The turbulent but structured nature of the flow is seen clearly in a volume rendering of enstrophy (vorticity squared) shown in figure 14.2 b. Vorticity concentrated at the converging interstices of the downflow network lanes is amplified by the rotational influence to produce prominent vortical structures which extend over the full depth of the layer. These are buffetted by the smaller-scale, more randomly-orientated vortex tubes of the turbulent interior. Rotation provides a pool of intrinsic vorticity as a source for enhancing all vortex structures and also a linear mechanism for mixing the small-scale vorticity components. The appearance of the interior turbulent motions is *not* due to the dependence of the diffusivities on the density stratification. The two completely independent codes of Paper 0 produce qualitatively similar results aeven though one code uses a constant dynamic viscosity (diffusivity $\sim 1/\rho$), whereas the other operates with (an effective) constant kinematic viscosity (diffusivity constant). The small-scale fluctuations in the interior are most likely produced by instabilities of the rapid downflows which are then forcefully returned to the body of domain by the impenetrable boundary. Preliminary results from simulations where this restrictive boundary condition is relaxed in favour of a stable layer of fluid below (Clune, Brummell and Toomre, 1996) show a similar but somewhat different scenario, with the role of the downflows enhanced but more decoupled from the return upflowing motions.

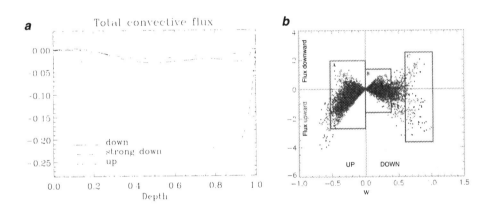

Figure 14.3: *a* Horizontally-averaged convective flux F_c as function of depth for turbulent convection ($R_a = 5 \times 10^5$, $T_a = 10^5$, $P_r = 0.1$ and $R_o = 7$), revealing that the near cancellation of enthalpy and kinetic fluxes within the downflows leads to the upflows dominantly carrying the convective flux. *b* Phase diagram of the convective flux F_c pointwise as function of the vertical velocity in a horizontal plane near midlayer for the turbulent convection in *a*. The boxes A, B and C delineate horizontal sites of upflow, downflow, and strong downflow within the horizontal sampling plane.

The transition to turbulence in the local models has major consequences for the heat transport. Despite the fact that the strong downflowing structures communicate over multiple scale heights and are coherent in time and space, surprisingly it is the small-scale turbulent motions that carry the majority of the total vertical convective flux F_c, the sum of the enthalpy ($F_e = \gamma \rho w T'/(\gamma - 1)$) and kinetic ($F_k = \rho w |\mathbf{u}|^2/2$) fluxes. This may be seen in figure 14.3 a which exhibits the upflow and downflow contributions to the horizontally-averaged F_c with depth. The contributions from the upflow regions obviously dominate in this typical example. There are major fluctuations in F_c from point to point across any given horizontal plane, as the phase diagram of vertical velocity w and convective flux F_c in figure 14.3 b reveals. Despite the scatter, in the upflows (A) the pointwise F_c values are sufficiently skewed that there is a net upward flux (negative) when averaged over all points, whereas the distribution is more nearly symmetrical for the moderate (B) and strong (C) downflows, with only a small value for the net transport emerging after averaging. Further examination of the composition of F_c reveals that the strong downflowing coherent structures are so organized that their strong kinetic flux directed downward just about counterbalances their enthalpy or heat flux upward (in a horizontal average), rendering them relatively inefficient and leaving the primary transport to the smaller turbulent motions of the interior. This convenient cancellation of F_c appears not to be a pointwise cancellation as was first thought in Paper 0. The original paper proposed a theory that outlines the cancellation of the constituent fluxes at each point in the downflows as being due to a strong correlation between the square of the vertical velocity and the temperature fluctuations. This could occur when the motion is nearly steady, the mean stratification is nearly adiabatic and viscous and thermal conduction are negligible. Whilst these conditions are reasonably well met in the turbulent simulations, it appears that this initial theory may be over-simplistic. It is clear from isection (C) of figure 14.3 b that points with strong downward velocities can have widely differing and large values of F_c. Instead of pointwise cancellation, there exists sizeable non-local patches in the downflows where the values of F_c are actually quite large and definitely non-zero, but which miraculously cancel in the horizontal average. The cancellation of the convective flux appears to be more a phenomena related to large-scale horizontal intermittency in time and space in the flux productions than by pointwise destruction as was first thought.

However, these local area models suggest a new picture of sufficiently turbulent convection in which the large-scale coherent motions serve primarily to organize the convective circulations, with the majority of the net transport being achieved by the small-scale turbulent motions. Interestingly, this is somewhat in the spirit of mixing-length approaches (see e.g. Gough and Weiss, 1976), which require all transport to be achieved by convective eddies which are only coherent over a fraction of a scale height. The inclusion of latent heat transport arising from ionisation effects may well alter this picture (see Rast, this issue, and Rast, 1993a,b), as may also the relaxation of a solid boundary at the bottom of the domain (Clune, Brummell and Toomre, 1996). However, these matters emphasize that turbulent motions can possess transport properties which are distinctly different from those of laminar flows.

14.2.2   Turbulent transport of (angular) momentum

Since the turbulent transport of convective energy is distinctly different from the laminar case, then a similar variance in the transport of momentum may also exist. Perhaps the missing ingredient from the earlier global simulations with regard to the solar differential rotation was the degree of nonlinearity and the freedom to transport momentum in a turbulent rather than laminar manner.

To understand the production of differential rotation in the local model, one must examine the mean horizontal flows $\bar{u}(z)$ and $\bar{v}(z)$. With the inclusion of rotation in the local compressible convection model described earlier, horizontally-averaged (mean) shearing flows in both the zonal and meridional directions are found to exist. Without the effects of rotation, even though the equations permit mean flow solutions, none have been found in simulations to date. With $\Omega = 0$, the creation of mean motions would require an internally-generated spontaneous (reflectional) symmetry-breaking by a mechanism like that of Howard and Krishnamurti (1986). With the effects of rotation included, and with the f-plane model positioned away from the pole ($\phi \neq 90°$), the symmetry of the flow is automatically broken and any solution must have definite handedness. Mean zonal and meridional flows therefore nearly always exist. Their amplitudes are dependent on the degree of nonlinearity of the solution and on the influence of the rotation, but are in general significant although not over-whelming, containing on the order of a few percent of the total kinetic energy. The mean flows are time-dependent, exhibiting inertial oscillations which can be large in amplitude. Averaging over a long time in the simulations (many inertial periods) removes the inertial oscillations and reveals the underlying persistent mean flows, $\langle u(z) \rangle$ and $\langle v(z) \rangle$.

The generation of the zonal and meridional flows is governed by the mean flow equations 14.15 and 14.16. The first term on the each of the right hand sides of these equations corresponds to a generation of the mean component via non-linear interactions of the fluctuating velocities. The second-order correlation of the velocities as seen here is often referred to as the Reynolds stress. The second, or Coriolis, term represents a transfer between the mean horizontal motions by inertial forces. The last terms in the equations correspond to viscous diffusion of the mean flows. The Reynolds stress terms are the only source terms in these equations. The vertical derivative of these stresses, retarded by the action of the diffusive terms, entirely determines the strength of the mean flows produced. The Coriolis terms serve only to swap energy between the two mean components and do no work themselves. In a steady state (or equivalently a long term time average, where the time derivatives also van-ish since the momenta must remain finite as the averaging time tends to infinity), equations 14.15, 14.16 reduce to

$$f_z \overline{\rho v} = -\partial_z \overline{\rho u w} + \sigma C_k \partial_z^2 \overline{u}, \tag{14.17}$$

$$f_z \overline{\rho u} = \partial_z \overline{\rho v w} - \sigma C_k \partial_z^2 \overline{v}. \tag{14.18}$$

The latter diffusive term here is generally small and the dominant balance is of production of the mean zonal and meridional flows in the Coriolis terms by correlations of the velocities in the Reynolds stresses. A strong correlation between w and u produces mean meridional momentum; a similar correlation between w and v produces mean zonal momentum. No-tice now that any mechanism that introduces a net tilted flow into the system produces a correlation between vertical and horizontal momenta and therefore generates mean flows. A rotation vector which is not parallel to gravity generally produces such tilted motions in the convection. Most notably, turbulent motions produce a very different form of tilt to laminar cellular convection, as was found in Papers I and II.

Paper I extends the non-rotating work in Paper 0 and examines the differences that rotation introduces to the flow topology and evolution. Briefly, the inclusion of rotation changes the characteristics of both the surface network and the turbulent interior. The surface network becomes more curvaceous, less connected and more time-dependent under the rotational influence. The overall mobility of the cellular pattern is due to inertial motions of the surface flows induced by the Coriolis forces. Rotational enhancement of a dynamical buoyancy mechanism operating at the junctions of the network leads to self-destruction of these interstices and thus a new method of cell creation.

More importantly, Paper I exhibits that tilted flow structure (i.e. non-vertical cellular boundaries, streamlines and coherent structures) can readily be built in compressible convec-tion under the influence of rotation, although the resultant correlated flows differ according to the circumstances. With the tilted rotation vector of the modified f-plane model utilized

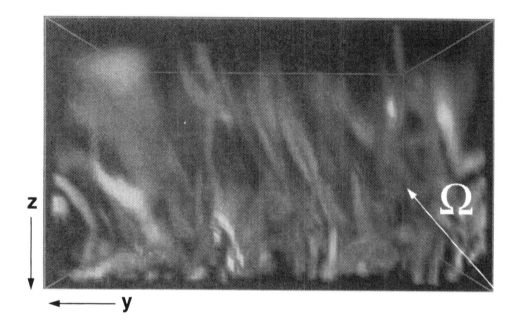

Figure 14.4: Enstrophy of a strongly rotationally constrained case showing turbulent alignment of coherent structures with the rotation vector. $(R_a = 5 \times 10^5, T_a = 10^7, P_r = 0.1, R_o = 0.71, \phi = 45°)$

in these simulations, both horizontal and vertical components of Ω are present. For laminar convection, the flows are smooth and cellular and the nonlinear production of vertical vorticity from the convectively-driven horizontal vorticity is weak. In this case, the action of the horizontal component of Ω dominates, tilting cellular streamlines in the zonal direction in a linear manner (Hathaway, Gilman and Toomre, 1979; Hathaway, Toomre and Gilman, 1980). This provides a strong correlation between vertical and zonal motions, producing a source for persistent meridional mean flows in the long-term time average.

When conditions are such that the convection is turbulent, the flow structure is different, as described earlier. The small-scale, fast overturning interior motions decorrelate from the inertial influence. Only structures with significant spatial and temporal coherence are left to sense the rotation. The strong vortical coherent plumes which pierce the convection over multiple scale heights tend to form naturally in alignment with the tilted rotation vector (see figure 14.4). These objects, though not space-filling, provide a weak (negative) correlation between vertical and meridional components of the momentum, thus providing a source term for modest steady mean zonal flows.

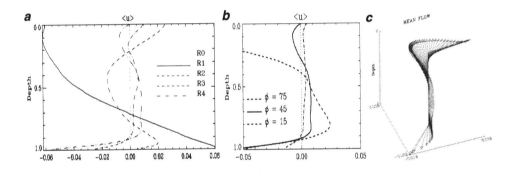

Figure 14.5: *a.* The zonal mean flows produced at various rotation rates for a turbulent case. The rotational influence increases from case R0 through R4 with R_o varying from ∞ to 0.71. ($R_a = 5 \times 10^5, P_r = 0.1, \phi = 45°$) *b.* The zonal mean flows produced at various latitudes. ($R_a = 5 \times 10^5, T_a = 10^7, P_r = 0.1, R_o = 0.71$) *c.* An example of spiralling with depth in the total mean velocity vector $\langle \mathbf{U}(z) \rangle = (\langle u(z) \rangle, \langle v(z) \rangle)$. ($R_a = 5 \times 10^6, T_a = 10^7, P_r = 0.1, R_o = 2.2, \phi = 45°$)

The mean velocity profiles for rotationally-influenced and turbulent flows are thus distinct from those for laminar and/or high R_o situations (see figure 14.5 a). When rotation is significant, the laminar means are strong, substantially sheared and are produced from positive Reynolds stresses generated by the laminar tilting of streamlines and cellular boundaries. The turbulent mean profiles are weak from the decorrelation of turbulent motions but can become dominated by the small contribution of the aligned vortical structures when the rotation is strong enough. The meridional Reynolds stress vw then changes sign in accordance with these structures and the zonal flow reverses, providing a small amplitude but constant value for $\bar{u}(z)$ across the interior, as in cases R3 and R4 in figure 14.5 a. The interplay between the laminar streamline alignment, which can generate both zonal and meridional mean flows, and the turbulent structural alignment, which favours constant zonal flows, is very delicate and strongly dependent on the parameters of the solution. However, these results are intriguing because the non-zero but constant-with-depth zonal flows in the interior of the solutions at certain parameters correspond (within this local model) to the 'constant-on-radii' differential profile of the solar convection zone inferred from helioseismic observations.

The comparison of simulations and obervations, whilst being tenuous due to the limitations of the local model, can be carried further! The strength of the zonal flow $\langle u(z) \rangle$ (the local equivalent of the differential rotation) decreases in the interior as the domain is moved from the equator to the pole (figure 14.5 b), as do the observed values of the zonal angular velocity. This can be understood in the turbulent scenario as the interplay between the two Reynolds stress generation mechanisms. At low latitudes both laminar and turbulent mechanisms would act to strengthen the mean zonal flow since the horizontal component of Ω is large. At high latitudes both mechanisms are absent as the horizontal component of Ω weakens, correspondingly weakening the mean flow. Also somewhat in accord with observations, the vector sum of the mean flow components $\langle \mathbf{U}(z) \rangle = (\langle u(z) \rangle, \langle v(z) \rangle)$ often displays a spiralling with depth, as is shown in figure 14.5 c. Surprisingly, this concurs with some findings elicited from ring diagrams of helioseismic data from the sun by Patron et al., 1995. These observations which span 9 different sites spread over a small section of latitude and longitude indicate that there exists a moderate mean flow below the solar surface whose vector seems to spiral with depth. The source of such spirals in the simulations may be related to the compressible nature of the calculations in conjunction with the Coriolis forces when the rotation is moderate and to Ekman-like effects in the balance of Reynolds stresses and Coriolis effects when the rotation is strong, although these mechanisms are still under investigation (Paper I; Brummell and Julien, 1996).

14.3   Conclusions

Numerical simulations of rotating, compressible convection in a local domain have allowed sufficient degrees of freedom in the model for turbulent convective flows to be studied. The topology of the high Reynolds number motions is significantly different from the laminar cellular overturnings. The highly nonlinear domain consists of small-scale turbulent motions punctuated by coherent columns of downflow emanating from a stratification-imposed laminar upper boundary layer. The transport of heat and momentum differ substantially in this turbulent regime when compared to the laminar. Whilst coherent structures spanning the multiple scale heights of the stratified layer exist, it appears that the small-scale turbulence of the broad upflows carries the majority of the convective energy, much in the spirit of mixing length theory. Even when the rotational influence is strong, the Reynolds stresses generated in the turbulent regime reflect a general decorrelation of the fast overturning motions. However, rotational alignment of the turbulent coherent structures provides a reversal in the meridional momentum transport and a subsequent change of sign of the zonal flows. This effect remains weak but constant throughout the interior for high Reynolds number quickly-rotating flows, providing a significant z-independent zonal flow equivalent to the constant-on-radii inferences of helioseismic investigations of the sun. Indeed, the balance of turbulent structural alignment and laminar streamline tilting provides a variance with latitude and a sub-surface structure that also agrees with the observations.

In final conclusion, whilst the simple local models cannot pretend to model solar convection, the variance of the transport properties with the transition from laminar to turbulent convective flows has added to our intuition with regard to the underlying dynamics of sun.

Acknowledgements

NHB would especially like to thank the other authors who have contributed to this body of work, namely Fausto Cattaneo, Neal Hurlburt, Andrea Malagoli and Juri Toomre. This work was partially supported by the National Science Foundation through grant ESC-9217394, and by the National Aeronautics and Space Administration through grants NAG 5-2218, NAG 5-2256, NAS 8-39747 and NAS 5-30386 and through Lockheed Independent Research Funds. Computations were performed mainly at the Pittsburgh Supercomputer Center on the Cray C-90 under MetaCenter grant MCA93S005P. NHB would also like to thank the organisers of the GAC95 meeting for the opportunity to express these opinions and ideas, and for their hospitality.

Bibliography

Baker, L., & Spiegel, E.A. "Modal analysis of convection in a rotating fluid" *J. Atmos. Sci.*, **32**, 1909 (1975).

Brummell, N.H., Cattaneo, F. & Toomre, J. "Turbulent dynamics in the solar convection zone", *Science* **269**, 1313 (1995).

Brummell, N.H., Hurlburt, N.E. & Toomre, J. "Turbulent compressible convection with rotation. I: Flow structure and evolution" *Astrophys. J.*, **473**, 494 (1996).

Brummell, N.H., Hurlburt, N.E. & Toomre, J. "Turbulent compressible convection with rotation. II: Mean flows and differential rotation" *Astrophys. J.*, **493**, 955 (1998).

Brummell, N.H., Julien, K. "Spiralling mean flows in rotating convection" preprint (1996).

Busse, F.H. "Differential rotation in stellar zones" *Astrophys. J.* **159**, 629 (1970).

Cattaneo, F., Brummell, N.H., Toomre, J., Malagoli, A., & Hurlburt, N.E. "Turbulent compressible convection" *Astrophys. J.* **370**, 282 (1991).

Clune, T., Brummell, N.H. & Toomre, J. "Penetrative compressible convection in three dimensions" preprint (1996).

Durney, B.R. "The generalization of mixing length theory to rotating convection zones and applications to the sun" in The Internal Solar Angular Velocity, ed. B.R. Durney and S. Sofia (Dordrecht: Reidel), 235 (1987).

Gilman, P.A. and Miller, J. "Nonlinear convection in a rotating spherical shell" *Astrophys. J. Suppl. Ser.*, **61**, 585 (1986).

Glatzmaier, G. A. "A review of what numerical simulations tell us about the internal rotation of the sun" in The Internal Solar Angular Velocity, ed. B.R. Durney and S. Sofia (Dordrecht: Reidel), 263 (1987).

Glatzmaier, G. A., & Toomre, J. "Global-scale solar turbulent convectionand its coupling to rotation" in ASP Conf. Ser. 76, GONG '94: Helio- and Astero-Seismology from the Earth and Space, ed. R.K. Ulrich, E.J. Rhodes, Jr., & W. Däppen (San Francisco: ASP), 200 (1995).

Gough D.O., & Toomre J. "Seismic observations of the solar interior" *Ann. Rev. Astron. Astrophys.*, **29**, 627 (1991).

Gough, D.O., & Weiss, N.O. "The calibration of stellar theories" *Mon. Not. Roy. Astron. Soc.*, **176**, 589 (1976).

Hathaway, D.H. "A convective model for turbulent mixing in rotating convection zones" *Astrophys. J.*,**276**, 316 (1984).

Hathaway, D. H., Gilman P.A., & Toomre J. "Convective instability when the temperature gradient and rotation vector are oblique togravity. I. Fluids without diffusion" *Geophys. Astrophys. Fluid Dyn.*, **13**, 289 (1979).

Hathaway, D. H., Toomre, J., & Gilman P.A. "Convective instability when the temperature gradient and rotation vector are oblique togravity. II. Real fluids with effects of diffusion" *Geophys. Astrophys. Fluid Dyn.*, **15**, 7 (1980).

Howard, L.N. & Krishnamurti, K. "Large scale flow in turbulent convection: A mathematical model" *J. Fluid Mech.*, **170**, 385 (1986).

Libbrecht, K.G. "Solar p-mode frequency splittings" *Astrophys. J.*, **336**, 1092 (1989).

Patron, J., Hill, F., Rhodes, E.J. Jr., Korzennik, S.G., Cacciani, A. "Velocity fields within the solar convection zone: Evidence from oscillation ring diagram analysis of Mount Wilson dopplergrams" *Astrophys. J.*, **455**, 746, (1995).

Rast, M.P., and Toomre, J. "Compressible convection with ionization. I. Stability, flow asymmetries, and energy transport" *Astrophys. J.* **419**, 224-239 (1993).

Rast M.P. Nordlund A. Stein R.F. Toomre J. "Ionization effects on solar granulation dynamics" *Astrophys. J.*, **408**, L53-56 (1993).

Rosner, R., & Weiss, N.O. "Differential rotation and magnetic torques in the interior of the sun" *Nature*, **317**, 790 (1985).

Rüdiger, G. Differential Rotation and Stellar Convection: Sun and Solar-type Stars, (New York: Gordon & Breach) (1989).

Chapter 15

POTENTIAL VORTICITY, RESONANCE AND DISSIPATION IN ROTATING CONVECTIVE TURBULENCE

PETER BARTELLO

Recherche en prévision numérique
Atmospheric Environment Service
2121, voie de Service nord
Route Transcanadienne
Dorval (Québec) H9P 1J3
Canada [1]

From observations of atmospheric energy spectra, Gage (1979) proposed an inverse (i.e. upscale) energy cascade from convective to larger scales in order to account for the -5/3 range observed below 300–400 km. Lilly (1983) put the suggestion on firmer theoretical grounds by examining stratified turbulence using the scaling arguments of Riley *et al.* (1981). Herring & Métais (1989) attempted to simulate such an inverse cascade in a numerical model of stratified Boussinesq turbulence with two-dimensional forcing. A -5/3 range was not observed. More recently, Métais, Riley & Lesieur (1994) repeated the simulations of Herring & Métais (1989) with the addition of solid-body rotation. Only in the case of rather rapid rotation ($N \approx f$, where N is the Brunt-Vaisala frequency and f is the Coriolis parameter) was an inverse cascade with a -5/3 range observed. It was proposed by Bartello (1995) that an application of the analysis, used by Warn (1986) on the shallow-water equations, to the Boussinesq system could explain a great deal of the behaviour observed in these numerical results. The analysis makes use of the existence of two disparate timescales as Rossby or Froude numbers tend to zero. In addition it exploits the conservation of potential vorticity, which is seen to prevent significant interaction between modes which possess PV and those that do not in the limit of low Rossby and Froude numbers. Two recent studies (Vallis *et al.*, 1997; Lilly, 1996) of convective simulations with realistic meteorological models have reported inverse cascades with little or no rotation. This apparent contradiction with previous work is examined by considering the robustness of the results as a function of Reynolds number.

---

[1]Now at: Department of Atmospheric and Oceanic Sciences, McGill University, Montreal (Québec) Canada, H3A 2K6

15.1   Background

This paper is perhaps a little different from the rest of this volume in that it concerns what happens after the convection is over. In the atmosphere, deep convection has the effect of increasing the stability of the layer over which it occurs and also of creating a kinetic energy disturbance in the form of a large outflow near the top of the convective plume. What happens to this energy? Observations show that large thunderstorms can interact with each other, merge and form 'mesoscale convective systems' with horizontal length scales on the order of 100 km (see McAnelly and Cotton 1986). In addition, convection has long been observed to boost synoptic-scale cyclogenesis in some circumstances, suggesting a large-scale organization of the kinetic energy created by the convection (Vincent and Schlatter 1979).

When observations of the atmospheric energy spectrum down to a few kilometers in horizontal scale became available (for a review see Gage & Nastrom, 1986), the -3 range of the synoptic-scale quasigeostrophic enstrophy cascade (Charney, 1971) was observed. It was naturally expected that this would become more shallow at smaller scales since fully three-dimensional flow becomes a possibility below $O(10)$ km. Surprisingly, the transition to a -5/3 spectrum occurred near 300–400 km, where the aspect ratio of the flow is still far from unity. On this basis Gage (1979) suggested that an inverse (or upscale) energy cascade, analogous to that observed in two-dimensional turbulence and forced by very much smaller scales, was responsible for the -5/3 range. An alternative, involving a spectrum dominated by inertial-gravity waves was advanced by Van Zandt (1982). Although this is the subject of current work, this paper will focus on the inverse cascade hypothesis. Lilly (1983) examined Gage's suggestion in light of the scaling arguments of Riley *et al.* (1981), employed in a study of stratified turbulence. There it was noted that with sufficient stratification, turbulence evolves to a state dominated by the horizontal components of velocity and with very little vertical coherence. It can be crudely modelled conceptually by a set of independent horizontal layers, each evolving according to 2D dynamics until their decoupled evolution leads to a local thermal instability and three-dimensional overturning. Lilly (1983) argued that an inverse cascade from convective scales was possible in such an environment.

Herring and Métais (1989) attempted to simulate such an inverse cascade in numerical simulations based on the Boussinesq equations. The maximum resolution of 3D Boussinesq simulations is still very low and obviously was even lower in 1989. It should be noted that the extent to which the flow is dominated by the dissipation is never particularly clear. In their simulations they were unable to produce an upscale cascade with a red spectrum. Instead, as $N \to \infty$ they observed that with 2D forcing the velocity field became increasingly horizontal (i.e. perpendicular to the mean density gradient) and a substantial fraction of the total dissipation was due to an interlayer drag that acts upon all horizontal scales. This had the effect of flattening the slope of the energy spectrum in the hypothetical inverse-cascade range.

Bartello *et al.* (1994) performed a numerical study of rotating homogeneous turbulence, investigating the asymmetry in the vertical vorticity component (i.e. along the rotation axis, $\mathbf{\Omega} = f\hat{\mathbf{z}}/2$) at Rossby numbers of order unity. As part of that study, simulations of decaying 3D turbulence with isotropic initial conditions were carried out at low Ro. It was found that the flow two-dimensionalized via a more efficient downscale cascade of energy with vertical structure ($k_z \neq 0$). By contrast, the energy of modes with $k_z = 0$ decayed less rapidly. As a result, the 2D flow emerged as it began to dominate the 3D modes which were decaying around it. As $t \to \infty$ it was characterized by intermittent 2D vortices and high kurtoses, much as in simulations of decaying 2D turbulence at similar resolutions (Fornberg 1977).

Since 2D turbulence certainly sustains an upscale cascade from forced scales, and since rotation can in certain circumstances lead to a Taylor-Proudman two-dimensionalization, Metais, Riley & Lesieur (1994) examined simulations of the sort performed by Herring and Métais (1989), except that rotation was now added. Energy spectra from a number of very similar simulations are shown below in Fig. 15.1. Since the goal was to determine the effect of rotation on low-Froude number flows, they organized the discussion in terms of the ratio N/f. This is also adopted here, where it is tacitly assumed that N is adjusted to yield a low Froude number, implying that the ratio N/f determines the Rossby number for an isotropic velocity. In summary, they found that in the presence of weak rotation ($N/f = 10$) a -5/3 range was not observed. However, with strong rotation, $N/f \approx 1$, a clear upscale cascade of geostrophic energy was noted. The latter was defined as the part of the flow in geostrophic and hydrostatic balance. Métais et al. (1994) reasoned that $N/f \approx 1$ is not the physically meaningful criterion, but rather the requirement is for both Ro and Fr to be small. Numerically, this is expensive to simulate explicitly unless $N \approx f$. However, in light of the theoretical work summarized here, it seems reasonable to expect that an inverse cascade can be simulated with any value of N/f provided both Ro and Fr are small. Returning to the original motivation of the study, N/f is of the order of 100 in the atmosphere, although the Brunt-Vaisala frequency is highly variable. However, a reasonable working definition of 'mesoscale' is motion with timescales between buoyancy and inertial periods, or equivalently with Rossby numbers of order unity or above, but with low Froude numbers. The $N/f \approx 1$ simulation of Metais et al. (1994) reflected rather the low-Ro low-Fr limit of synoptic-scale quasigeostrophic turbulence discussed by Charney (1971). Paralleling this study, decaying simulations of homogeneous rotating stratified flows were performed by Bartello (1995) and analysed in terms of theory formulated in a study of the shallow-water equations by Warn (1986). The theory was adapted to the Boussinesq equations and was found to explain a great deal of the behaviour of these simulations and of those of Métais et al. (1994). It is summarized in the next section. A number of the same conclusions were reached in independent studies by Embid and Majda (1996, 1997), using a more mathematical approach.

More recently, Vallis et al. (1997) and Lilly (1996) have presented simulations of convection with meteorological models containing such complications as the lower boundary, moisture variables, variable $N(z)$, etc. These simulations were dissipated by meteorological subgridscale models, dependent on the local Richardson number, which have the advantage of being able to turn themselves off where appropriate. Using reasonable meteorological parameters (e.g. $N/f \approx 100$) the simulations did show some evidence of an inverse cascade of both rotational and divergent energy (defined on a horizontal surface near the maximum outflow). However, these models employed very few grid points in the vertical. It would seem from previous work that the inverse cascade is arrested by the convective overturning, which is the inevitable result of decoupled layers of 2D turbulence. It is necessary to determine the lengthscale associated with this vertical motion and then to compare it to the scales where model dissipation acts. Dimensionally, the former can be associated with the Ozmidov scale, L_b, obtained by equating the eddy turnover time at wavenumber k with the buoyancy frequency. Traditionally this is done with a Kolmogorov spectrum, yielding $L_b \sim (\epsilon/N^3)^{1/2}$, where ϵ is the energy dissipation. Typical values of ϵ inside cumulus clouds were measured by MacPherson and Isaac (1977) to be near $\epsilon \approx 10^{-2} \mathrm{m}^2 \sec^{-3}$, implying $L_b \approx 10^2 \mathrm{m}$. For the intense thunderstorms simulated by Lilly (1996), ϵ is much greater and L_b is initially $O(10^3)\mathrm{m}$. Lilly (personal communication, 1996) suggests that the Ozmidov scale can be associated with the vertical scale of the thunderstorm's outflow anvil, which spreads out horizontally and whose thickness decreases with time. Although the model of

Vallis *et al.* (1997) used a grid with resolution concentrated near the ground, the upper grid was much coarser than L_b. The grid used by Lilly (1996) had $\Delta z = 500$m, implying that L_b was initially resolved, but soon fell into the model dissipation range. If the overturning is on a length scale much smaller than the Kolmogorov scale imposed by the model grid, then we should not expect to see it in a simulation. If there is no overturning, should we then expect an inverse cascade? These numerical results would suggest that we should. It is interesting to note here that Maxworthy also observed an inverse cascade in laboratory experiments of decaying stratified turbulence. His measurements of the vertical scale of the vorticity layers indicated that it was determined by viscosity, implying that the Ozmidov scale was below the dissipation scale in his experiment, much as in the simulations of Vallis *et al.* (1997) and Lilly (1996).

In order to clarify matters, this report presents simulations investigating the sensitivity of the inverse cascade to vertical dissipation. The next section summarizes the theory presented in Bartello (1995) to complete the discussion of the role of potential vorticity conservation. Then a number of simulations imitating the work of Metais *et al.* (1994) are presented. Once again, an inverse cascade is not observed at low rotation and strong stratification. The focus here is on the effect of adding a supplementary diffusion term acting on the momentum and temperature, which smooths out vertical second derivatives. The operator was devised to mimic a reduced vertical resolution in a numerical model. It is demonstrated that this has the effect of producing an inverse cascade with very little rotation ($N/f = 100$). The implications are discussed in the final section.

15.2 Normal mode equations, conservation and resonance

Since the mathematics are laid out in Bartello (1995), only the major results are outlined here. The departure point is the normal mode equations of the rotating stratified Boussinesq system, linearized about a state of rest,

$$\frac{\partial}{\partial \tau} \mathbf{W_k} = i\mathcal{L}(\mathbf{W_k}), \tag{15.1}$$

where

$$\mathbf{W_k} = \begin{pmatrix} \zeta_{\mathbf{k}}/k_h \\ -ik\hat{w}_{\mathbf{k}}/k_h \\ \hat{b}_{\mathbf{k}}/N \end{pmatrix} \quad \text{and} \quad \mathcal{L} = \begin{pmatrix} 0 & ifk_z/k & 0 \\ -ifk_z/k & 0 & -Nk_h/k \\ 0 & -Nk_h/k & 0 \end{pmatrix}, \tag{15.2}$$

where $b = g\theta/\theta_o$ is the buoyancy, (u, v, w) is the velocity and $\zeta = \hat{\mathbf{z}} \cdot \nabla \times \mathbf{u}$ is the vertical vorticity. In (2) it is assumed that $k_h = (k_x^2 + k_y^2)^{1/2} \neq 0$ and that $k_z \neq 0$, although removing these restrictions poses no particular difficulty to the discussion below. With this choice of variables $\mathcal{L}$ is hermitian and its eigenvectors can therefore form an orthonormal set of basis functions. The eigenvalues are zero (the geostrophic mode) and plus and minus the gravity-inertial wave dispersion relation

$$\sigma_{\mathbf{k}} = \left(f^2 \cos^2 \theta_{\mathbf{k}} + N^2 \sin^2 \theta_{\mathbf{k}} \right)^{1/2}, \tag{15.3}$$

where $\theta_{\mathbf{k}}$ is the angle between wavevector $\mathbf{k}$ and $\hat{\mathbf{z}}$.

The energy is

$$E = \frac{1}{2} \sum_{\mathbf{k}} |G_{\mathbf{k}}|^2 + |\mathcal{A}_{\mathbf{k}}^{(+)}|^2 + |\mathcal{A}_{\mathbf{k}}^{(-)}|^2, \tag{15.4}$$

where $G_\mathbf{k}$ is the geostrophic-mode amplitude and $\mathcal{A}_\mathbf{k}^{(\pm)}$ are those of the ageostrophic modes. Energy can easily be divided into slowly-varying geostrophic and gravity-inertial wave contributions for fields far from balance. Central to the study of Warn (1986) and to the discussion here is the potential vorticity. For the Boussinesq system it is

$$\Pi = \left(\nabla \times \mathbf{u} + f\hat{\mathbf{z}}\right) \cdot \left(N^2 \hat{\mathbf{z}} + \nabla b\right). \tag{15.5}$$

It can be shown (Bartello 1995) that if either Ro or Fr tends to zero,

$$\langle \Pi^2 \rangle \to N^4 f^2 + N^4 \left\langle \left(\zeta + \frac{f}{N^2}\frac{\partial b}{\partial z}\right)^2 \right\rangle, \tag{15.6}$$

where angle brackets denote an ensemble average, implying that

$$V = \frac{N^2}{2}\left\langle \left(\zeta + \frac{f}{N^2}\frac{\partial b}{\partial z}\right)^2 \right\rangle = \frac{1}{2}\sum_\mathbf{k} k^2 \sigma_\mathbf{k}^2 |G_\mathbf{k}|^2 \tag{15.7}$$

is conserved. Two things are apparent about V. First, it is quadratic in this limit, implying that conservation in the full nonlinear set survives Fourier truncation since each triad of interacting wavevectors conserves V separately. Second, it involves only the geostrophic modes as Ro or Fr tend to zero. Subsequent reasoning assumes that E and V are the *only* invariants. For this reason, it is necessary to avoid the limit Fr $\to \infty$, since there the thermodynamics decouples from the dynamics and entropy becomes a passive scalar, thereby modifying the list of invariants.

Restoring the nonlinear terms (N and M) to the normal-mode equations yields

$$\frac{\partial G_\mathbf{k}}{\partial \tau} = N_\mathbf{k} \qquad \text{and} \qquad \left(\frac{\partial}{\partial \tau} \mp i\sigma_\mathbf{k}\right)\mathcal{A}_\mathbf{k} = M_\mathbf{k}. \tag{15.8}$$

The equations are then nondimensionalized using the maximum gravity-inertial wave frequency, $\sigma_{max} = max[f, N]$, and isotropic velocity and length scales U and L. The latter are chosen in order to focus on the effects of stratification and rotation on isotropic turbulence resulting from small-scale atmospheric convection. Using

$$\omega_\mathbf{k} = \pm\frac{\sigma_\mathbf{k}}{\sigma_{max}}, \qquad\qquad t = \tau\,\sigma_{max}, \tag{15.9}$$

and nondimensionalizing G, $\mathcal{A}$, M and N by appropriate combinations of U and L, we have

$$\frac{\partial G_\mathbf{k}}{\partial \tau} = \epsilon N_\mathbf{k} \qquad \text{and} \qquad \left(\frac{\partial}{\partial \tau} - i\omega_\mathbf{k}\right)\mathcal{A}_\mathbf{k} = \epsilon M_\mathbf{k}, \tag{15.10}$$

where $\epsilon = U/max[f, N]L = min[\text{Ro}, \text{Fr}]$. Here, all quantities are order unity with the exception of ϵ. As is well known in large-scale meteorology, this is a two-timescale problem as $\epsilon \to 0$. In this case take

$$\mathcal{A}_\mathbf{k} = A_\mathbf{k}(\epsilon t)\,e^{i\omega_\mathbf{k}t}. \tag{15.11}$$

Finally, the ensemble-average normal-mode variance (energy) equations can be written

$$\frac{\partial}{\partial(\epsilon t)}\langle |G_\mathbf{k}|^2\rangle = \sum_\Delta N_1\langle G_\mathbf{k}^* G_\mathbf{p}^* G_\mathbf{q}^*\rangle + N_2\langle G_\mathbf{k}^* G_\mathbf{p}^* A_\mathbf{q}^*\rangle e^{i\omega_\mathbf{q}t} + N_3\langle G_\mathbf{k}^* A_\mathbf{p}^* A_\mathbf{q}^*\rangle e^{i(\omega_\mathbf{q}+\omega_\mathbf{p})t} \tag{15.12}$$

and

$$\frac{\partial}{\partial(\epsilon t)}\langle|A_{\mathbf{k}}|^2\rangle = \sum_{\Delta} M_1 \langle A_{\mathbf{k}}^* G_{\mathbf{p}}^* G_{\mathbf{q}}^*\rangle e^{i\omega_{\mathbf{k}}t} + M_2 \langle A_{\mathbf{k}}^* A_{\mathbf{p}}^* G_{\mathbf{q}}^*\rangle e^{i(\omega_{\mathbf{k}}+\omega_{\mathbf{p}})t} + M_3 \langle A_{\mathbf{k}}^* A_{\mathbf{p}}^* A_{\mathbf{q}}^*\rangle e^{i(\omega_{\mathbf{k}}+\omega_{\mathbf{p}}+\omega_{\mathbf{q}})t},$$

$$(15.13)$$

(Warn 1986) where the sums are over wavenumber triads such that $\mathbf{k}+\mathbf{p}+\mathbf{q}=0$. The most efficient interactions are those which remove the fast timescale t by a judicious selection of modes with appropriate frequencies. For these interactions the modal amplitudes are 'forced' on their own slow timescale ϵt. Each of the terms are now considered, keeping in mind the restrictions imposed by the two conserved quantities and the relative ease of finding interactions that are resonant.

15.2.1   The $\langle GGG\rangle$ interactions

The $\langle GGG\rangle$ term appears only in the geostrophic-mode equation (12). It is always resonant and would yield the geostrophic equations if the $A_{\mathbf{k}}$ were all set to zero. In this case it is well known that an inverse cascade of energy and a downscale cascade of potential enstrophy is the result (Charney 1971).

15.2.2   The $\langle AAA\rangle$ interactions

The $\langle AAA\rangle$ term appears in the ageostrophic-mode equation and can be efficient if triple wave resonances can be found such that $\omega_{\mathbf{k}} + \omega_{\mathbf{p}} + \omega_{\mathbf{q}} = 0$ (for details see McComas and Bretherton 1977). If $N = f$, resonance is impossible. Since computer simulations with two very different timescales are expensive, numerical results concerning these interactions are somewhat lacking.

15.2.3   The $\langle GAA\rangle$ interactions

These interactions are found in both $A_{\mathbf{k}}$ and $G_{\mathbf{k}}$ equations and therefore represent a possible exchange mechanism. Resonating them is trivial since it suffices to have two wave modes with the same $\theta_{\mathbf{k}}$ (possibly with very different $|\mathbf{k}|$) and completing the triad with the appropriate geostrophic mode. Most interesting here is the potential vorticity constraint. Since only one member of the triad contributes to V, its conservation by the interaction requires that the geostrophic mode remain unaffected by the exchange. In other words these triads are catalytic in nature. The $G_{\mathbf{k}}$ mode allows two $A_{\mathbf{k}}$ modes to interact, but remains unaffected, implying that $N_3 = 0$. Lelong and Riley (1991) noted that the resulting interaction between ageostrophic modes isotropizes the wave field and Bartello (1995) employed simulations and statistical mechanical arguments (analogous to Warn 1986) to argue that it acted to cascade wave energy downscale. It is speculative but perhaps reasonable to suggest that it can be associated with gravity wave breaking, in that a velocity shear imposed by geostrophically-balanced flow can cause a freely-propagating gravity wave to break and give its energy to smaller-scale waves. This point is currently under investigation.

15.2.4   The $\langle GGA\rangle$ interactions

Finally, we come to the only interaction that can induce exchange between geostrophic and ageostrophic modes. Here the conservation arguments are not very restrictive. However, since there is only a single gravity-inertial wave mode, resonance is impossible if all frequencies are large. This is obviously the case if $N = f$, since then all frequencies are the same, implying that the highly disparate timescales would make this interaction weak, resulting in no significant rotational-wave modal echanges. In addition, triple-wave resonances are also ruled out when $N = f$, making the dynamics particularly simple. The inevitable result here is that the geostrophic interaction $\langle GGG \rangle$ would cascade geostrophic energy upscale and the $\langle AGG \rangle$ interaction would cascade wave energy downscale. This is exactly what Métais *et al.* (1994) (forced) and Bartello (1995) (decaying) observed when $N \approx f$. It corresponds to dynamics in the largest atmospheric scales where both Ro and Fr are low.

More interesting to the atmospheric mesoscale is the case $N \gg f$. There, the frequencies get much smaller as $\theta_{\mathbf{k}} \to 0$ and rotational-wave modal exchanges become a possibility. The exchange takes place through the lowest-frequency (vertically propagating) waves, although the interaction coefficient is zero precisely at the resonance at $\theta_{\mathbf{k}} = 0$ (Embid and Majda 1996). Significant off-resonant interaction presumably takes place for wavectors with $\theta_{\mathbf{k}} < f(\text{Fr})$, where $f(\text{Fr})$ is yet to be determined. It is the presence of low frequencies at small $\theta_{\mathbf{k}}$ that makes the low-Froude number limiting dynamics quite different when the Rossby number is fixed, as compared to the quasigeostrophic limit where Ro $\sim$ Fr, as stressed by Embid and Majda (1997). The present analysis can say little about the case $N \gg f$ without some further closure assumption. The geostrophic energy feels the opposite effects of the $\langle GGG \rangle$ triad, which attempts to move variance upscale and the $\langle AGG \rangle$ triad, which according to the statistical mechanical arguments does not. The simulations of Herring and Métais (1989) show that an inverse cascade with a -5/3 range is not observed with strong stratification without rotation, implying that $\langle AGG \rangle$ wins. More recent simulations by Vallis *et al.* (1997) and Lilly (1996) appear to contradict these results. Since it remains to be seen how robust they are and how they depend on the Reynolds number, a number of simulations focussing on this issue are presented in the next section.

15.3   Numerical Results

15.3.1   Métais *et al.* (1994) revisited

In order to obtain a few control runs for comparison with the more viscous results presented below, simulations were set up along the lines of those presented in Métais *et al.* (1994). The spectral transform model employed a dealiased grid of 60^3, implying an isotropic wavenumber range of only 20. Although this is very small by today's standards it suffices to reproduce the results of Métais *et al.* (1994), which are shown to be very robust. Both the forcing and dissipation terms are different from that study. The former was inspired by Sullivan *et al.* (1994). It simply holds constant the contribution to the variance of the vertical vorticity component from all modes with $k_z = 0$, over the range $k_f - \frac{1}{2} \le k < k_f + \frac{1}{2}$, where $k_f = 10$ is the forcing wavenumber. The forcing is therefore two-dimensional and horizontally isotropic. Although it holds the waveband-integrated vertical vorticity variance constant, individual modal phases and amplitudes within the forced waveband vary considerably.

In order to take full advantage of the small grid, band-limited dissipation terms were used for both momentum and temperature. In Fourier space

$$\frac{\partial}{\partial t} Q_{\mathbf{k}} = \cdots - \nu(k)\, Q_{\mathbf{k}}, \qquad \text{where} \quad \nu(k) = \begin{cases} 0 & \text{if } k \le k_f \\ \nu_o k_t^2 \left(\frac{k - k_f}{k_t - k_f} \right)^2 & \text{if } k > k_f, \end{cases} \qquad (15.14)$$

and $Q_{\mathbf{k}}$ is potential temperature or a velocity component, implying that there is no dissipation at k_f, nor is there any over the hypothetical inverse cascade range. The coefficient was set to $\nu_o = 4.5 \times 10^{-4}$ and $k_t = 20$ is the truncation wavenumber.

The model was spun up from rest with $N = f = 0$ until statistical stationarity was reached by $t = 80$. The temperature field (which becomes a passive scalar as $N \to 0$) was initialized to zero, where it remained. By the end of the run the total energy had levelled off at around $E = 0.043$, implying $u_{rms} = 0.29$. Although the forcing injected energy only into modes with $k_z = 0$, the spectra indicate isotropic fields (see Fig. 15.1 below), with modal energy equipartition for $k < k_f$. These fields at $t = 80$ served as initial conditions to all the runs discussed below.

In order to investigate the inverse cascade in meteorological models a strong stratification with low Froude number was desired. A few runs were performed at various N and with the meteorological value of $N/f = 100$ in order to obtain such a baseline. The value $N = 20$ was chosen as it resulted in a simulation achieving statistical stationarity at $\mathrm{Fr} = [\partial \mathbf{u}_h / \partial z]_{rms}/N = 0.075$ and $\mathrm{Ro} = \zeta_{rms}/f > 10$ with $E \approx 0.090$. An inverse cascade was not observed. From this meteorological case, we approach the régime where Métais *et al.* (1994) observed an inverse cascade via two other simulations, each with the same value of N, but with $N/f = 10$ and with $N/f = 1$.

Figure 15.1 shows the three-dimensional, horizontal and vertical wavenumber spectra of the three simulations, as well as that of the non-rotating unstratified case, which served to initialize the others. The results are as in Métais *et al.* (1991) in that an inverse cascade is only apparent when N/f is of order unity. More detailed study of the data also shows that only the geostrophic energy cascades upscale with $N/f = 1$ as in the discussion of section 2. For the other cases, the spectra show little significant difference from a state of modal equipartition over the range $k < k_f$. Note that modal equipartition implies $E_3(k) \sim k^2$, $E_h(k) \sim k$ and $E_z(k) \sim \mathrm{const}$.

15.3.2   Simulations with large vertical dissipation

In the introduction it was stated that the meteorological simulations of convection by Vallis et al. (1997) and Lilly (1996) displayed an inverse energy cascade while employing vertical grids which did not resolve the Ozmidov scale. In this section we investigate the effect of vertical resolution, or equivalently the effect of increased vertical dissipation since the numerical grid must resolve the dissipation. To do this the above simulation with $N = 20$ and $N/f = 100$ was rerun with an extra dissipation term,

$$\frac{\partial}{\partial t} Q_{\mathbf{k}} = \cdots - \nu(k) \, Q_{\mathbf{k}} - \nu_z k_z^2 \, Q_{\mathbf{k}}, \tag{15.15}$$

where ν_z was 0.001, 0.005 and 0.010.

The values of ν_z were chosen according to a rather crude argument, which in hindsight seems to have had at least some validity. Since the initial conditions of all of these runs are isotropic, assume we can estimate the eddy turnover time at k as

$$T_e(k) = \left[\int_{k_{min}}^{k} p^2 E_3(p) \, dp \right]^{-1/2}. \tag{15.16}$$

The wave timescale varies between the shortest and longest wave periods,

$$N^{-1} \le T_w(k) \le f^{-1}, \tag{15.17}$$

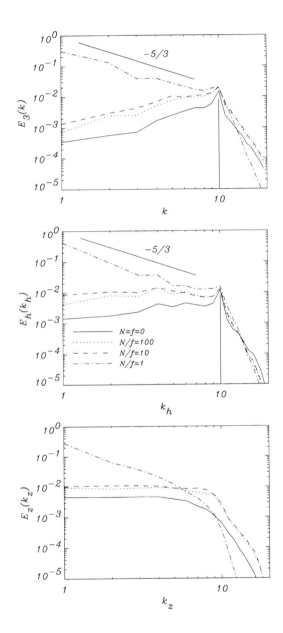

Figure 15.1: Total energy spectra at $t = 140$ for forced simulations at various rotation rates. Three-dimensional wavenumber spectra, $E_3(k)$ (top); horizontal wavenumber spectra, $E_h(k_h)$ (middle) and vertical wavenumber spectra, $E_z(k_z)$ (bottom). Vertical line indicates forced wavenumber, k_f.

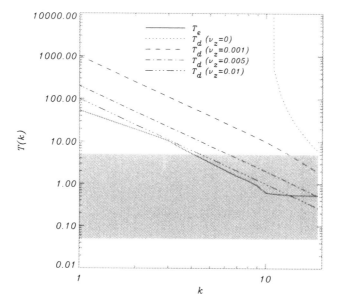

Figure 15.2: Timescales, $T(k)$, as a function of wavenumber for the initial conditions of the $N/f = 100$ runs. $T_e(k) = [\int^k p^2 E(p)\ dp]^{-1/2}$ is the eddy turnover time, while the $T_d(k)$ are the dissipative timescales for the various runs. The shaded region is bounded by minimum (N^{-1}) and maximum (f^{-1}) wave periods.

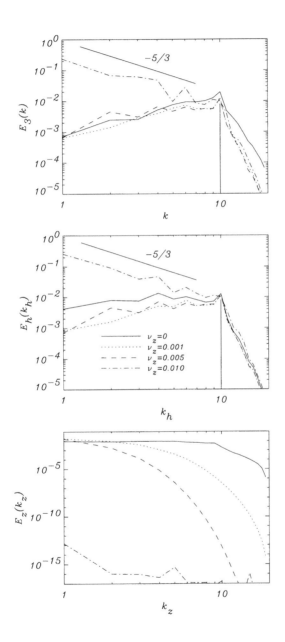

Figure 15.3: Total energy spectra as in Fig. 15.1 at $t = 160$ for forced simulations with $N/f = 100$ at various vertical dissipations.

and finally, the dissipative timescale (at least along the $\theta_{\mathbf{k}} = 0$ axis, where it is a minimum) is

$$T_d(k) = [\nu(k) + \nu_z k^2]^{-1}. \tag{15.18}$$

All of these curves are plotted in Fig. 15.2, including the three different values of ν_z considered. It is apparent that wave periods are of the order of the eddy timescale at high wavenumbers, implying that to within order-one factors the Ozmidov scale is at least close to being resolved by the model. Convective overturning was visibly taking place in the real-space fields with $\nu_z = 0$. Since $T_e \sim k^{-5/2}$ for $k < k_f$ if $E(k) \sim k^2$, it is clear that $T_e(k)$ approximately parallels the dissipative timescale $T_d(k) \approx \nu_z k^{-2}$ over this range. The value $\nu_z = 0.01$ is singled out since it yields $T_e(k) \approx T_d(k)$ over the entire range $k < k_f$. Of course these quantities are defined only to within order-one factors and stratification soon renders the flow anisotropic such that these crude arguments are of questionable validity. They served here merely as a point of departure for numerical experimentation.

The spectra are displayed in Fig. 15.2, where it can be seen that with the largest value of $\nu_z = 0.01$, there is a clear inverse cascade with a -5/3 range. However, it is simply a low Reynolds number effect resulting from an inhibition of energy in all but the forced $k_z = 0$ modes. It is argued here that this is perhaps connected to the behaviour observed in meteorological models with only $O(10)$ vertical grid points. There are differences with respect to Vallis et al. (1997) and Lilly (1996) however, in that the low Reynolds number cascade shows very little divergent energy. It remains to be seen how this varies with the nature of the forcing. It is interesting to note the abrupt change in behaviour between the runs with $\nu_z = 0.005$ and $\nu_z = 0.01$. With only a factor of two difference in the coefficient, the effect on the vertical wavenumber spectrum is enormous (11 orders of magnitude). Clearly, there is a virtual extinction of the transfer of energy out of the forced modes with $k_z = 0$. (The vertical wavenumber spectrum at $\nu_z = 0.01$ is of the order of the machine error in Fig. 15.3.) This abrupt change of behaviour is in contrast to what one would expect in unstratified 3D turbulence, although there too one could presumably obtain quasi-two-dimensional flow with ν_z sufficiently large.

15.4   Conclusions

After a brief summary of the history of Gage's (1979) proposal of an inverse cascade from convective scales in mesoscale meteorology, some analysis of the interaction between inertial-gravity waves and rotational modes was presented. While both types of modes contribute to the energy, only the latter have potential vorticity. In addition, the most significant nonlinear interactions are the resonant ones when the timescales characterizing the two types of modes are very different. Following Warn's (1986) analysis of the shallow-water equations, some properties of these interactions for the 3D Boussinesq system were summarized.

While it is clear that inverse cascades and an approach to quasigeostrophic flow are possible when both Ro and Fr are small, the atmospheric mesoscale is characterized by low Fr and Ro $\geq O(1)$. In this régime Métais et al. (1994) were unable to simulate an inverse cascade, while Vallis et al. (1997) and Lilly (1996) were. It was stated in the introduction that it was not clear how these results depend on a number of things. Clearly, the numerical models differed considerably in their geometry, dynamics (e.g. moisture) and forcing mechanisms. However, the latter two models differed from the former in having only $O(10)$ grid points in the vertical direction. For this reason, simulations similar to those of Métais et al. (1994) were undertaken. After reproducing those results concerning the lack of an inverse cascade at Fr$\approx$ 0.1 and $N/f = 100$, increased vertical viscous and diffusive terms were added. It was found that, when the viscous timescale became of the same order as the nonlinear eddy timescale over the range $k < k_f$, there was an abrupt change of behaviour, with very little transfer out of the 2D forced modes and an obvious inverse cascade. In this case, the cascade is a spurious low Reynolds number effect.

It is not clear if the inverse cascades observed by Vallis *et al.* (1997) and Lilly (1996) were simply due to insufficient resolution of the convective overturning at the Ozmidov scale. There are significant differences between the low Reynolds number case simulated here and their results. The forcing mechanisms and model configurations abound with differences in the two types of studies. It will be necessary to explore these differences one at a time. This paper is intended as a preliminary step in that process.

Acknowledgements I would like to thank Robert M. Kerr, Peter Fox and Chin-Hoh Moeng for inviting me to this meeting. I am also grateful to Jack Herring, Greg Holloway, Pascale Lelong, Marcel Lesieur, Doug Lilly, Andrew Majda, Olivier Métais, Jim Riley, Geoff Vallis, and Tom Warn for many interesting exchanges on this topic.

Bibliography

Bartello, P., "Geostrophic adjustment and inverse cascades in rotating stratified turbulence", *J. Atmos. Sci.* **52**, 4410–4428 (1995).

Bartello, P., Métais, O. and Lesieur, M., "Coherent structures in rotating three-dimensional turbulence", *J. Fluid. Mech.* **273**, 1–29 (1995).

Charney, J. G., "Geostrophic Turbulence", *J. Atmos. Sci.* **28**, 1087–1095 (1971).

Embid, P. F. and Majda, A. J., "Averaging over fast gravity waves for geophysical flows with arbitrary potential vorticity", *Commun. Partial Diff. Eqns.* **21**, 619-658 (1996).

Embid, P. F. and Majda, A. J., "Low Froude Number Limiting Dynamics for Stably Stratified Flow with Small or Fixed Rossby Numbers", submitted to *Geophys. Astrophys. Fluid Dyn* 1997.

Fornberg, B., "A Numerical Study of 2-D Turbulence", *J. Comput. Phys.* **25**, 1–31 (1977).

Gage, K. S., "Evidence for a $k^{-5/3}$ Law Inertial Range in Mesoscale Two-Dimensional Turbulence", *J. Atmos. Sci.* **36**, 1950–1954 (1979).

Gage, K. S. and Nastrom, G. D., "Theoretical Interpretation of Atmospheric Wavenumber Spectra of Wind and Temperature Observed by COmmercial Aircraft During GASP", *J. Atmos. Sci.* **43**, 729–740 (1986).

Herring, J. R. and Métais, O., "Numerical experiments in forced stably-stratified turbulence", *J. Fluid Mech.* **202**, 97–115 (1989).

Lelong, M.-P. & Riley, J. J., "Internal wave-vortical mode interactions in strongly stratified flows", *J. Fluid Mech.* **232**, 1–19 (1991).

Lilly, D. K., "Stratified Turbulence and the Mesoscale Variability of the Atmosphere", *J. Atmos. Sci.* **40**, 749–761 (1983).

Lilly, D. K., "Stratified and Geostrophic Turbulence in the Atmospheric Mesoscales", preprints of the Stratified and Rotating Turbulence Workshop, 30 July – 2 Aug., Boulder, Colorado (1996).

MacPherson, J. I. and Isaac, G. A., "Turbulent Characteristics of Some Canadian Cumulus Clouds", *J. Appl. Meteor.* **16**, 81–90 (1977).

Maxworthy, T., "Experiments on Stratified Turbulence", preprints of the Stratified and Rotating Turbulence Workshop, 30 July – 2 Aug., Boulder, Colorado (1996).

McAnelly, R. L. and Cotton, W. R., "Meso-β-scale characteristics of an episode of meso-α-scale convective complexes", *Mon. Wea. Rev.* **114**, 1740–1770 (1986).

McComas, C. H. and Bretherton, F. P., "Resosant interaction of oceanic internal waves", *J. Geophys. Res.*, **82**, 1397–1412 (1977).

Métais, O., Riley, J. J. & Lesieur M., "Numerical simulations of stably-stratified rotating turbulence". In *Stably-Stratified Flows - Flow and Dispersion over Topography* (eds. I. P. Castro & N. J. Rockliff), Clarendon Press-Oxford, pp. 139–151 (1994).

Riley, J. J., Metcalfe, R. W. and Weissman, M. A., "Direct numerical simulations of homogeneous turbulence in density stratified fluids". In *Proc. AIP Conf. Nonlinear Properties of Internal Waves* (ed. B. J. West), 79–112 (1981).

Sullivan, N. P., Mahalingam, S., and Kerr, R. M., "Deterministic forcing of homogeneous, isotropic turbulence". *Phys. Fluids*, **6**, 1612–1614 (1994)

Vincent, D. G. and Schlatter, T. W., "Evidence of deep convection as a source of synoptic-scale kinetic energy", *Tellus*, **31**, 493-504 (1979).

Vallis, G. K., Shutts, G. J. and Gray, M. E. B., "Balanced Mesoscale Motion and Stratified Turbulence Forced by Convection", *Quart. J. Roy. Meteor. Soc.* in press.

Van Zandt, T. E., "A Universal Spectrum of buoyancy waves in the atmosphere", *Geophys. Rev. Lett.* **9**, 575–578.

Warn, T., "Statistical mechanical equilibria of the shallow water equations", *Tellus*, **38A**, 1–11 (1986).

Chapter 16

NUMERICAL SIMULATIONS OF CONVECTION IN PROTOSTELLAR ACCRETION DISKS

W. CABOT[1]

[1] *Center for Turbulence Research*
Stanford University, Bldg. 500
Stanford, CA 94305-3030, USA

Thermal convection has been thought to play a role in the evolution of low-mass disks composed of gas and dust orbiting young stars through the outward radial transport of angular momentum generated by turbulence in the presence of differential rotation. This would cause disk material in convectively unstable regions to spiral inwards to be processed by the central star. Standard phenomenological models of Reynolds stresses used in the previous decade give reasonable evolutionary time scales compared with those deduced from observational surveys of young stellar systems. However, results from recent hydrodynamical simulations of this flow situation contradict the standard models, indicating that net angular momentum transport due to convection tends to be directed *inward* and that the mean flow gains energy at the expense of smaller scale convective motions. The character of disk convection and the implications for disk structure suggested by the hydrodynamic simulation results are discussed in this paper.

16.1   Introduction

16.1.1   What are protostellar accretion disks?

In the currently accepted paradigm of star formation (Shu *et al.*1993), there is a stage after a young star has contracted from its parent molecular cloud in which it is orbited by material with excess angular momentum that naturally takes the form of a flattened disk or "nebula" (Fig. 16.1). This material is 99% primordial hydrogen and helium gas with small amounts of dust and ice grains. Eighteenth century scientists Kant and Laplace hypothesized that it was from just such a disk about our own young sun ("the Solar Nebula") that the planets and other objects in the solar system formed. Many recent observations have confirmed that not only do disks occur frequently about young stars (Strom, Edwards & Skrutskie, 1993; Beckwith & Sargent, 1993), but that planets can form as well (Mayor & Queloz, 1995; Marcy

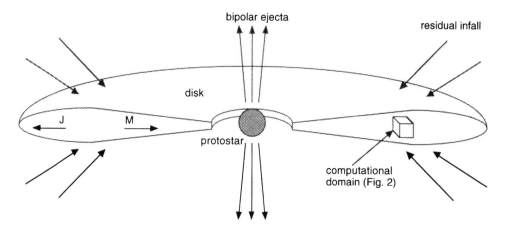

Figure 16.1: Idealized thin, low-mass protostellar disk orbiting a young star with mass (M) spiraling inward and angular momentum (J) being transported outward. Turbulence is simulated in localized region (box).

& Butler, 1996; Butler & Marcy, 1996), so that the situation in our own planetary system is probably not at all unique.

One of the keys to understanding the very complex process of planet formation is to establish the structure of and conditions in primordial circumstellar disks, and the way they evolve and eventually disperse all of the gaseous and dust components not trapped in planets and smaller solid bodies. In order to remove most of the material in the circumstellar disk, which is rotating nearly in centrifugal balance at keplerian velocities, the angular momentum of the material must be changed; then material that has lost angular momentum will spiral inward toward the central star, while material that has gained angular momentum will spiral outward away from the star. Because angular momentum increases with radius, only a small fraction of the total mass needs to spiral outward (to very large radius) to carry off most of the excess angular momentum, while most of the mass spirals inward to be accreted[1] or ejected poleward in energetic regions near the stellar surface.

16.1.2 Why is convection (potentially) important?

In early stages, the disk has a mass comparable to the central star and is subject to gravitational instabilities; the resulting large-scale spiral density waves have been shown to produce a nonlocal (long-range) transfer of substantial amounts of angular momentum or even lead to a binary fission of the system (Adams & Lin, 1993). In later stages, surviving disks contain much less mass than their central stars, and gravitational instabilities are not thought to be very important for transferring angular momentum, so other mechanisms are required. Although the frictional torque from molecular viscosity is a natural way to cause disk accretion (Pringle, 1981), the time scale to remove a circumstellar disk is far too long (longer, in

---

[1]Hence the name "accretion disk" in analogy to the more energetic disks that orbit white dwarfs and blackholes; they are "protostellar" because they orbit young stars, and they may be "preplanetary" or "protoplanetary" if there is incipient planet formation, although the latter can also refer to disks orbiting young planets.

fact, than the age of the universe). The time scale for young, solar-mass stars to lose their disks is deduced from observational surveys to be only a few million years (Strom, Edwards & Skrutskie, 1993).

An alternative to molecular viscosity is turbulent Reynolds stresses, which can act like a viscosity in gas or fluid flows, but at much higher levels. However, a mechanism for driving the turbulence is needed, which is where convection comes in. It was demonstrated by Lin & Papaloizou (1980) that a radiatively cooling protostellar disk, contracting vertically (along the rotation axis), would eventually become convectively unstable due to the blockage of escaping radiation by dust opacities. Turbulent convection motions in the presence of a strong radial rotational shear could then possibly generate the Reynolds stresses needed to transport angular momentum in protostellar disks. In order for the convection to be self-sustaining in a large region of a disk over long times ($\sim$ 1 Myr), gravitational energy made available from the material spiraling inward must roughly balance the heat generated by dissipated turbulent motions, which in turn maintains the convective instability; the energy is ultimately radiated away through the disk surfaces. The key to this rather intricate balance is *an outward transport of angular momentum in the disk.*

Lin and coworkers constructed phenomenological Reynolds stress models based on convective speeds estimated from stellar mixing length models; a high positive correlation of radial and azimuthal turbulent velocity components (u and v) was assumed such that the Reynolds stress component ($-\langle \rho uv \rangle$, where ρ is the density and $\langle \cdots \rangle$ denotes a suitable ensemble average) was downgradient to the rotational shear and angular momentum was transported outward *by design.* This Reynolds stress component is often parametrized by "α" as

$$-\langle \rho uv \rangle = \alpha \langle \rho c_s^2 \rangle \, , \qquad (16.1)$$

where c_s is the sound speed and S is the rotational shear rate. This standard model gave $\alpha \sim 10^{-2}$ and evolutionary time scales in disks comparable to those deduced from observations (Ruden & Lin, 1986). Another ad hoc model by Cabot et al.(1987), which also featured an outward angular momentum balance by fiat but used a different eddy viscosity formulation (Canuto & Goldman, 1985), gave a wide range of Reynolds stresses ($\alpha \sim 10^{-2}$–10^{-4}). This disparity prompted us to turn to simplified hydrodynamic simulations of turbulence as a "numerical experiment" to provide a physical basis for Reynolds stress modelling of convection in rapid differential rotation.

The original goals of these simulations were to determine convective heat transport rates and, more importantly, to determine angular momentum transport rates in order to calibrate Reynolds stress models, estimate time scales for mass transfer, and establish the conditions needed for the convection to sustain itself. The results from hydrodynamic simulations by myself and others, which largely contradict the phenomenological models, will be the main focus of this paper. In Sect. 16.2 the general properties of disks deduced from observations and models, and some simplifications employed to study them, are discussed. In Sect. 16.3 the equations, physics, and some of the numerics that go into the hydrodynamic simulations are presented. In Sect. 16.4 some of the salient results from the simulations to date are presented, and in Sect. 16.5 the implications of these results with regard to disk structure are discussed.

16.2   Properties of Protostellar Disks

Before describing the conditions that we expect to encounter in protostellar disks orbiting solar mass stars, it should be emphasized that most of what is known about disk structure has not been directly observed, but rather has been indirectly deduced from analyses of spectra at various wavelengths (cf. Adams, Lada & Shu, 1987; Bertout, Basri & Bouvier, 1988; Skrutskie et al.1990; Beckwith et al.1990). Except for our own planetary system, star-disk systems have been heretofore too remote for conventional instruments to resolve spatially; only recently has the Hubble Space Telescope and new ground-based interferometry begun to resolve marginally the nearest star-disk systems. Our own solar system offers a scant (and sometimes enigmatic) thermochemical record of the original primordial disk in planetary, asteroidal and cometary compositions.

16.2.1   General disk properties

The disk is flattened by rotation and gravitational forces, primarily from the central star, are balanced radially by centrifugal forces with little radial pressure support, so disk material has near-keplerian orbits. This means that the gas and dust rotates very rapidly and differentially. Because angular momentum increases radially outward in these disks, they are dynamically stable to the differential rotation. The residual gravity, and hence convective buoyancy, acts primarily in the direction along the disk's rotation axis. In this sense, the geometry of the disk resembles the polar regions of differentially rotating planet. The effective gravity points toward the midplane from either side; the midplane region has almost no net gravity (or buoyancy), and disk material is in virtual free fall there.

The disks of young solar-mass stars extend out to about 100 astronomical units (AU). The vertical scales in the disk are likewise "astronomical", about 0.01 to 0.1 of the distance from the central star: the thickness of the disk at jovian orbits is typically greater than 1 AU, and the vertical sound-crossing time is about a year (one orbital period). Local length and time scales vary over many orders of magnitude from inner to outer radii, making global simulations extremely difficult at present. Temperature and pressures also vary over many orders of magnitude; e.g., temperatures range from a few tens of degree Kelvin at outer radii to about 2000K at inner orbits near the central star. Although the molecular thermal diffusivity is comparable to the molecular viscosity, the radiative thermal diffusivity is about 8 orders of magnitude larger due to the extremely low densities, so that the Prandtl number based on the net thermal diffusivity is minuscule. Reynolds numbers can also be constructed that are meaninglessly immense due to the small molecular viscosity and large length scales. It is often assumed therefore (perhaps fallaciously) that the flow in disks must be turbulent, even in the absence of any demonstrable fluid instability.

16.2.2   Under what conditions do disks become convective?

Convection in protostellar disks probably occurs over limited radial extents at certain stages of development. (This, of course, poses problems for disk models requiring convection to drive all of the angular momentum and mass transfer globally.) Disks will be convectively unstable in regions where thermodynamic conditions give high enough opacities to block radiation trying to escape vertically through the disk's surfaces. The surfaces of the disk are always capped by convectively stable layers due to the very rapid thermal diffusion in these rarefied, optically thin regions. The heat source that drives convection must be generated internally or derived from the release of gravitational energy as the disk contracts vertically.

The opacity in disks is due predominantly to ice grains at temperatures $T \sim 100K$ and small dust (mostly silicate) grains at $T \sim 1000K$ (Pollack *et al.*1994). In this temperature range, most of the hydrogen gas is molecular and provides little opacity (in contrast to much hotter stellar interiors). A low-mass disk orbiting a solar mass central star will have appreciable water ice opacities at jovian orbits (where the Rosseland mean opacity κ varies roughly as T^2), and silicate grain opacities at terrestrial orbits ($\kappa \propto T$). If an appreciable amount of the grains joins together into larger bodies and/or settles to the midplane, there may be too little opacity in the bulk of the disk material to sustain convectively unstable temperature gradients. Heating of the disk surface by radiation from the central star, shock heating by residual infall of material from the parent molecular cloud, or other sources will tend to suppress convection by reducing the midplane to surface temperature difference. Because vertical optical depths are moderate in these disks (~ 100 to 1000), convection is not very efficient at transporting heat, and in this sense the conditions resemble those in stellar envelopes and atmospheres.

16.2.3   Simplifying assumptions

A number of simplifying assumptions are generally made when discussing the structure of protostellar disks: 1) The disk orbits an isolated, single central star. Protostars that form in dense star-forming regions like the Taurus-Auriga or Orion nebulæ are not isolated, being subject to the gravitational pull of neighbouring stars and the sometimes intense ultraviolet irradiation by high-mass OB stars. Perhaps as much as half of these young stars are in fact in binary or multiple star systems. 2) The masses of the disks (M_d) under consideration here are low compared with that of the central star ($M_\star$), $M_d \lesssim 0.1 M_\star$, making self-gravity negligible, and they are thin (their vertical thickness being much less than the distance from the central star). For the purpose of convection studies, we also assume 3) that the gas and small grain dust in the disk are perfectly coupled so that the dust opacity is evenly distributed, 4) that heating of the disk's surface by stellar radiation or shocks from vertically infalling material is negligible, and 5) that magnetohydrodynamic effects are negligible. Many of these approximations are likely to be invalid in real protostellar disks at any given stage, and provide fertile ground for current and future research endeavours beyond what is presented here.

16.3   Numerical Hydrodynamic Simulations

In this section, I will present some details of hydrodynamic simulations primarily by Cabot & Pollack (1992) and Cabot (1996), and also by Stone & Balbus (1996) where they differ significantly.

16.3.1   Further simplifying assumptions

More simplifications are necessary to reduce the computational expense of the three-dimensional hydrodynamic simulations. It is assumed that one is treating a small, localized region of the disk, with dimensions much less the distance from the star (Fig. 16.1). In this case, one can neglect curvature terms and use local cartesian coordinates (Fig. 16.2). Radial pressure gradients and any other radial forces are neglected, and gravity, solely from the central star, is assumed to be balanced exactly at the midplane by centrifugal forces, giving exact keplerian rotation there ($\Omega^2 = GM_\star/r^3$, where r is the distance from the central star).

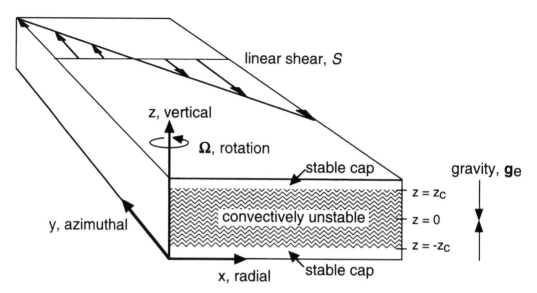

Figure 16.2: The computational geometry for disk convection, featuring uniform rotation about the z-axis aligned with the linear effective gravity and a linear mean shear in the horizontal (x-y) plane.

Because the gravitational force from the star is spherically symmetric while the centrifugal force is cylindrically symmetric, there is a residual vertical gravity,

$$\boldsymbol{g}_e = -\Omega^2 \boldsymbol{z} \tag{16.2}$$

to leading order in a thin disk, which confines material about the midplane ($z = 0$). The keplerian differential rotation can be treated in the cartesian limit as a linear, lateral shear superposed on uniform mean rotation about the vertical axis (Cabot & Pollack, 1992; Cabot, 1996), as sketched in Fig. 16.2. Simple (though sometimes unphysical) boundary conditions are also employed (see Sect. 16.3.3).

The physics that goes into the governing equations for the simulations is also considerably simplified, since we are mostly interested in the hydrodynamical aspects of this problem. The gas-dust mixture is assumed to be governed by the compressible Navier-Stokes equations with constant specific heats and mean molecular weight, and a simple power-law opacity relation is used. In early studies (Cabot & Pollack, 1992), incompressible "Boussinesq" equations were used to gain insight into the basic hydrodynamics of disk convection. Prandtl numbers are usually chosen artificially high and Reynolds numbers artificially low in order to keep velocity and thermal spatial scales comparable and well resolved in the simulation. This is a critical issue with regard to making physical interpretations of the simulation results, and I will return to it later in the discussion (Sect. 16.5.3).

In the simulations of Cabot & Pollack (1992) and Cabot (1996), a distributed internal heat source is imposed at a level sufficient to maintain a superadiabatic temperature gradient and drive the convection. In the simulations of Stone & Balbus (1996), an unsteady heat source is also used to keep the interior temperature at a set, superadiabatic level. This artifice allows the turbulence driven by convection to reach a statistical steady state in the simulations when the imposed heat and any heat generated by viscous dissipation balances the thermal energy radiated from the surfaces. Stationary turbulence statistics are then calculated with time and space averages. In compressible simulations, if the viscous heat dissipation supplies a sufficient amount of the internal heating, it can in principle drive convection without the aid of an imposed heat source, deriving its energy from the imposed keplerian shear (which is maintained ultimately at the expense of orbital energy). In fact, the realistic amount and spatial distribution of frictional heating due to convection motions is unknown *a priori*, and it is something we hope to learn from compressible simulations. The more centrally concentrated the heat release is, the steeper the superadiabatic temperature gradients will be.

16.3.2 Governing equations

The perfectly coupled gas and dust in the disk is governed by the Navier-Stokes equation,

$$\frac{\partial \rho \boldsymbol{v}}{\partial t} + \boldsymbol{\nabla}\cdot(\rho \boldsymbol{v}\boldsymbol{v}) = \rho\boldsymbol{g} - \boldsymbol{\nabla}p + \boldsymbol{\nabla}\cdot\boldsymbol{T} \,, \tag{16.3}$$

where $\boldsymbol{v}$ is the velocity, $\boldsymbol{g}$ is the gravity, and p is the pressure. The (Stokes) viscous stress tensor $\boldsymbol{T} = 2\mu\boldsymbol{S}^*$ (where μ is the kinematical viscosity coefficient and $\boldsymbol{S}^*$ is the trace-free velocity strain tensor) is retained in our simulations to dissipate the turbulence. With the aforementioned simplifying assumptions for the geometry and physics, this equation is cast in a local cartesian coordinate (x, y, z) system rotating about the vertical axis at a mean uniform rate of Ω; (x, y, z) correspond to the radial, azimuthal and vertical directions (Fig. 16.2). The velocity $\boldsymbol{u} = (u, v, w)$ is now measured with respect to an imposed, uniform mean shear $S = -3\Omega/2$ in the azimuthal velocity component, hence in the rotating frame the velocity is $(u, v + Sx, w)$. Eq. (16.3) becomes

$$\frac{\delta \rho \boldsymbol{u}}{\delta t} + \boldsymbol{\nabla}\cdot(\rho \boldsymbol{u}\boldsymbol{u}) + 2\boldsymbol{\Omega} \times \rho\boldsymbol{u} + \rho u S\hat{\boldsymbol{y}} = \rho\boldsymbol{g}_e - \boldsymbol{\nabla}p + \boldsymbol{\nabla}\cdot\boldsymbol{T} \,. \tag{16.4}$$

Note the usual addition of the Coriolis term in the rotating frame, as well as the shear term $\rho u S$ in the azimuthal direction ($\hat{\boldsymbol{y}}$); the effective gravity $\boldsymbol{g}_e$ is given by (16.2). The modified time derivative

$$\frac{\delta}{\delta t} = \frac{\partial}{\partial t} + Sx\frac{\partial}{\partial y} \tag{16.5}$$

includes the effect of the lateral shear. The continuity equation is given by

$$\frac{\delta\rho}{\delta t} + \boldsymbol{\nabla}\cdot(\rho\boldsymbol{u}) = 0 \,. \tag{16.6}$$

We sometimes use a high-order hyperviscosity term with this equation to damp two-delta waves during violent transient events (Cabot, Thompson & Pollack, 1995).

The total energy $E = e + k + \rho\Phi_e$ (where e is the internal energy, $k = \rho u^2/2$ is the kinetic energy, and $\Phi_e = \Omega^2 z^2/2$ is the effective gravitational potential) is governed by

$$\frac{\delta E}{\delta t} + \boldsymbol{\nabla}\cdot[(E + p)\boldsymbol{u} - \boldsymbol{u}\cdot\boldsymbol{T} + \boldsymbol{F}] = Q + (2T_{12} + \mu S)S - \rho u v S \,. \tag{16.7}$$

A perfect gas is assumed, such that the pressure and internal energy are related by $p = (\gamma - 1)e = \mathcal{R}\rho T$, where $\gamma = 7/5$ is the constant ratio of specific heats, and $\mathcal{R}$ is the gas constant divided by a constant mean molecular weight. The right-hand side of (16.7) is a collection of putative source terms: the imposed internal heating rate Q, much smaller frictional heating terms, and a Reynolds stress term $-\rho uvS$ that extracts orbital energy. From the kinetic and internal energy equations,

$$\frac{\delta k}{\delta t} + \boldsymbol{\nabla}{\cdot}(k\boldsymbol{u} - \boldsymbol{u}\cdot\boldsymbol{T}) = \rho\boldsymbol{u}\cdot\boldsymbol{g}_e - \boldsymbol{u}\cdot\boldsymbol{\nabla}p - \boldsymbol{T}:\boldsymbol{\nabla}\boldsymbol{u} - \rho uvS \ , \tag{16.8}$$

$$\frac{\delta e}{\delta t} + \boldsymbol{\nabla}{\cdot}[(e + p)\boldsymbol{u} + \boldsymbol{F}] = Q + \boldsymbol{u}\cdot\boldsymbol{\nabla}p + \boldsymbol{T}:\boldsymbol{\nabla}\boldsymbol{u} + (2T_{12} + \mu S)S \ , \tag{16.9}$$

it is seen that the Reynolds stress source term $-\rho uvS$ contributes to kinetic energy production, which is dissipated as frictional heat through the $\boldsymbol{T}:\boldsymbol{\nabla}\boldsymbol{u}$ term. The internal heating rate Q was chosen by Cabot (1996) to resemble the α-model version of $-\rho uvS$,

$$Q = \alpha_s p|S| \ , \tag{16.10}$$

with $\alpha_s \approx 10^{-2}$; however, much lower values ($\alpha_s \sim 10^{-3}$–10^{-4}) are probably more realistic for disk convection.

The radiative flux $\boldsymbol{F}$ is usually treated with a simple heat diffusion equation (Cabot & Pollack, 1992; Stone & Balbus, 1996). Cabot (1996) used a somewhat more realistic mean grey intensity equation (Mihalas, 1978),

$$\boldsymbol{F} = -\frac{4\pi}{3\kappa\rho}\boldsymbol{\nabla}J \ , \quad \frac{1}{3\kappa\rho}\boldsymbol{\nabla}{\cdot}\frac{1}{\kappa\rho}\boldsymbol{\nabla}J = J - B \ , \tag{16.11}$$

where J is the mean grey intensity, and $B = \sigma_r T^4/\pi$ is the Planck blackbody function with σ_r being the Stefan-Boltzmann constant. An ice opacity $\kappa \propto T^2$, appropriate for the $T \sim 100K$ conditions at jovian orbits, is used. Eq. (16.11) reduces to the standard radiative diffusion equation,

$$\boldsymbol{F} = -K_r\boldsymbol{\nabla}T \ , \quad K_r = 16\sigma_r T^3/3\kappa\rho \ , \tag{16.12}$$

in the optically thick interior where $J \approx B$. In the optically thin surface regions, (16.11) realistically limits the radiative diffusion rate, which is also advantageous in numerical codes with explicit time advancement, allowing longer time steps.

In the incompressible simulations of Cabot & Pollack (1992), the internal energy equation (16.9) is replaced with a simple scalar diffusion equation for the potential temperature. Information about acoustics and the absolute level of internal energy (assumed to be very much larger than the kinetic energy) is discarded. The potential temperature equation is coupled to the incompressible form of the Navier-Stokes equation (16.4) through the usual linear buoyancy term derived from the fluctuating component of $\rho\boldsymbol{g}_e$. And the continuity equation (16.6) is replaced by the isochoric condition, $\boldsymbol{\nabla}\cdot\boldsymbol{u} = 0$. The internal heating Q was taken to be uniform or modified to include near-wall sinks to give convectively stable caps.

16.3.3   Boundary conditions

The boundary conditions are usually periodic in the azimuthal direction and shear-periodic in the radial direction (i.e., periodic in the cosheared frame). In the vertical direction, impermeable, no-stress boundary conditions are imposed on the velocity field; either fixed temperature boundary conditions or low optical depth boundary conditions for the mean grey intensity (Mihalas, 1978) are used. The vertical boundaries are placed at a vertical optical depth of about 0.1 to 0.01 in Cabot (1996). The vertical velocity condition probably generates the worst errors by blocking upward flow and by spuriously reflecting various waves from the surfaces, and could stand some improvements. However, the statistics of the interior convection are not found to be very sensitive to the placement of the walls.

The geometry of the numerical domain is shown in Fig. 16.2. The radial-vertical aspect ratio is typically 2–3 to allow several large-scale convective structures to coexist, and the azimuthal-radial aspect ratio is around 3 to allow for the elongation of the structures by the shear.

16.3.4   Parameters

Convective parameters are based on a measure of the convective buoyancy proportional to the thermal lapse rate:

$$-N^2 = \boldsymbol{g}_e \cdot \boldsymbol{\nabla} s / c_p = \boldsymbol{g}_e \cdot \boldsymbol{\nabla} \ln \theta \;, \tag{16.13}$$

where N is the Brunt-Väisälä frequency, s is the entropy, c_p is the specific heat at constant pressure, and θ is the potential temperature. Stellar astrophysicists often use dimensionless temperature gradients, $\nabla \equiv d \ln T / d \ln p$. In these terms,

$$-N^2 = (\nabla - \nabla_a) \boldsymbol{g}_e \cdot \boldsymbol{\nabla} \ln p \approx (\nabla - \nabla_a) \rho g_e^2 / p \;, \tag{16.14}$$

using hydrostatic equilibrium ($\nabla p \approx \rho \boldsymbol{g}_e$), and where the adiabatic temperature gradient $\nabla_a = 1 - 1/\gamma$. Values of $(\nabla - \nabla_a)$ are usually less than 0.5. Note that $-N^2$ vanishes at midplane ($\propto z^2$) and peaks near the outer parts of the convection zone.

Parameters can be constructed on the basis of a convective buoyancy rate N_c, given by $N_c^2 \approx \max(-N^2)$ in the initial base state (i.e., without convection). A measure of the convection that does not involve the viscosity is the Péclét number Pe, which is the ratio of N_c to the large-scale thermal diffusion rate, $K_r / c_p \rho z_c^2$, evaluated at midplane (where z_c is the height of the convection zone); we find that $Pe \approx 1/\alpha_s \sim 10^2$. The Prandtl number $Pr = c_p \mu / K_r$ is chosen to be about 0.1 near midplane in the simulations. A convective Reynolds number based on N_c can also be defined, $Re_c = Pe/Pr \sim 10^3$.

The epicyclic frequency ω_e, which measures the restorative force due to rotation, is given by $\omega_e^2 = 2\Omega(2\Omega + S)$, which for keplerian rotation is just $\omega_e = \Omega$. The ratio ω_e/N_c measures the effect rotation has on convection (its inverse being a Rossby number of sorts). Because N_c itself scales with Ω through $\boldsymbol{g}_e$, it turns out that $\omega_e/N_c \sim (\nabla - \nabla_a)^{-1/2} \gtrsim 1.5$ in the convective core, indicating that rotation *always* has a strong influence on disk convection. Because the rotational shear rate is the same order as the rotation rate, it too is expected to have a strong influence on the convective structure.

A measure of the strength of developed turbulence is the Taylor Reynolds number (Tennekes & Lumley, 1972), $Re_\lambda \approx \langle k \rangle / (\mu \langle \epsilon \rangle)^{1/2}$, where $\langle k \rangle$ is the mean turbulence kinetic energy and $\langle \epsilon \rangle = \langle \boldsymbol{\mathcal{T}} : \boldsymbol{\nabla u} \rangle$ is the mean dissipation rate. Simulations by Cabot (1996) have values of Re_λ between 20 and 80, while values appreciably greater than 100 are really required for production and dissipation spatial scales to be well separated with an intermediate inertial (cascade) range.

16.4   Simulation results

16.4.1   Incompressible simulations

Three-dimensional Boussinesq (incompressible) simulations were performed by Cabot & Pollack (1992) using a pseudospectral method in the cosheared frame. These simulations were performed for $Pe \approx 100$ and different values of Pr (0.2 to 0.03), ω_e/N_c (0 to 0.8), and shear rates (uniform rotation, $S = 0$, to uniform angular momentum, $S = -2\Omega$).

Rapid uniform rotation was observed to change the morphology of convection from one featuring irregular cells of broad, hot upwellings with narrow, cool downdrafts to one with narrower columns of hot rising fluid between broad, cool downwellings (Cabot *et al.*1990). The convective heat flux was also reduced by almost an order of magnitude. When rotational shear was added, the convective structures were elongated in the azimuthal direction. At the maximal keplerian rotation rate simulated by Cabot & Pollack (1992), which is about half of the minimum actually found in convective disks, the flow began to transition to a two-dimensional state (as evident in the contours in Fig. 16.3) in which the shear wipes out most of the azimuthal structure. The convective takes more the form of alternating lanes of warm upflow and cold downflow.

At low keplerian rotation rates ($\omega_e/N_c \ll 1$), the flow retains a large degree of three-dimensionality and $\langle \rho uv \rangle$ is found to be positive, corresponding to the the situation anticipated by disk modellers in which angular momentum is transported outward and a cascade of turbulent kinetic energy to smaller scales is dissipated by viscosity. However, it was found that increasing the keplerian rotation rate close to the two-dimensional regime ($\omega_e/N_c \sim 1$) caused $\langle uv \rangle$ and the shear production $-S\langle uv \rangle$ to become negative (Fig. 16.4), contrary to the standard convective disk models. The outer regions of the convection zone, where convection weakens, develop large negative values of $\langle uv \rangle$ that overwhelm smaller, positive values of $\langle uv \rangle$ in the midplane region. This transition occurs at higher rotation rates the higher (lower) the Reynolds (Prandtl) numbers, therefore not excluding the possibility that three-dimensional turbulence and outward angular momentum transport occurs at very high Reynolds numbers at realistic Reynolds numbers ($\omega_e/N_c \gtrsim 1.5$). The transition to two-dimensionality was observed to occur roughly when the shear rate $|S|$ times the eddy turnover time, $\tau = \langle k \rangle / \langle \epsilon \rangle$, is greater than about 10, which has also been observed in other turbulent shear flows.

16.4.2   Compressible simulations

Eqs. (16.4), (16.6), and (16.11) are used in the fully compressible simulations of Cabot (1996) for realistic keplerian rotation rates ($\omega_e/N_c \approx 1.6$) using a fourth-order central finite difference scheme and fourth-order Runge-Kutta time advancement. The outer quarter to third of the flow domain is convectively stable (beyond a vertical optical depth of about 3). Before reaching statistical equilibrium, the flows often exhibited violent transients with transsonic flow and mild shocks. Supersonic flow was also observed near walls (Cabot, Thompson & Pollack, 1995) as in earlier studies of compressible convection (Malagoli *et al.*1990), but equilibrium flow with convectively stable caps (which acts like "soft walls") usually remains subsonic. The stable cap is insufficient to brake all of the outward momentum of the convective flow from the interior, but the vertical momentum blocked by impermeable walls is quite low due to the low near-wall densities. The flows, after about ten rotation periods, become very nearly two-dimensional with virtually no azimuthal variation.

Statistics are shown in Figs. 16.5 and 16.6 for a case with $\alpha_s = 0.01$, $Pr = 0.1$ at midplane, and $\omega_e/N_c = 1.6$. These are averaged in time and over horizontal planes above

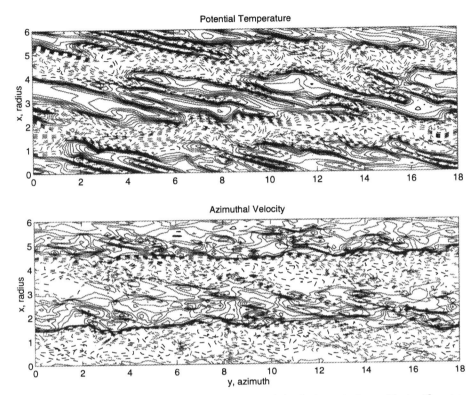

Figure 16.3: Planforms of Boussinesq convection with keplerian rotation with significant two-dimensional structure, $Pe \approx 100$, $Pr = 0.06$, and $\omega_e/N_c \approx 0.7$. Variations from the means of (a) potential temperature and (b) azimuthal velocities are shown, with solid/dashed contours represent positive/negative values.

and below the midplane. Many of the statistical results of the interior convection from the compressible simulations agree quite well with those from the Boussinesq simulations. This is because the compressible convective flow has mean rms Mach numbers M_t of only about 0.2 and peak instantaneous values about 2.5 times that, so that acoustic terms, which scale as M_t^2, are not very important. And even though the density falls by 2 orders of magnitude (Fig. 16.5), this occurs mostly near the optically thin surface, where it behaves like a gaussian in z due to the linearly increasing gravity. The density stratification is much more modest in the interior convection zone, falling by a factor of 3.

Horizontal velocity fluctuations are seen to peak just above the mean convectively neutral boundary at z_c in Fig. 16.6 at about a tenth of the midplane sound speed. However their correlation is seen to be very small in the convective interior, and significant but negative in the outer regions. This leads again to the result that $\langle \rho uv \rangle$ is negative on average; values of α defined in (16.1) are typically found to be $\sim -\text{few} \times 10^{-4}$. Note that the computed α in Fig. 16.6 bears no resemblance to the value of $\alpha_s \approx +0.01$ used to generate the internal

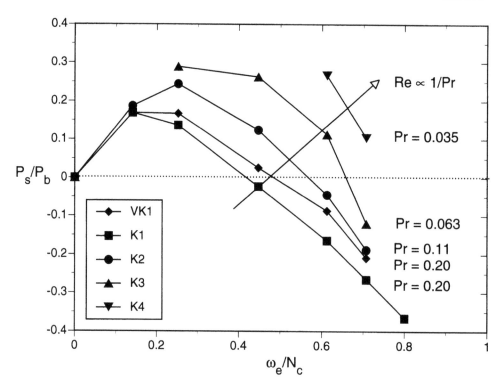

Figure 16.4: Net shear production of kinetic energy $P_s = -S\langle uv \rangle$ divided by buoyancy production P_b in Boussinesq simulations (Cabot & Pollack, 1992) with different Prandtl numbers and keplerian rotation rates. Negative shear production corresponds to an inward transport of angular momentum.

heating! The consequences of $\overline{\rho uv} < 0$ in a convective flow are discussed subsequently.

The instantaneous convective flow structure in the disk is illustrated in Fig. 16.7, showing contours of entropy and velocity for a flow with $\alpha_s = 0.0064$, $Pr = 0.01$, and $\omega_e/N_c = 1.6$. The convective pattern is dominated by large-scale vertical motions that fully penetrate the midplane, closely resembling the dominant linear mode found by Ruden, Papaloizou & Lin (1988). The convective velocity is organized in azimuthally counterflowing lanes (Fig. 16.7). Flow away from midplane tends to lag the mean flow ($v < 0$), while toward the midplane tends to lead the mean flow ($v > 0$). The flow contrives to move material radially in surface regions where the angular momentum gradient is sharply reduced (Fig. 16.8). There is a general "east-west" asymmetry, in that the radial transition from lagging (westward) flow to leading (eastward) flow is quite sharp, while the converse transition is more gradual. This results in narrow regions where $\partial v/\partial x > 0$ largely cancels the rotational shear (S) and broader regions where $\partial v/\partial x < 0$ cancels the angular momentum gradient ($\propto 2\Omega + S$). The convective flow therefore sets up a stepwise rotation curve with a keplerian mean. It is even possible for transient regions with negative angular momentum gradient to arise at higher Reynolds numbers, which may lead to secondary instabilities.

Instantaneous contours of ρuv are also shown in Fig. 16.8 scaled by the mean midplane

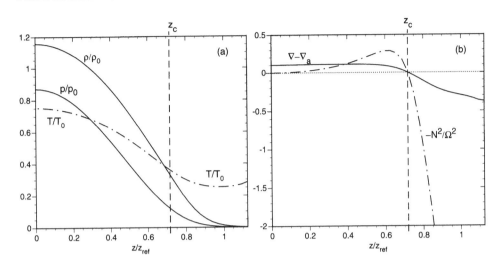

Figure 16.5: Typical (a) mean thermodynamic profiles (with respect to initial base state values at midplane) and (b) superadiabatic gradient $\nabla - \nabla_a$ and convective buoyancy $-N^2$ scaled with Ω^2 in compressible simulations (Cabot, 1996). The mean convectively neutral boundary occurs at z_c where $N^2 = 0$.

value of ρc_s^2; one can view this as contours of the disk model parameter α. Relatively large magnitudes (0.01) are attained in alternating positive and negative regions. Horizontal plane averages yield values an order of magnitude less, which can be either positive or negative at any given time; the regions near the convectively stable boundaries predominately have large, negative values, while the midplane region generally has smaller positive and negative values that cancel to a high degree over time.

The vertical thermodynamic structure is still dominated by hydrostatic equilibrium,

$$\frac{\partial p}{\partial z} \approx -\rho \Omega^2 z \ , \tag{16.15}$$

while a "geostrophic balance" holds in the radial direction:

$$\frac{\partial p}{\partial x} \approx 2\Omega \rho v \ . \tag{16.16}$$

Because of this, large-amplitude pressure and density ridges form in the surface regions (Fig. 16.9). About half of the heat is carried outward by convection in the midplane region, the other half by radiation. Very little net turbulent kinetic energy transport occurs, in contrast to nonrotating compressible convection, because of the more symmetric upflow-downflow structure.

Because these simulations approach a near two-dimensional, axisymmetric state, one can in fact reproduce the flow statistics with a strictly two-dimensional simulation. Ironically, Kley, Papaloizou & Lin (1993) performed two-dimensional simulations of convection in an annular section in spherical coordinates including curvature terms. They also reported an inward transport of angular momentum (as we will see in Sect. 16.5.1, they had no choice),

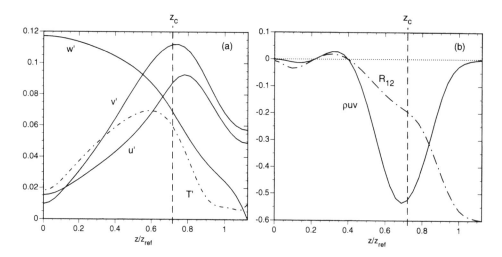

Figure 16.6: Profiles from compressible simulations by Cabot (1996) of (a) rms velocity components scaled by the midplane sound speed, and rms temperature fluctuations relative to horizontal means, and (b) the mean angular momentum transport $\overline{\rho uv}$ scaled by $10^{-3}\overline{\rho c_s^2}$ at midplane and its correlation coefficient $R_{12} = \overline{\rho uv}/(\overline{\rho u^2}\,\overline{\rho v^2})^{1/2}$.

but dismissed the result in the belief that nonaxisymmetric (azimuthal) variations would come to the rescue.

Stone & Balbus (1996) have also recently carried out compressible hydrodynamic simulations of disk convection, and they obtain similar results. Convection is sustained with an unsteady internal heat source designed to maintain the interior temperature at a certain level. The thermal diffusion coefficient is assumed constant. They used a piecewise-parabolic method for a nominally inviscid flow; their numerical scheme however has a numerical dissipation from which they can estimate an effective Reynolds number for their simulations (Balbus, Hawley & Stone, 1996) that is probably about an order of magnitude larger than that treated by Cabot (1996). They use values of $\omega_e/N_c \sim 3$ and $Pe \sim 20$. Their convective flow structure is again highly elongated in the azimuthal direction, but still retains a fair degree of three-dimensional structure. A time history of the volume average of ρuv scaled by the midplane pressure shows positive and negative values between $+1 \times 10^{-4}$ and -2×10^{-4}, with predominantly negative values; the long-time average yields a value of about -4×10^{-5}. Thus they reach the conclusion that disk convection tends to produce a very small net *inward* transport of angular momentum. Ryu & Goodman (1992) found similar results in a linear analysis of rotating shearing sheets.

16.5 Discussion

The simulations indicate that disk convection may generate, if anything, an inward transport of angular momentum. Here I discuss why we think this occurs, what it means for the disk structure, and what additional things we need to do to convince ourselves that this is the situation in real disks.

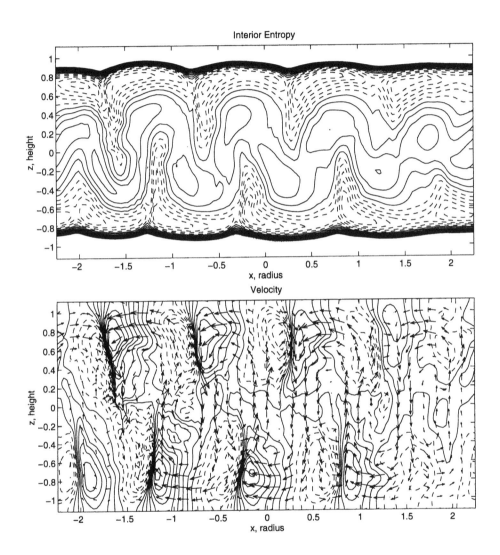

Figure 16.7: Radial-vertical cross-section of compressible convective flow: (a) interior entropy with high surface values surpressed, where dashed contours denote low values; and (b) azimuthal velocity contours, where solid/dashed contours denote positive/negative values, overlaid with velocity vectors in the plane.

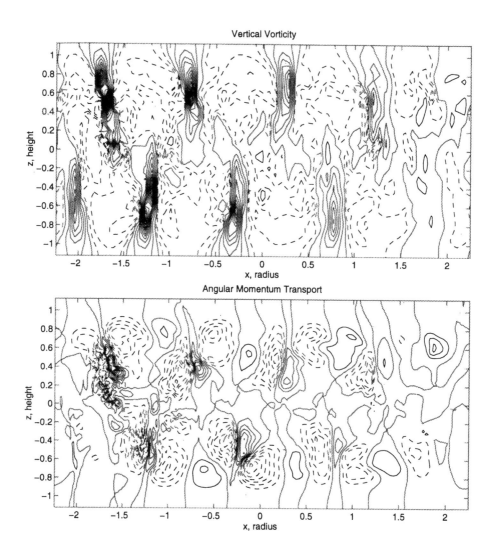

Figure 16.8: Radial-vertical cross-section of compressible convective flow: (a) vertical vortic-ity $\omega_z \approx \partial v/\partial x$, where negative (dashed) regions have very low angular momentum gradients, and (b) instantaneous ρuv, scaled by midplane ρc_s^2, with positive and negative (solid and dashed) peaks of about ± 0.01.

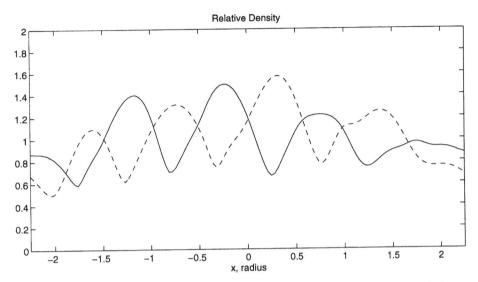

Figure 16.9: Radial density profiles in compressible flow near the optical surfaces ($z/z_{\rm ref} \approx$ ± 0.85 in Figs. 16.7 and 16.8), relative to the average values at those heights.

16.5.1   Why does disk convection generate inward transport of angular momentum?

In the standard disk models it was always assumed that turbulence (from any source) generates Reynolds stresses that are downgradient with respect to the angular velocity gradient (i.e., the rotational shear), e.g., $\overline{\rho uv} \sim -\mu_t S$ in an eddy viscosity form. This is certainly the case in standard (nonrotating) turbulent shear flow. Because the angular velocity gradient is directed radially inwards for keplerian rotation and the angular momentum gradient is directed outward, this would result in an outward transfer of angular momentum.

But if material mixes from different orbits in a way that at least approximately conserves angular momentum, then angular momentum transport is *inward*, downgradient with respect to angular momentum, not angular velocity (cf. Ryu & Goodman, 1992). This situation can occur when the generation of azimuthal motions by some forcing term (which is absent in vertically driven convection) or the transfer of energy to azimuthal modes through the velocity-pressure term is strongly inhibited. The governing equation for the azimuthal velocity component v from (16.4) is

$$\frac{\delta \rho v}{\delta t} + \boldsymbol{\nabla}{\cdot}(\rho v \boldsymbol{u}) = -(2\Omega + S)\rho u - \frac{\partial p}{\partial y} + (\boldsymbol{\nabla} \cdot \boldsymbol{T})_2 \ . \tag{16.17}$$

The governing equation for the azimuthal component of kinetic energy ρv^2 is

$$\frac{1}{2}\left[\frac{\delta \rho v^2}{\delta t} + \boldsymbol{\nabla}{\cdot}(\rho v^2 \boldsymbol{u})\right] = -(2\Omega + S)\rho uv - v\frac{\partial p}{\partial y} + v(\boldsymbol{\nabla} \cdot \boldsymbol{T})_2 \ . \tag{16.18}$$

For a statistical steady state with no-stress wall conditions, the terms on the left-hand side only redistribute ρv^2. In a nearly two-dimensional state where $\partial/\partial y \to 0$, the velocity-pressure gradient term becomes negligible, and the global balance is primarily between the positive-definite dissipation $-\langle v(\boldsymbol{\nabla} \cdot \boldsymbol{T})_2 \rangle$ and the Reynolds stress term, $-(2\Omega + S)\langle \rho uv \rangle$, which too must be positive. However, $2\Omega + S$, proportional to the angular momentum gradient, is positive for keplerian rotation, so ρuv must be negative on average (though it may be positive locally). A similar Reynolds stress term, $2\Omega\langle \rho uv \rangle$, appears as a net sink in the governing equation for radial normal stress. The sum of radial and azimuthal terms yields a net sink term, $-S\langle \rho uv \rangle$, in both the kinetic energy equation (16.8) and total energy equation (16.7).

Both the rapid shear and the rapid rotation contribute to this situation. The rapid shear is responsible for stretching and aligning convective motions in the mean flow direction, thereby suppressing azimuthal pressure gradients. The rapid rotation not only stabilizes the base flow against the shear, it inhibits the convective instability. Cabot & Pollack (1992) found that values of the shear rate $|S|$ times the eddy turnover time $\tau = \langle k \rangle / \langle \epsilon \rangle$ (inversely measuring the vigour of convective motions) greater than about 10 lead to highly two-dimensionalized flow; for $|S|\tau < 10$, net angular momentum transport is outward. For an outward transport of angular momentum, more vigorous turbulence is required which can maintain a healthy degree of three-dimensionality in the turbulence such that a steady supply of eddies are formed that turn into (oppose) the mean shear (Dowling, 1995; Pedlosky, 1987). Balbus, Hawley & Stone (1996) argue that only hydrodynamic instabilities that are directly unstable to the mean shear flow, like MHD instabilities, can generate outward angular momentum transport like this.

In summary, let me again emphasize that it is the combination of *weak* convection with a stable differential rotation that leads to an inward transport of angular momentum. What may make this the universal situation in protostellar disks is the fact that the rotation (and its shear) dominate the dynamics of disks, so that, except for unusual transient events, thermal convection *is* always weak with respect to the rotation and shear.

16.5.2   What are the consequences?

If the results of these (essentially low Reynolds number) numerical simulations remain valid in more realistic (very high Reynolds number) situations, then convection cannot be a mechanism by which protostellar disks transport angular momentum outward and mass inward, and other mechanisms, like MHD instability or spiral gravity waves, are needed.

In terms of energy, internal heating from some dissipative mechanism or the gravitational energy from the vertical contraction of the disk is converted to kinetic energy via the convective instability. Some of this kinetic energy may cascade to small scales to be dissipated as heat, but some if it will also be converted to large-scale mean motions. Convection cannot then be self-sustaining, because instead of converting orbital energy to heat, it returns thermal energy to the mean flow.

How an inward transport of angular momentum actually affects the large-scale disk structure is a still unexplored, though intriguing, avenue of research. One can speculate that the whole convection region, or some parts thereof, may approach a state of uniform angular momentum, which would in fact *reduce* the dynamic stability of the flow and perhaps lead to other instabilities. Or a system of radial jets may form, analogous to zonal jets in planetary atmospheres, a possibility suggested by the structure in the hydrodynamic simulations. Angular momentum transport varies greatly with height in the disk, which could drive large-scale meridional circulation; it could also enhance baroclinicity by driving the flow away from rotation on cylinders.

On the other hand, it is still not clear from the present simulation results if disk convection generates enough angular momentum transport (in any direction) to alter the disk structure significantly on the relevant evolutionary time scale for planet formation and disk dispersal. If not, then the role of convection in disks would seem to be limited simply to augmenting the vertical heat transport.

16.5.3   What more needs to be done?

At this point, some researchers have abandoned convection as a useful mechanism for transporting any significant amount of angular momentum in protostellar disks. (Worse, it may transport angular momentum "the wrong way"!) Though this judgment may prove to be correct, it may be premature. There are still some major flaws with the hydrodynamic simulations that need to be addressed. And, if angular momentum *is* transported inward in disk convection, the structural consequences certainly need to be investigated.

Cabot & Pollack (1992) showed that the Reynolds stress has a significant Reynolds (or Prandtl) number dependency, suggesting that very high Reynolds number flows may retain enough three dimensionality to produce outward angular momentum transport. The most important shortcoming of current numerical simulations (even the nominally inviscid ones) are that they essentially treat low Reynolds number flows, while astrophysical turbulence has enormously larger Reynolds numbers. Realistic disk convection has a huge range of spatial (and temporal) scales that numerical simulations cannot hope to treat directly with current computer resources. Large-eddy simulations (e.g., Mason & Thomson, 1992; Schumann, 1991; Moin *et al.*1991) can in principle fill the gap by using models of the effect of unresolved small scales on the large scale motions in a more realistic way than high molecular viscosities or artificial numerical dissipation. On the other hand, one must trust the subgrid-scale models for stress and heat flux to capture properly the complicated physics of this problem and the high degree of flow anisotropy. We ideally need to perform large-eddy simulations of more vigorous disk convection at Péclét numbers one or two orders of magnitude larger than present, requiring a much larger dynamic range of temperature scales (or an adequate model for subgrid-scale heat flux). The Reynolds number should be effectively infinite; the subgrid-scale model for the stress should be responsible for transporting all of the energy to unresolved scales, where it is presumably dissipated as heat at much smaller viscous scales. The numerical scheme one employs is also critical, since low-order finite difference schemes and dissipative schemes can seriously degrade the representation of turbulence at small resolved scales (cf. Ghosal, 1996; Kravchenko & Moin, 1997).

There are several improvements to the numerical treatment that should be implemented. The impermeable wall boundary conditions should be modified to allow the transmission of waves. The thin-disk and pseudocartesian approximations should be relaxed since these "thin disks" are not all that thin and the numerical domains actually have an appreciable radial extent; this means performing simulations on a annular section in spherical coordinates and including curvature terms. This would admit baroclinicity in the mean flow, and it would perhaps allow meridional circulation patterns to be studied, provided appropriate boundary conditions can devised. A detailed picture of convection and larger circulation patterns may clarify some issues about compositional migration and mixing in preplanetary disks. An interesting addition to the physics of the disk convection would be the use of dust opacities that vary with flow conditions, which could in principle approximate the effects of turbulent sweeping, settling, or changes in the size distribution of dust grains (cf. Weidenschilling & Cuzzi, 1993; Champney, Dobrovolskis & Cuzzi, 1995).

An analysis of the convective flow structure is needed, especially to explain the formation of accelerated and decelerated lanes of azimuthal flow that form broad regions of low angular momentum gradient. The observed "east-west" asymmetry calls to mind a similar phenomenon in planetary atmospheres. An analysis based on the conservation of a potential vorticity in this flow might be fruitful; Adams & Watkins (1995) have already employed this method in protostellar disks in the context of maintaining coherent, large-scale vortices in analogy to the atmospheres of jovian planets.

16.5.4 Conclusions

Hydrodynamic simulations of convection in a protostellar disk environment were originally conceived as numerical experiments to calibrate and refine the standard phenomenological models for disk convection by providing detailed turbulence statistics and realistic levels at which angular momentum is transported outward. This would have given us, to lowest order, the time scale on which mass spirals inward and disks evolve. Instead, a widening consensus of numerical simulation and analytical work suggests that convection in the presence of rapid keplerian (dynamically stable) differential rotation generates small net amounts of *inward* angular momentum transport. So, instead of clarifying the role the convection in protostellar disks, this work has, in a sense, raised more questions. For example, if significant inward angular momentum transport occurs over times short compared with evolution time scales, one can imagine unsteady disruptions of the disk structure occurring in the vicinity of extended convective regions. It should be emphasized, however, that the simulations results need to be verified at much higher Reynolds numbers.

There are, of course, several other viable local and global mechanisms for transporting angular momentum outward in disks (e.g., magnetohydrodynamic instabilities and spiral density waves) that are being actively investigated, even if disk convection is relegated to secondary status (cf. Hawley, 1995). One of the most attractive alternatives to convection in low-mass disks is magnetohydrodynamic (MHD) instability. If the disk is able to has support even a small magnetic field, it will be unstable to a MHD instability due to the differential rotation (Balbus & Hawley, 1992). This requires the disk to maintain a sufficient level of ionization, e.g., from stellar irradiation or cosmic ray bombardment. If these conditions are realized, Brandenburg et al.'s (1995) MHD simulations show that a substantial outward transport of angular momentum (with $\alpha \approx 0.003$) can be generated.

Acknowledgements

Support for this work was provided in part by NASA's Origins of Solar Systems Program under grant RTOP 452-22-94-02.

Bibliography

Adams, F.C., Lada, C.J., & Shu, F.H., "Infrared spectra of rotating protostars," *Astrophys. J.* **308**, 836–853 (1987).

Adams, F.C., & Lin, D.N.C., "Transport processes and the evolution of disks," in *Protostars and Planets III*, ed. E.H. Levy & J.I. Lunine (University of Arizona Press, Tucson), pp. 721–748 (1993).

Adams, F.C., & Watkins, R., "Vortices in circumstellar disks," *Astrophys. J.* **451**, 314–327 (1995).

Balbus, S.A., & Hawley, J.F., "A powerful local shear instability in weakly magnetized disks. IV. Nonaxisymmetric perturbations," *Astrophys. J.* **400**, 610–621 (1992).

Balbus, S.A., Hawley, J.F., & Stone, J.M., "Nonlinear stability, hydrodynamical turbulence, and transport in disks," *Astrophys. J.*, **467**, 76–86 (1996).

Bertout, C., Basri, G., & Bouvier, J., "Accretion disks around T Tauri stars," *Astrophys. J.* **330**, 350–373 (1988).

Beckwith, S.V.W., Sargent, A.I., Chini, R.S., & Güsten, R., "A survey for circumstellar disks around young stellar objects," *Astron. J.* **99**, 924–945 (1990).

Beckwith, S.V.W., & Sargent, A.I., "The occurrence and properties of disks around young stars," in *Protostars and Planets III*, ed. E.H. Levy & J.I. Lunine (University of Arizona Press, Tucson), pp. 521–541 (1993).

Brandenburg, A., Nordlund, Å., Stein, R.F., Torkelsson, U., "Dynamo-generated turbulence and large-scale magnetic fields in a Keplerian shear flow," *Astrophys. J.* **446**, 741–754 (1995).

Butler, R.P., & Marcy, G.W., "A planet orbiting 47 Ursae Majoris," *Astrophys. J. Letters* **464**, L153–L156 (1996).

Cabot, W., "Numerical simulations of circumstellar disk convection," *Astrophys. J.* **465**, 874–886 (1996).

Cabot, W., Canuto, V.M., Hubickyj, O., & Pollack, J.B., "The role of turbulent convection in the primitive solar nebula. II. Results," *Icarus* **69**, 423–457 (1987).

Cabot, W., & Pollack, J.B., "Direct numerical simulations of turbulent convection: II. Variable gravity and differential rotation," *Geophys. Astrophys. Fluid Dyn.* **64**, 97–133 (1992).

Cabot, W., Thompson, K.W., & Pollack, J.B., "Numerical simulations of stratified compressible convection with internal heating and disklike variable gravity," *Astrophys. J. Suppl.* **98**, 315–343 (1995).

Canuto, V.M., & Goldman, I., "Analytical model for large scale turbulence," *Phys. Rev. Lett.* **54**, 430–433 (1985).

Champney, J.M., Dobrovolskis, A.R., & Cuzzi, J.N., "A numerical model for multiphase flows in the protoplanetary nebula," *Phys. Fluids* **7**, 1703–1711 (1995).

Dowling, T.E., "Dynamics of Jovian Atmospheres," *Ann. Rev. Fluid Mech.* **27**, 293–334 (1995).

Ghosal, S., "An analysis of numerical errors in large-eddy simulations of turbulence," *J. Comp. Phys.* **125**, 187–206 (1996).

Hawley, J.F., "Keplerian complexity: Numerical simulations of accretion disk transport," *Science* **269**, 1365–1370 (1995).

Kley, W., Papaloizou, J.C.B., & Lin, D.N.C., *Astrophys. J.* **416**, 679–688 (1993).

Kravchenko, A.G., & Moin, P., "On the effects of numerical errors in large eddy simulations of turbulent flows," *J. Comp. Phys.*, **131**, 310–322 (1997).

Lin, D.N.C., and Papaloizou, J., "On the structure and evolution of the primordial solar nebula," *Mon. Not. R. Astron. Soc.* **191**, 37–48 (1980).

Malagoli, A., Cattaneo, F., & Brummell, N.H., "Turbulent supersonic convection in three dimensions," *Astrophys. J. Letters* **361**, L33–L36 (1990).

Marcy, G.W., & Butler, R.P., "A planetary companion to 70 Virginis," *Astrophys. J. Letters* **464**, L147–L151 (1996).

Mason, P.J., & Thomson, D.J., "Stochastic backscatter in large-eddy simulations of boundary layers," *J. Fluid Mech.* **242**, 51–78 (1992).

Mayor, M., & Queloz, D., "A Jupiter-mass companion to a solar-type star," *Nature* **378**, 355–359 (1995).

Mihalas, D., *Stellar Atmospheres*, 2nd ed. (San Francisco: W.H. Freeman, 1978).

Moin, P., Squires, K., Cabot, W., & Lee, S., "A dynamic subgrid-scale model for compressible turbulence and scalar transport," *Phys. Fluids A* **3**, 2746–2757 (1991).

Pedlosky, J., *Geophysical Fluid Dynamics*, 2nd ed. (Springer-Verlag: New York, 1987).

Pollack, J.B., Hollenbach, D., Beckwith, S., Simonelli, D.P., Roush, T., & Fong, W., "Composition and radiative properties of grains in molecular clouds and accretion disks," *Astrophys. J.* **421**, 615–639 (1994).

Pringle, J.E., "Accretion disks in astrophysics," *Ann. Rev. Astron. Astrophys.* **19**, 137–162 (1981).

Ruden, S.P., Papaloizou, J.C.B., & Lin, D.N.C., "Axisymmetric perturbations of thin gaseous disks. I. Unstable convective modes and their consequences for the solar nebula," *Astrophys. J.* **329**, 739–763 (1988).

Ruden, S.P., & Lin, D.N.C., "The global evolution of the primordial solar nebula," *Astrophys. J.* **308**, 883–901 (1986).

Ryu, D., & Goodman, J., "Convective instability in differentially rotating disks," *Astrophys. J.* **388**, 438–450 (1992).

Schumann, U., "Subgrid length-scales for large-eddy simulation of stratified turbulence," *Theor. & Comp. Fluid Dyn.* **2**, 279–290 (1991).

Shu, F., Najita, J., Galli, D., Ostriker, E., & Lizano, S., "The collapse of clouds and the formation and evolution of stars and disks," in *Protostars and Planets III*, ed. E.H. Levy & J.I. Lunine (University of Arizona Press, Tucson), pp. 3–45 (1993).

Skrutskie, M.F., Dutkevitch, D., Strom, S.E., Edwards, S., & Strom, K.M., "A sensitive 10μm search for emission arising from circumstellar dust associated with solar-type pre-main-sequence stars," *Astron J.* **99**, 1187–1195 (1990).

Stone, J.M., & Balbus, S.A., "Angular momentum transport in accretion disks via convection," *Astrophys. J.* **464**, 364–372 (1996).

Strom, S.E., Edwards, S., & Skrutskie, M.F., "Evolutionary time scales for circumstellar disks associated with intermediate- and solar-type stars," in *Protostars and Planets III*, ed. E.H. Levy & J.I. Lunine (University of Arizona Press, Tucson), pp. 837–866 (1993).

Tennekes, H., & Lumley, J.L., *A First Course in Turbulence* (Cambridge: MIT Press, 1972).

Weidenschilling, S.J., & Cuzzi, J.N., "Formation of planetesimals in the solar system," in *Protostars and Planets III*, ed. E.H. Levy & J.I. Lunine (University of Arizona Press, Tucson), pp. 1031–1060 (1993).

Chapter 17

A NEW MODEL FOR TURBULENCE: CONVECTION, ROTATION AND 2D

V.M. CANUTO, M.S. DUBOVIKOV
A. DIENSTFREY and D.J. WIELAARD

NASA Goddard Institute for Space Studies
2880 Broadway, New York, NY 10025

We apply a recent turbulence model to convective, rotating and 2D turbulence. The model has no adjustable parameters. Agreement with the data is good.

17.1 Turbulent Convection

17.1.1 New Stochastic Equations

The renewed interest in thermal convection stems from the Chicago experiments (10, 37) which have provided a new scaling law $Nu \sim Ra^{\gamma}$ with $\gamma = 2/7$, as well as other turbulent statistics. A model was proposed in (10) and shortly thereafter a second one was proposed ((28). The latter assumes that the thermal boundary layer is within the viscous one, which holds true for values of Ra sufficiently low. Since recent DNS data (18) show a deviation from this assumption even at $Ra \sim 10^5$, extending (28) to large Ra may lead to incorrect conclusions (Siggia, private communication).

To deal with convection, we extend the basic model (9) to incorporate inhomogeneous turbulence and derive a closed set of dynamic equations governing the basic turbulent statistics of turbulent convection: turbulent kinetic energy e, z-component of the turbulent kinetic energy e_z , temperature variance e_{θ} and temperature-velocity correlation j. The equations are solved numerically for the case of large aspect ratio A and without a coherent flow. The restriction is not required by the our approach, it is a matter of simplicity in this first comparison with the data. The predictions are checked against both laboratory data and recent DNS results (18) which provide statistics not measured in laboratory. The model explains the data in spite of the absence of free parameters.

One of the key points of the method developed in (9) is that the non-linear Navier-Stokes equations and those for temperature fields $\theta(k,t)$, be substituted with the equivalent stochastic, Langevin-like, equations which are linear in the fields $\vec{u}$ and θ but with non-linear coefficients. The general form is (a dot represents time derivative)

$$\dot{\vec{u}} = \vec{f}^{ext} + \vec{f}^t - \nu_d(k)k^2\vec{u}, \qquad \dot{\theta} = f_\theta^{ext} + f_\theta^t - \chi_d(k)k^2\theta \qquad (17.1)$$

Here, f^{ext}'s represent the external forces which are problem dependent and the f^t 's are the turbulent (stochastic) forces acting on an eddy k. Although equations of the type (17.1) are not new in turbulence, the novelty of the approach (9) is the way the dynamical $\nu_d(k)$ and $\chi_d(k)$, as well as the work of the stochastic forces, are related to the energy spectra $E(k)$ and $E_\theta(k)$. To evaluate $\nu_d(k)$ and $\chi_d(k)$, we used the general Renormalization Group (RNG) technique and treated the case of stirring forces of arbitrary form. The results are

$$\nu_d(k) = (\nu^2 + \frac{2}{5}\int_k^\infty E(p)p^{-2}dp)^{1/2}, \qquad \frac{\partial\chi_d}{\partial\nu_d} = \frac{10}{3}\nu_d(\nu_d + \chi_d)^{-1} \qquad (17.2)$$

with the initial condition $\chi_d(\nu) = \chi$, where ν and χ are the molecular viscosity and conductivity. The structure of (17.2) indicates that $\nu_d(k)$ and $\chi_d(k)$ are ultraviolet properties since the effect on an eddy k is due to all the eddies with wavenumbers larger than k. In (18) we discussed the fact that while RNG technique is appropriate to compute correctly UV properties like Eq.(17.2), it cannot give the spectrum $E(k)$ since by construction it neglects infrared contributions such as the work of the turbulent forces $a_t \sim< f_i^t(\vec{k}',t)u_i(k,t) >$. In [(10, 37) 4], use was made of the well established fact that energy transfer among eddies is essentially a local phenomenon. This consideration alone allowed the evaluation of the a^t's. The results are (18)

$$a^t = -(4\pi k^2)^{-1}r(k)E'(k), \qquad a^\theta = -(4\pi k^2)^{-1}r_\theta(k)E_\theta'(k) \qquad (17.3)$$

where $r(k)$ is the rapidity (velocity of energy transfer in k-space) given by

$$r(k) = 2\int_0^\infty [\nu_d(p) - \nu]p^2 dp \qquad (17.4)$$

To identify $\vec{f}^{ext}$, we use the Navier-Stokes and temperature equations to obtain $f_i^{exp} = -\alpha P_{ij}(\vec{k})g_j\theta(k)$, and $f_\theta^{ext} = \beta w(k)$, where $w = u_3, P_{ij}$ is the projection operator, α is the volume expansion coefficient, $\vec{g}$ is the gravitational acceleration and $\beta = -\frac{\partial T}{\partial z}$. The governing dynamical equations can be obtained from Eqs. (17.1). They are:

$$\dot{e} = a^t + \alpha gj - 2k^2\nu_d e + \frac{\partial}{\partial z}\left[\nu_d(k_0)\frac{\partial e}{\partial z}\right] \qquad (17.5)$$

$$\dot{e_\theta} = a_\theta^t + \beta j - 2k^2\chi_d e_\theta + \frac{\partial}{\partial z}\left[\chi_d(k_0)\frac{\partial e_\theta}{\partial z}\right] \qquad (17.6)$$

$$\dot{j} = 2\alpha g P_{zz}e_\theta + 2\beta e_z - k^2(\nu_d + \chi_d)j + 1/2\frac{\partial}{\partial z}\left\{[\nu_d(k_0) + \chi_d(k_0)]\frac{\partial j}{\partial z}\right\} \qquad (17.7)$$

$$\dot{e_z} = 1/2P_{zz}a^t + \alpha g P_{zz}j - 2k^2\nu_d e_z + 1/2\frac{\partial}{\partial z}\left[\nu_d(k_0)\frac{\partial e_z}{\partial z}\right] \qquad (17.8)$$

The equations must be supplemented by the equation governing for the mean temperature field. Calling $J(z,t)$ the integral over all k's of $J(z,k,t)$, we have

$$\frac{\partial}{\partial t}T(z,t) = -\frac{\partial}{\partial z}J(z,t) + \chi\frac{\partial^2}{\partial z^2}T(z,t) \qquad (17.9)$$

The equations are fully prognostic and have no adjustable parameters.

17.1.2   Numerical results

In Fig. 17.1, panel a), we present the experimental (37) (dash dot), numerical (squares) and predicted (solid line) Nu vs. Ra relationship for He($\sigma = 0.7$). The discrepancy of about 25% may be an aspect ratio effect since it has been observed that Nu decreases with increasing A. The predicted Nu for Hg, $\sigma = 0.025$, is also plotted. Although the laboratory data are available only for small values of A's (11) and thus, a direct comparison is not possible, two quantitative features are noteworthy. First, the slope changes with the Prandtl number, being somewhat smaller for smaller σ's and the data (37, 11) confirm the trend. Second, Nu decreases with decreasing σ, a feature recently confirmed by the laboratory data (37, 11). In panel b), we present the scaling law for the horizontal Peclet number evaluated at the center of the cell. The agreement with DNS data (18) is good. In panel c), we present the temperature variance vs. Ra. The two upper curves correspond to data near the wall (subscript w). No laboratory data are available for this case. Although the exponents appearing in the power laws are roughly the same in our model and DNS data, the coefficients disagree by some 40%. The discrepancy can be attributed to the presence of a coherent flow in (18). In fact, the discrepancy disappears as one moves to the center of the cell (lower three curves) for which case we also have laboratory data. In Fig.17.2 we present the z-profiles of several statistics for the largest $Ra = 2 \times 10^7$ attainable with DNS (18). In panel a), we exhibit the temperature variance: the agreement with the DNS data is good. The vertical Pe number, panel b), is in good agreement with the DNS data. As for the horizontal Pe, panel c), we obtain good agreement at the center of the cell but not at the boundaries. In Fig. 17.3 we plot several spectra vs. the horizontal wavenumber k_h . We compare the theoretical spectra averaged over z with the correspondent DNS data. We plot the normalized spectra since the integrated values are in good agreement with the data, as we have seen. The agreement is quite satisfactory with the exception of a small region of small wavenumbers which does not contribute significantly to the total integrals. The effect is caused by boundary layer shears (18).

17.1.3   Conclusions

Although the largest Ra presently attainable via DNS is not sufficient for the spectra to exhibit an extended inertial region, the DNS data provide a large and important set of data to check the validity of any model of convection. Our model yields results that overall agree with the data. The only discrepancies are the values of Pe(horizontal) which are due to the effect of coherent flows which we have not included. It is important to stress that the formulation we employ here was not tailored to convection: it was previously shown to reproduce quite accurately a host of other turbulent statistics ranging from free decay, shear generated turbulence, etc. (9). Convective turbulence is a further, very important, test and validation of the physical assumptions underlying our approach.

17.2   Rotating turbulence

17.2.1   Basic results

If one were to apply the Taylor-Proudman theorem to turbulence, as Ω increases, rotating turbulence should tend toward a 2D state characterized by vertical L_v and horizontal L_h length scales such that $L_v \gg L_h$. DNS and LES, however, do not confirm such expectations.

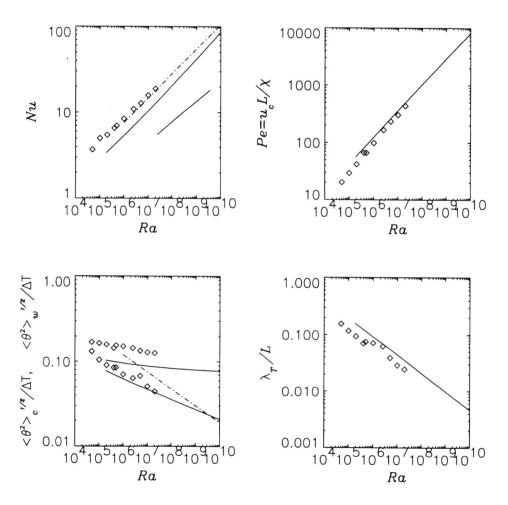

Figure 17.1: Scaling laws. Comparison of theoretical results (solid lines) with experimental (37) data (dash-dotted lines) and DNS (18) data (squares).

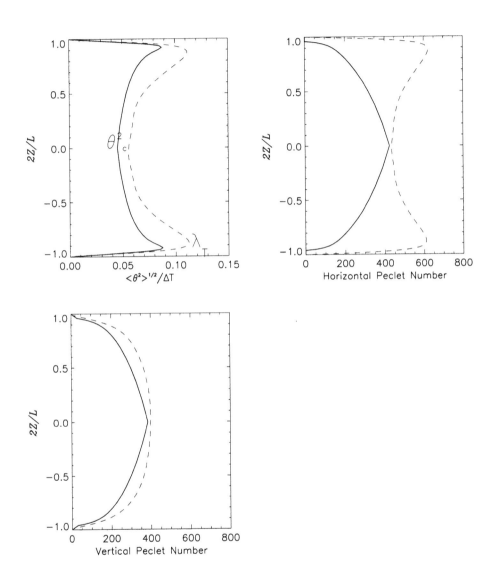

Figure 17.2: z-profile of statistics. Solid lines, theoretical results; dashed-lines, DNS data (18).

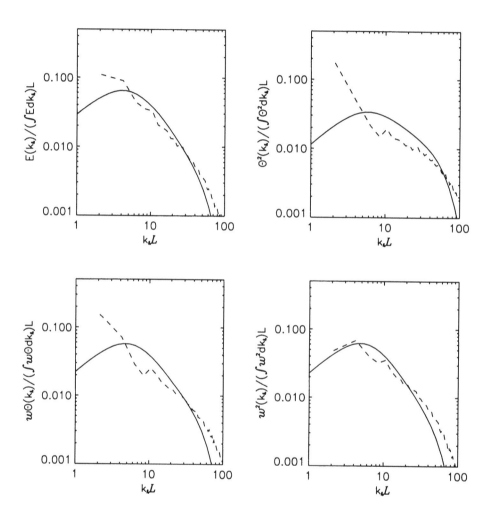

Figure 17.3: Spectra averaged over z. Same notation as in Fig.17.2.

Early DNS (1) and experiments (34) confirmed the trend toward $2D$ but further DNS work (22) with larger Ω yielded the opposite result: L_v/L_h first grows with Ω but then decreases returning toward a $3D$ state. Using LES, the tendency toward $2D$ was seen and it was thought that lateral $L_{11,3}$ and longitudinal $L_{33,3}$ vertical length scales would *both* be larger than L_h . It was, however, found (3· 32) that $L_{11,3} \gg L_h$, but $L_{33,3} \sim L_h$. It was stated that (32) "the decoupling was unexpected especially considering the strong coupling between vertical and horizontal fluctuations", and that (3) "most striking is the large growth rate in $L_{11,3}$ which attains values between 5-10 larger than $L_{33,3}$".

Here, we show that both DNS-LES results can be reproduced and understood on the basis of (9). For large Ω, we show that there exist two quite different regimes separated by the new number $N = K/\nu\Omega$, where K is the kinetic energy. For $N < 1$, rotation is so strong as to suppresses the energy cascade altogether. No inertial regime, defined by the constancy of the energy flux, develops. In a freely decaying case, viscosity remains the only operating mechanism and, in the absence of energy transfer, an initially isotropic 3D turbulence remains thus and never tends towards a 2D state. This explains the DNS data. For $N > 1$, the energy cascade is restored and the flow consists of mutually interacting $2D$ and $3D$ states. We further show that $L_{33,3}$ belongs to a $3D$ state (where all lengths are of the same order), while $L_{11,3}$ belongs to a $2D$ state and thus, $L_{11,3} \gg L_{33,3} \sim L_h$. This explains the different behavior vs. Ω found in LES (3· 32). We also compute the power law exponents for energy and length scales for freely decaying rotating turbulence and show that the results reproduce LES data (31· 32). The non-linear interactions, though weakened, are the main cause of the $2D$ state which is not of the Proudman-Taylor type since the latter requires negligible non-linearities. The equations relevant to the present case are (9)

$$\frac{\partial}{\partial t}E(k) = T(k) - 2\nu k^2 E(k) + A_s(k) \tag{17.10a}$$

where the last term represents the work done by the stirring forces and $T(k)$ is the transfer,

$$T(k) = -\frac{\partial}{\partial k}\Pi(k), \qquad \Pi(k) = E(k)r[k, E(k)] \tag{17.10b}$$

where the rapidity r is given by Eq. (17.4). How can one include rotation? There is ample experimental, numerical and model evidence (36· 17· 8) that the transfer $T(k)$ is inhibited by rotation. In our model, the inhibition affects the rapidity $r(k)$ (we do not write the $E(k)$ dependence for ease of notation) and write

$$r(k) \rightarrow r_\Omega(k) = f(k, \Omega)r(k) < r(k) \tag{17.10c}$$

Using the helical formalism (35), we have determined that $\Pi(k)$ is reduced by the factor

$$\frac{1}{\Omega_\tau} \sim k^2\nu_t(k)/\Omega < 1, \qquad f(k, \Omega) = \frac{k^2}{\Omega}\nu_t(k) \tag{17.10d}$$

Let us consider a steady state and assume that the stirring forcing is concentrated in the low k region. For a high Reynolds number flow, there is an extended inertial region in which $\Pi(k) = \epsilon$. We obtain

$$E(k) = (\frac{45}{8})^{1/2}(\epsilon\Omega)^{1/2}k^{-2}, \qquad r_\Omega(k) = (\frac{8}{45})^{1/2}(\epsilon/\Omega)^{1/2}k^2 \tag{17.11a}$$

Eq.(17.11a) was first derived in (39) where it was also shown why (17.11a) differs from $E(k) \sim (\epsilon \Omega^2)^{2/5} k^{-11/5}$ (38). The spectrum (17.11a) has been recently obtained (13) by solving the model (7). In the presence of Ω, the turbulent viscosity $\nu_\Omega(k)$ is given by

$$\nu_\Omega(k) = \frac{1}{2k^2} \frac{\partial}{\partial k} r_\Omega(k) = (\frac{8}{45})^{1/2} (\frac{\epsilon}{\Omega})^{1/2} k^{-1} = \frac{K^2}{\epsilon} g \quad , g \equiv \frac{8}{45} \frac{\epsilon}{K\Omega} \tag{17.12}$$

The last equality, which corresponds to $k = k_0 \sim L$, coincides with that of (39) in the limit of large Ω. The condition that $f(k, \Omega) < 1$ is equivalent to

$$k < k_\Omega, \qquad k_\Omega \equiv 2\epsilon^{1/2} \Omega^{3/2} \tag{17.13}$$

Thus, we envisage a spectrum that in the interval $k_0 < k < k_\Omega$ is given by (7a), while for $k_\Omega < k < k_d$, is given by Kolmogorov. Here, k_d has the usual expression $k_d \equiv (\epsilon \nu^{-3})^{1/4}$. However, since k_Ω increases with Ω, it may become larger than k_d. In that case, Kolmogorov is no longer obtained and one goes directly from (17.11a) into a dissipation region which begins at a wavenumber k_d^* defined as the location where $\nu = \nu_\Omega(k_d^*)$. This gives

$$k_d^* \equiv 2(\epsilon \Omega^{-1})^{1/2} \nu^{-1} \tag{17.14}$$

The condition for a Kolmogorov spectrum to exist, $k_\Omega < k_d^*$, translates into $\nu \epsilon^{-1} \Omega^2 < 1/4$, or $Ro^\omega > 1$, where $Ro^\omega \equiv \omega/2\Omega$ is the Rossby micro-number and $\omega \equiv (\epsilon/\nu)^{1/2}$. The condition for the spectrum (17.11a) to exist is instead (L is the size of the system) $k_\Omega > k_0 \sim L^{-1}$, or $Ro^L < 1$, where $Ro^L \equiv K^{1/2}/\Omega L$ and from (17.11a), $K \sim (\epsilon \Omega)^{1/2} L$. In the case of strong rotation, the new dissipation wavenumber k_d^* can become even smaller than $k_0 \sim L^{-1}$. This will happen when

$$N \equiv \frac{K}{\nu \Omega} < 1 \tag{17.15}$$

When this occurs, there is no inertial region where $\Pi(k) = \epsilon$. The spectrum is no longer universal and it depends on the specific type of external forcing. We suggest that N is a new number to characterize rotating turbulence.

17.2.2 $2D - 3D$ states in rotating turbulence

The Coriolis force, $2\vec{\Omega}_{\vec{k}} \times \vec{u}(\vec{k})$, $\vec{\Omega}_{\vec{k}} = k^{-2}(\vec{\Omega} \cdot \vec{k})\vec{k}$, affecting an eddy $\vec{u}(\vec{k})$, vanishes if $\vec{k} \perp \vec{\Omega}$. This implies that eddies with $k_h \gg k_z$ form a quasi $2D$ mode which is weakly affected by rotation. Since in $3D$ energy cascades mostly forward, while it is mostly backward in $2D$, the energy flux from $3D \to 2D$ occurs mostly at large k's, whereas the $2D \to 3D$ transition occurs mostly at lows k's where the inhibition of energy transfer factor f^{-1} is the largest. Thus, if the initial energy densities $e(2D) = e(3D)$, flux $(3D \to 2D) >$ flux $(2D \to 3D)$, which leads to a flow of energy from $3D$ to $2D$ until equilibrium is reached in which $e(2D) \gg e(3D)$. As Ω increases, so does $e(2D)$, while there is a corresponding decrease of the volume of the mode whose boundary is defined by the condition $\Omega_k \sim k^2 \nu_\Omega(k)$. Using the above expression for $\vec{\Omega}_{\vec{k}}$ and the second equality (17.12), this yields

$$\frac{k_z}{k} \sim \Omega^{-1} (\epsilon \Omega^{-1})^{1/2} k \tag{17.16}$$

As a result of the two opposing tendencies, the total $2D$ energy density can either increase, decrease or tend to a fixed asymptotic value. LES data exhibits a symmetry of the Reynolds stress tensor for quite a long decay time during which the Rossby number decreases quite significantly. We can thus assume that the ratio of the energies in the $2D - 3D$ modes tends to a finite value as $\Omega \to \infty$. In addition, local interactions among eddies of either mode limit the value of the ratio $E(2D)/E(3D)$ at each k so as to prevent an infinite energy flux from one mode to the other. This, in turn, can be viewed as an indication that the two energy spectra must have similar shapes.

17.2.3   Decaying turbulence

The very different behavior of $L_{33,1}$ and $L_{11,3}$ found by LES has not yet received an explanation. The hierarchy of regimes introduced above and the basic relation (17.16) provide a natural explanation. We begin by assuming an initial small k behavior of $E(k)$ of the form $E(k, t = 0) = B_{s-2}k^s$, where $s = 2, 4$ (20). Carrying out a similarity analysis similar to the one in (20), but with the spectrum given by Eq.(17.11a), one obtains the following results

$$K(t, s = z) \sim \Omega^{3/5}t^{-3/5}, \qquad K(t, s = 4) \sim \Omega^{5/7}t^{-5/7} \tag{17.17}$$

The results agree with LES data (32). Furthermore, k_0 , the value at which the spectra (17.11a) and $E(k, t) = 0$ coincide, is computed to be

$$k_0(t, s = 2) \sim (t/\Omega)^{-1/5}, \qquad k_0(t, s = 4) \sim (t/\Omega)^{-1/7} \tag{17.18}$$

which are used to evaluate the length scales

$$L_{\alpha\alpha,\beta}(t) = <u_\alpha^2>^{-1} \int <u_\alpha(\vec{r})u_\alpha(\vec{r} + s\vec{e}_\beta)> ds \tag{17.19}$$

(no summation over α); $\vec{e}_\beta$ is the unit vector along the x_β axis and $\vec{r}$ is the radius-vector. Since turbulence is axially symmetric (with respect to the x_3 axis, $\vec{\Omega} = (0, 0, \Omega)$), there are three horizontal $L_{\alpha\alpha,1}(\alpha = 1, 2, 3)$ and two independent vertical scales, $L_{11,3}$ and $L_{33,3}$. In a $3D$ mode, all five length scales are of the same order of magnitude k_0^{-1}

$$L_{\alpha\alpha,1} \sim L_{33,3} \sim k_0^{-1} \sim t^{1/5}, t^{1/7} \tag{17.20}$$

depending on $s = 2, 4$. The results are in agreement with LES data (32). In a $2D$ mode, there is an horizontal $L_{11,1} \sim k_0^{-1}$ and a vertical length scale, $L_{11,3}$, defined by the $2D/3D$ boundary value (17.16) of k at $k_z = k_0$

$$L_{11,3} \sim (k_z^0)^{-1} = k_0^{-2}(\Omega^3\epsilon^{-1})^{-1/2} \tag{17.21}$$

Using Eqs. (17.18), we obtain (for $s = 2, 4$)

$$L_{11,3} \sim (\Omega^4 t^6)^{1/5}, \qquad L_{11,3} \sim (\Omega^6 t^8)^{1/7} \tag{17.22}$$

which grow much faster than $L_{33,3}$. The results agree with LES data and explain the decoupling of the vertical length scales $L_{33,3}$ and $L_{11,1}$ that surprised the authors of (22, 32).

17.2.4   Conclusions

We have constructed a new hierarchy of regimes of rotating turbulence with which we have recovered laboratory, DNS and LES data that were called "unexpected". We also show that the $2D$ state of rapidly rotating turbulence is intrinsically different from the $2D$ state arrived at via the Taylor-Proudman mechanism: instead of a pure $2D$ state, there is equilibrium between $2D$ and $3D$ states. The latter occurs because of the non-linear interactions which are neglected in the Taylor-Proudman mechanism.

17.3 $2D$ Tubulence

17.3.1   Basic features

The distinctive feature of $2D$ turbulence is the existence of two conservation laws, one representing energy and the other enstrophy

$$\int_0^\infty T(k)dk = 0, \qquad \int_0^\infty k^2 T(k)dk = 0 \qquad (17.23)$$

In $3D$, case there is only "energy cascade" (from low to high k), while in $2D$, energy cascades from large to small k while enstrophy goes from small to large k's (19, 15, 26).

In the case of freely decaying $2D$ turbulence, there are two quite different stages. A first stage characterized by non-coherent structures (if, of course, the initial velocity field is also incoherent) and a second stage in which, separated, almost axisymmetric, coherent vortices emerge from the structureless velocity field (25, 2, 27). The final stage consists always of a small number (~ 2) of vortices which are maximally separated. In this second stage, the decay of enstrophy slows down considerably. Thus, the energy spectrum $E(k)$ is a representative measure only during the first stage, while in the subsequent stage, one must also consider the characteristics of a "dilute vortex gas" such as density of vortices, energy, enstrophy, average radii of vortices, etc. It is now clear that the early theory of $2D$ turbulence (19, 15), as well as the EDQNM and TFM models (26) and the model presented here, can only deal with the first stage of development. A description of the second stage (from where vortices begin to appear) must incorporate both the turbulence processes (selective decay) as well as the vortices. With this proviso, we suggest a model for $2D$ turbulence, derive the relevant equations, solve them and compare the results with a host of DNS data. The overall agreement is quite satisfactory.

17.3.2   Basic equations. Time evolution of the energy spectrum.

In the $2D$ case, Eqs. (17.10c) and (17.3) are are no longer valid because there are now two cascade processes. Because of the conservation of enstrophy, in addition to (17.10a), we also have the dynamic equation for the enstrophy spectrum $H(k)$

$$\frac{\partial}{\partial t}H(k) = \tau(k) + k^2 A^{ext}(k) - 2\nu k^2 H(k) \qquad (17.24)$$

where $H(k) = k^2 E(k)$, and $\tau(k) = k^2 T(k)$. The function $\tau(k)$

$$\int_0^\infty \tau(k)dk = 0, \qquad \tau(k) = \frac{\partial}{\partial k}\Pi_H(k) \qquad (17.25)$$

where $\Pi_H(k)$ is the enstrophy flux, $\Pi_H(0) = \Pi_H(\infty) = 0$. In (19, 15) it was suggested that enstrophy propagates mainly from low to high k's. Suppose that $\Pi_H(k)$ can be represented in manner analogous to (17.10c), namely $\Pi_H(k) = r(k)H(k)$, with $r(k)$ expressed by Eq. (17.4) and $\nu_d(k)$ (k) by Eq.(17.2) with $1/2$ instead of $2/5$ for 2D solution. Since $r(k)$ and $\Pi_H(k)$ are positive definite, this ensures the uniqueness of the direction of the enstrophy flux from low to high wavenumbers for any spectral function $H(k)$. If we substitute Π_H in (17.24), we find

$$\frac{\partial}{\partial t}E(k) = -2\frac{r}{k}E(k) - r(k)\frac{\partial E}{\partial k} - 2k^2(\nu_d + \nu)E(k) + A^{ext} \qquad (17.26)$$

which does not satisfy the first of (17.23). We will replace the expression $\Pi_H(k) = r(k)H(k)$ with the following enstrophy flux

$$\Pi_H(k) = rH + ak\frac{\partial}{\partial k}(rH) \qquad (17.27)$$

With $a = -1/2$, we can satisfy (17.23). The two fluxes are now given by

$$\Pi(k) = -rE - 1/2k\frac{\partial}{\partial k}(rE), \qquad \Pi_H(k) = -rH - 1/2k\frac{\partial}{\partial k}(rH), \qquad (17.28)$$

Numerical solution of Eq. (17.10a,17.10b) and (17.29a) show that if there exists a stationary external forcing of the type $A^{ext}(k) = \epsilon\delta(k - k_i)$, the reverse energy cascade is realized at $k < k_i$. In the inertial regime, $r(k)E(k) = \epsilon$, the solutions are

$$E(k) = C_E\epsilon^{2/3}k^{-5/3}, \qquad C_E = \tfrac{1}{3}\cdot 10^{2/3} \qquad (17.29a)$$

and

$$H(k) = C_H h^{2/3}k^{-1}, \qquad C_H = 2^{1/3} \qquad (17.29b)$$

where h is rate of enstrophy dissipation, defined by (P is the palinstrophy)

$$h = 2\nu P, \qquad P \equiv \int k^4 E(k)dk \qquad (17.29c)$$

17.3.3   Numerical results

In Fig. 17.4 we compare the compensated spectra $k^4E(k)$ (panel a) and $k^3E(k)$ (panel b) with the results of the DNS work (5, 6). Two times were chosen, $t = 13$ and $t = 40$. The initial spectrum and molecular viscosity were $E(k,0) = 2Ckk_0^{-2}\exp[-(k/k_0)^2], \nu = 7.5 \times 10^{-5}$. The agreement is good. In Fig. 17.5 we compare the total enstrophy dissipation rate (panel a) and skewness S (panel b) 2

$$\frac{\partial\Omega}{\partial t} = -2\nu P, \quad \Omega \equiv \int_0^\infty H(k)dk = \int_0^\infty k^2E(k)dk, \quad S = P^{-1}\Omega^{-1/2}\int_0^\infty k^4T(k)dk$$
$$(17.30)$$

as function of time. The DNS data are from (5, 6). The model performs quite well. The discrepancy at early times has the same origin as in the $3D$ case (18). In the DNS case, one starts with Gaussian statistics at $t = 0$ which results in $T(k) = 0$, so that "natural statistics" are achieved only at later times. By contrast, our model entails natural statistics

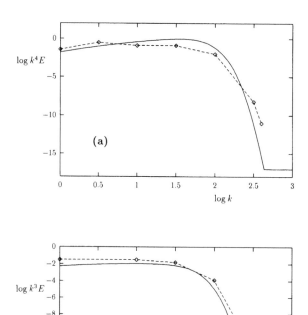

Figure 17.4: The compensated energy spectrum $k^4 E(k)$ at time $t = 13$ (panel a) and $k^3 E(k)$ at time $t = 40$ (panel b). Solid curve, present model; dashed curve, DNS (5). The molecular viscosity is 7.5×10^{-5} and the initial spectrum is taken to be $E(k, 0) = 2Ckk_0^{-2} \exp[-(k/k_0)^2]$, with $k_0 = 2$ and $\int E(k)dk = C = 4 \times 10^{-2}$.

from the very beginning, that is $T(k) \neq 0$ even at $t = 0$ and this implies that S is also non-zero. A comparison of the time dependence of the integrated variables $(\Omega/K)^{1/2}, (P/\Omega)^{1/2}$ is presented in Fig. 17.6. The DNS data are from (23, 24). The overall agreement is quite good up to a time $t \approx 1$, after which the emergence of coherent structures changes the qualitative behavior of the system. In particular, vortices slow down the decay of the enstrophy. If we describe the decay law by a power law $\Omega \sim t^{-a}$, DNS data (23, 24) suggest $a \approx 1.5 - 1.6$ toward the end of the non-vortex stage while when vortices are fully formed $a \approx 0.35$. As we have already mentioned, the present model can only describe the first stage as it yields $a \approx 1.5 - 1.6$ which agrees with DNS data but not with Batchelor's exponent of $a = 2$. Next, we discuss the case of a flow stirred by a forcing $A^{ext}(k) = \epsilon\delta(k - k_i)$. In the inverse cascade region when $k < k_i$, the energy spectrum follows Kolmogorov law $E(k) \sim k^{-5/3}$ quite closely with a Kolmogorov constant of $C_E \approx 1.5$, in agreement with the analytic solution (17.29b). In the region of direct cascade, $k > k_i, E(k) \sim k^{-x}$ where $x \geq 3$. To compare with DNS data, one should recall that these data never encompass both regions $k < k_i$ and $k > k_i$. This is so because in order to achieve the largest inertial interval, the stirring force is usually located near the boundary between the inertial and the dissipation regions. For the same reason, a hyperviscosity is often employed in DNS calculations. The inverse energy cascade regime $E(k) \sim k^{-5/3}$ has been recovered in several DNS calculations (4, 21, 14, 12, 29). The values of the corresponding Kolmogorov constant C_E exhibit a wide dispersion $2 \leq C_E \leq 10$. All of these calculations, with the exception of (4), refer to the regime before the formation of vortex structures. By contrast, in (4) DNS calculations were carried out for sufficiently long time until a stationary state was reached thanks to the adoption of a dissipative mechanism at the lowest k. The final, stationary stage consists of a large number of vortices which are in equilibrium with a background field of non-vortex structures. A stirring force of the above type was located at the inertial-dissipation boundary and thus, almost the whole spectrum is located at values of $k < k_i$. The resulting energy spectrum did not exhibit the expected Kolmogorov form with exception of case C) in Fig. 1 of (4) where such behavior was observed in half a decade in k-space. In each of the three cases, A), B) and C) the spectrum $E(k) \sim k^{-3}$ was observed in more than half a decade in k-space. Since the volume occupied by the vortices does not exceed $\sim 2\%$ of the total volume while the energy injected by the above force is distributed uniformly over the entire volume, almost the whole external energy is injected in the structureless background which exhibits a $k^{-5/3}$ spectrum. The Kolmogorov constant is about 2 (see Fig. 17.4 and Fig. 17.1 for case C) which agrees with our numerical and analytical results.

17.3.4   Conclusions

We have shown that the locality of energy transfer, together with the conservation laws characterizing the flow under consideration, has a remarkable predictive power since it naturally leads to dynamical equations governing the energy spectra without the need to introduce free parameters. In the $3D$ case, the energy flux $\Pi(k)$ is expressed as

$$\Pi(k) = r(k)E(k) + \sum_{1}^{\infty} a_n k^n \left(\frac{\partial}{\partial k}\right)^n [r(k)E(k)] \tag{17.31}$$

where the a_n's are constants. Since $r(k)$ is a functional of $E(k)$ and not a simple function of k and $E(k)$, (17.32) cannot be interpreted as a diffusion model for $\Pi(k)$. To assure a local nature of the transfer, the number of terms in (17.32) must be finite, since derivatives of infinite order would result in a dependence of $\Pi(k)$ not only on k and $E(k)$ computed at

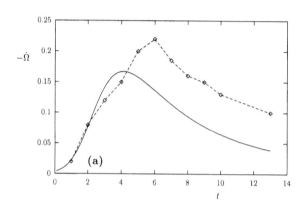

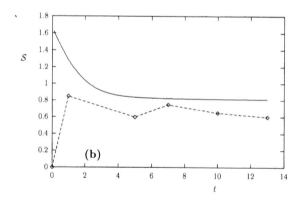

Figure 17.5: Time evolution of the enstrophy dissipation rate $-\dot{\Omega}$ (panel a) and of skewness S (panel b). Solid curve, present model, dashed curve, DNS (5, 6). The molecular viscosity is 2.33×10^{-5} and the initial spectrum is taken to be $E(k,0) = 2Ckk_0^{-2}\exp[-(k/k_0)^2]$, with $k_0 = 5$ and $\int E(k)dk = C = 6 \times 10^{-2}$.

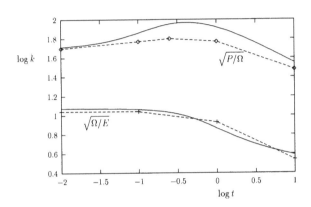

Figure 17.6: The time evolution of the logarithm of the mean wavenumbers $(\Omega/E)^{1/2}$ and $(P/\Omega)^{1/2}$. Solid curve, present model, dashed curve, DNS (23, 24). The initial energy spectrum and the value of viscosity are as in (23, 24).

k, but also on those functions computed at $k + \kappa$, which would imply a non-local character of the transfer. In particular, if we adopt a "minimal model" corresponding to $a_n = 0$ for $n \geq 1$, which automatically ensures a positive $\Pi(k)$, we obtain the dynamical equations that we have shown yield quite accurate predictions for a variety of turbulent flows, encompassing freely decaying, shear and buoyancy generated turbulence, as well as rotating turbulence. In the 2D case, we assume that the enstrophy flux $\Pi_H(k)$ can be represented in an analogous fashion

$$\Pi_H(k) = r(k)H(k) + \sum_1^\infty b_n k^2 (\frac{\partial}{\partial k})^n [r(k)H(k)] \tag{17.32}$$

In contrast with the 3D case, in 2D turbulence the presence of vortices implies non-locality and thus derivatives of infinite order must be included in (17.32). Since we limit our considerations to the pre-vortex stage, we can assume as a working hypothesis that locality does operate. This allows us to consider only a limited number of terms in (17.32). Specifically, we have assumed a minimal number of terms consistent with the energy conservation law. This in turn required that we keep at least the $n = 1$ term and take $b_1 = -1/2$. This provides an adequate model of 2D turbulence as comparison with DNS data shows.

Bibliography

Bardina, J., Ferziger, J. H., and Rogallo, R. S., "Effect of rotation on isotropic turbulence: computation and modeling", *J. Fluid Mech.* **154**, 321 (1985).

Benzi, R., Paternello, S., and Santangelo, P., "On the statistical properties of two-dimensional turbulence", *Europhys. Lett.* **3**, 811 (1987); Benzi, R., Patarnello, S., and Santangelo, P., "Self-similar coherent structures in two-dimensional decaying turbulence", *J. Phys. A: Math. Gen.* **21**, 1221 (1988).

Bertello, P., Metais, O., and Lesieur, M., "Coherent structure in rotating 3D turbulence", *J. Fluid Mech.* **273**, 1 (1994).

Borue, V., "Inverse energy cascade in stationary two-dimensional homogeneous turbulence", *Phys. Rev. Lett.* **72**, 1475 (1994).

Brachet, M. E., Meneguzzi, M. and Sulem, P. L., "Small scale dynamics of high Reynolds number two-dimensional turbulence", *Phys. Rev. Lett.* **57**, 683 (1986).

Brachet M. E., Meneguzzi, M., Politano, H., and Sulem, P. L., "The dynamics of freely decaying two-dimensional turbulence", *J. Fluid Mech.* **194**, 333 (1988).

Cambon, C., Bertoglio, J., and Jeandel, D., "Spectral closure of homogeneous turbulence", in AFOSR-Stanford Conf. on Complex Turb. Flows, (1981).

Cambon, C. and Jacquin, L., "Spectral approach to non-isotropic turbulence subjected to rotation", *J.Fluid. Mech.* **202**, 295 (1989).

Canuto, V. M. and Dubovikov, M. S., "A dynamical model for turbulence:I, General formalism", *Phys. of Fluids* **8**, 571 (1996); ibid., "A dynamical model for turbulence:I, shear Driven flows", **8**, 587 (1996); Canuto, V. M., Dubovikov, M. S., Cheng, Y., and Dienstfrey, A., "A dynamical modelfor turbulence: III, Numerical results", *Phys. of Fluids* **8**, 599 (1996).

Castaing, B., Gunaratne, G., Heslot, F., Kadanoff, L., Libchaber, A., Thomae, S., Wu, X. Z., Zaleski, S., and Zanetti, G., "Scaling of hard thermal turbulence in Rayleigh-Benard convection", *J. Fluid Mech.* **204**, 1 (1989).

Cioni, S., Ciliberto, S., and Sommeria, J., "Temperature structure functions in turbulent convection at low Prandtl number", *Europhysics Lett.* **32**, 413 (1995); "Experimental study of high Rayleigh numnber convection in mercury and water", *Dynamics of Atm. and Oceans* **24**, 117 (1996).

Frish, U. and Sulem, P. L., "Numerical simulation of the inverse cascade in two-dimensional turbulenc", *Phys. Fluids* **27**, 1921 (1984).

Godeferd, F. S., private communication.

Herring, J. R. and McWilliams, J. C., "Comparison of direct numerical simulation of two-dimensional turbulence with two-point closures: the effect of intermittency", *J. Fluid Mech.* **153**, 229 (1984).

Herring, J. R., Orzsag, S. A., Kraichnan, R. H., and Fox, D. G., "Decay of two-dimensional homogeneous turbulence", *J.Fluid Mech.* **66**, 417 (1974).

Hossain, M., "Reduction in the dimensionality of turbulence due to slow rotation", *Phys. Fluids* **A6**, 1077 (1994).

Jacquin, L., Leucther, O., Cambom, C., and Mathieu, J., "Homogeneous turbulence in the presence of rotation", *J. Fluid Mech.* **220**, 1 (1990).

Kerr, R. M., "Rayleigh number scaling in numerical convection", *J. Fluid Mech.* **310**, 139 (1996).

Kraichnan, R. H., "Inertial transfer in two and three dimensional turbulence", *J.Fluid Mech.* **47**, 525 (1971).

Lesieur, M., "Turbulence in Fluids", Kluwer Ac. Publ. (1991).

Lilly, D. K., "Numerical simulation of two-dimensional turbulence", *Phys. Fluids Suppl. 11* **12**, 240 (1969).

Mansour, N. N., Cambon, C., and Speziale, C. G., "Theoretical and computational study of rotating isotropic turbulence", in Studies in Turbulence, edited by Gatski T. B., Sarkar, S. and Speziale, C. G., Springer-Verlag, 59 (1992).

Matthaeus, W. H., Stribbling, W. T., Martinez, D., and Oughton, S., "Selective decay and coherent vortices in two-dimensional incompressible turbulence", *Phys. Rev. Lett.* **66**, 2731 (1991).

Matthaeus, W. H., Stribbling, W. T., Martinez, D., Oughton, S, and Montgomery, D., "Decaying two-dimensional, Navier-Stokes turbulence at very long times", *Physica* **D 51**, 531 (1991).

McWilliams, J. C., "The emergence of isolated coherent vortices in turbulent flow", *J. Fluid Mech.* **146**, 21 (1984); "The vortices of two-dimensional turbulence", *J. Fluid Mech.* **219**, 361 (1990).

Pouquet, A., Lesieur, M., Andre, J. C., and Basdevant, C., "Evolution of high Reynolds number two-dimensional turbulence", *J.Fluid Mech.* **72**, 305 (1975).

Robert, R. and Sommeria, J., "Statistical equilibrium states for two-dimensional flows", *J. Fluid Mech.* **229**, 291 (1991).

Shraiman, B. I. and Siggia, E. D., " Heat transfer in high Rayleigh number convection", *Phys. Rev.* **A42**, 3650 (1990).

Smith, L. M. and Yakhot, V., "Bose condensation and small-scale structure generation in a random force driven 2D turbulence", *Phys. Rev Lett.* **71**, 352 (1993).

Speziale, C. G., Mansour, N. N., and Rogallo, R. S., "The decay of isotropic turbulence in a rapidly rotating frame", Proc. 1987 Summer Program, Report CTR-S87, Center for Turbulence Research, NASA Ames Res. C., Moffet Field, Ca., (1987).

Squires, K. D., Chasnov, J. R., Mansour, N. N., and Cambon, C., "Investigation of the asymptotic state of rotating turbulence using LES", Ann. Res.Briefs-1993, Center for turbulence research, NASA Ames Res. Center, Stanford University, 157-170 (1993).

Squires, K. D., Chasnov, J. R., Mansour, N. N., Cambon, C., "The asymptotic state of rotating homogeneous turbulence at high Reynolds numbers", AGARD, 4-1, 4-9, 1994, Proc. of the 74th Fluid Dynamics Symposium "Applications of DNS and LES to transition and turbulence", Chania, Crete, Greece, (1994).

Teissedre, C. and Dang, K., "Anisotropic behavior of rotating homogeneous turbulence by numerical simulation", AIAA 87-1250, AIAA 19th Fluids Dynamics, Plasma Dynamics and Lasers Conference, Hawaii, (1987).

Veravalli, S. V., "An experimental study of the effects of rapid rotation on turbulence", Ann. Res. Briefs, NASA-Stanford Center for Turbulence Research, (1991).

Welfelle, F., "Inertial transfer in the local decomposition", *Phys. Fluids* **A5**, 677 (1993).

Wiegeland, R. D. and Nagib, H. M., "Grid generated turbulence with and without rotation about the streamwise direction, IIT Fluids and Heat Transfer", Report No R78-1, Ill. Inst. tech., (1978).

Wu, X. Z. and Libchaber, A. "Scaling relations in thermal turbulence: the aspect ratio dependence", *Phys. Rev.* **A45**, 842 (1992).

Zeman, O., "A note on the spectra and decay of rotating homogeneous turbulence", *Phys. Fluids* **6**, 3221 (1994).

Zhou, Y., "A phenomenological treatment of rotating turbulence", *Phys. Fluid* **7**, 2092 (1995).

Chapter 18

TRANSPORT USING TRANSILIENT MATRICES

ROLAND B. STULL

Atmospheric Science Programme
Dept. of Geography
The University of British Columbia
Vancouver, Canada V6T 1Z2

and

JERZY BARTNICKI

Norwegian Meteorological Institute (DNMI)
Postboks 43, Blindern , N-0313 Oslo, NORWAY

Convective turbulence is nonlocal, advective and vertically anisotropic. Such nonlocal vertical mixing can be described by a matrix of mixing coefficients, called a transilient matrix. Each matrix element indicates the fraction of air at a destination height that comes from any source height. These matrices can be parameterized as a function of mean flow conditions, and then used as a nonlocal, first-order closure in numerical weather forecast, pollutant transport, and other flow models. Such a transilient parameterization is described here, using atmospheric pollutant transport to illustrate mixing for some idealized scenarios.

18.1 Introduction

Convective turbulence is nonlocal, advective and vertically anisotropic. It is nonlocal because the undiluted cores of large convective thermals can transport fluid across large vertical distances such as the depth of the whole boundary layer. Unlike diffusion, this transport is like advection, involving the movement of large parcels of air. This is not surprising, because the turbulent transport terms of the Reynolds-averaged equations of motion come from the advection terms, not the diffusion terms. The resulting vertical motions can cause greater upward transport from a heated surface than downward, causing 1-D vertical anisotropy at any height.

Such nonlocal vertical mixing can be described by a matrix called a transilient matrix, where the name "transilient" has a Latin root meaning "jump over" or "leap across". The transilient matrix describes for every destination height the portion of air that comes from each possible source height. To conserve air mass and air state (heat, tracer, momentum, etc), each row and each column of the matrix must sum to one. This doubly-stochastic matrix results in a turbulence description that is absolutely numerically stable for any time step and any grid spacing (Stull, 1986). Like other first-order closures, it describes the vertically-resolved effects of horizontally unresolved turbulent eddies. However, unlike local schemes such as eddy-diffusivity theory, transilient matrices quantify the mixing between all source-destination height pairs, including nonlocal as well as local mixing. A comprehensive review of transilient turbulence is given by Stull (1993).

For free convection, the transilient matrix is typically asymmetric, which corresponds to a skewed vertical-velocity distribution. This matrix can be estimated at great computational expense using large-eddy simulation models. From the matrix can be calculated many other nonlocal measures of turbulence: true mixing length, mixing intensity, transport spectra (the partitioning of fluxes by the transport distance), vertical anisotropy, and source or destination effects (Ebert et al 1989).

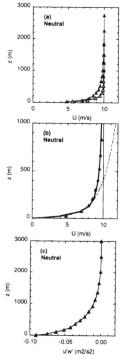

Figure 18.1: (a) Mean wind profile at 1 h (open triangles) and 6 h (solid triangles) into the simulation of a neutral boundary layer. Solid vertical line without data points shows the initial condition and imposed geostrophic wind. (b) The simulated wind profile (data points) at 6 h is compared to the best-fit surface-layer similarity theory (dashed line) and the best-fit neutral boundary-layer (eq 20) similarity theory (solid curved line). (c) Kinematic vertical turbulent flux of horizontal momentum.

The matrix can be parameterized at trivial computational expense for use as a turbulence closure scheme in models of fluid flow. The best parameterization schemes assume that turbulence is not a state of the flow, but a response to instabilities in the mean flow. By relating the off-diagonal matrix elements to the degree of flow instability, a parameterization is generated where turbulence undoes the mean instability. To date, parameterizations have been based on nonlocal representations of the Richardson number (Zhang and Stull 1992), and of the turbulence kinetic energy (TKE) equation (Stull and Driedonks 1987).

The latter approach will be discussed in sections 2 to 4, and will be illustrated in sections 5 and 6 in the framework of a prototype long-range pollutant transport model. The goal of this paper is to present parameterization details for transilient matrices, and to demonstrate their applicability to widely varying atmospheric conditions. Verification against atmospheric observations has been demonstrated elsewhere (e.g., Alapaty et al, 1997; Zhang and Stull, 1992), and will not be discussed here. Also not shown in this paper are the wide range of new nonlocal turbulence statistics that can be defined using transilient matrices (Ebert et al, 1989; Stull, 1993).

18.2   A transilient turbulence parameterization

Subgrid turbulent mixing in the vertical is illustrated using a modification of the transilient turbulence theory (T3) parameterization described by Stull and Driedonks (1987; hereafter abbreviated SD). This approach is absolutely conservative of pollutant concentration within the round- off errors of the computer, maintains the positive-definite nature of concentrations (i.e., negative concentrations are not possible), and includes both the diffusive action of small turbulent eddies as well as the advective-like transport of the larger coherent eddy structures.

Eddy structures cause turbulent transport across finite distances during finite time intervals. Transilient coefficient $c_{ij}(t, \Delta t)$ is the fraction of air in destination grid cell i that came from source cell j during the time interval from t to $t + \Delta t$. The fraction of air that stays in grid cell i during the same time interval is $c_{ii}(t, \Delta t)$. These coefficients are the elements of a transilient matrix, $\mathbf{c}(t, \Delta t)$, that describes mixing between all pairs of grid cells in a column. This matrix changes with time as the boundary layer and other turbulence regions evolve.

In a turbulent atmosphere with a spectrum of different eddy sizes, mixing can simultaneously occur between many pairs of points, resulting in a transilient matrix that is not sparse. Small-size eddies causing diffusion result in nonzero elements near the main diagonal of the matrix, while the effects of larger eddies are contained in elements further from the diagonal. Conservation of air mass and state require that each row and each column of the transilient matrix sum to 1, and each element must be non-negative for our universe where entropy (as a measure of randomness) increases. If there is no turbulent mixing to or from any grid cell i, then $c_{ii} = 1$, and all other matrix elements in the same row as c_{ii} are zero, as are all other elements in the same column.

Turbulence is dissipative, and exists only as a flow response to instabilities. The action of turbulence reduces the amount of instability, which is analogous to LeChatelier's principle of chemistry. In the absence of continued destabilizing forcings, turbulence would cease after it caused sufficient mixing to eliminate the instability (Stull and Hasagawa, 1984). The parameterization illustrated here follows the same principle, whereby the potential for vertical mixing is related to the amount of dynamic and static instability in the mean flow. Namely, greater instability should cause more mixing. To this end, a matrix of mixing potentials, Y, is computed as the first step in a parameterization that eventually gives the transilient matrix. Dispersion of tracers and boundary-layer evolution are coupled via this same physical parameterization of mixing.

18.2.1 Mixing potential, Y, first estimate

The following steps except the last are performed on only one triangle of the Y matrix. The first estimate of potential for mixing between grid points i and j can be found from:

$$Y_{ij} = \left(\frac{DU}{Dz}\right)^2 + \left(\frac{DV}{Dz}\right)^2 - \frac{g}{R_c \cdot \overline{\theta_v}}\left(\frac{D\theta_v}{D'z}\right) - Dis \qquad \text{for} \quad i \neq j \qquad (18.1)$$

where θ_v is mean virtual potential temperature in a grid cell, $\overline{\theta_v}$ is the average at cells i and j, $g = 9.8$ m/s^2 is the gravitational acceleration, R_c is the critical Richardson number, and Dis is a dissipation parameter. Nonlocal difference operator $D()$ represents $()_j - ()_i$, and $D'()$ is defined later by (18.2).

As an approximation to the TKE equation, the first two terms on the right give the mechanical production of turbulence by the mean wind shear, the next term represents the buoyant production or consumption, and the last term describes dissipation of turbulence. If wind shears are great enough to overpower the stabilizing effects of static stability and dissipation, then the flow is dynamically unstable and Y_{ij} is positive. If the air is statically unstable, then the buoyancy term changes sign and contributes positively to potential for turbulence. However, if the destabilizing effects are insufficient to overcome the stabilizing factors, then the resulting negative value of Y_{ij} is truncated to zero to indicate there is no potential for mixing. Along the main diagonal, mixing potentials of $Y_{ii} = 0$ are set as a first guess.

18.2.2 Influences of nonlocal static stability

For each element of Y_{ij} is associated a nonlocal static stability defined by the sign of $(\theta_{vj} - \theta_{vi}) / (z_j - z_i)$. That is, colder air above warmer air is unstable. Parameters in (18.1) are modified element by element, depending on the static stability between locations i and j. For unstable flow, mixing potentials are modified to account for undiluted rise of convective thermals. For stable flow, the critical Richardson number is modified to account for subgrid gravity waves. Details of both modifications are given next.

For unstable conditions, Crum and Stull (1987) have noted that rising thermals carry moisture and virtual potential temperature from the surface to the top of the mixed layer, with little dilution or lateral entrainment. The amount of buoyancy is not reduced by the vertical distance across which the thermal rises; that is, the buoyancy is not reduced by the separation distance between source and destination locations.

In (18.1), the normal distance increment Dz is defined in the usual way as $Dz = z_j - z_i$. However, in the buoyancy term a special distance increment $D'z$ is defined as follows to incorporate the non-reducing buoyancy in rising thermals:

$$D'z = Dz = z_j - z_i \quad \text{for} \quad \text{statically} \quad \text{stable} \quad \text{air} \quad (D\theta_v/D'z > 0) \qquad (18.2)$$

$$= \Delta z_1 \qquad \text{for} \quad \text{statically} \quad \text{unstable} \quad \text{air} \quad (D\theta_v/D'z < 0) \qquad (18.3)$$

where Δz_1 is the thickness of the bottom grid cell. In other words, for statically stable conditions $D\theta_v/D'z$ is a nonlocal approximation to the temperature gradient, causing the temperature difference to become less effective across greater distances. However, in unstable conditions, the temperature difference in the buoyancy term is parameterized to not depend on separation distance between i and j.

This distance-increment parameterization has the effect of making the mixing potentials equally large at those grid points thru which thermals might rise or sink, thereby creating well-mixed layers in the flow. This parameterization is equally effective for warm thermals rising from the ground, and for cold thermals sinking from radiatively or advectively-cooled cloud tops. Also, it need not be associated solely with the boundary layer.

In statically stable conditions, propagating gravity (buoyancy) waves in the real atmosphere can oscillate wind shears and static stabilities, causing the Richardson number Ri to oscillate closer and further from its constant critical value Ri_c (Finnigan et al 1984; Nappo 1992). To simulate the resulting oscillations in turbulence intensity, the modeled critical Richardson number (Ri_c) is caused to oscillate closer and further from the modeled flow Richardson number (Ri), which doesn't oscillate because gravity waves are not resolved. The net effect in both cases is an oscillation of the difference between the Richardson number and its critical value, thereby oscillating turbulence characteristics including onset and the intensity.

For statically stable conditions, the critical Richardson number is repeatedly cycled through the values of 0.25, 0.5, 1.0, and 0.5 . These values give reasonable balance between mechanical production and buoyant consumption of TKE (Mahrt 1981b, Stull 1988), and allow bursts of turbulence at night (Blackadar 1957). Superficially, this causes an oscillation in depth of turbulence and intensity of mixing, both near the ground in statically stable boundary layers, and at the top of the convective mixed layer where the air is again statically stable.

More significantly, such a scheme aids the modeled development of the stable boundary layer, which could otherwise evolve into one of two extreme states: (a) elimination of turbulence and extreme cooling of the lowest grid point, or (b) continuation of turbulence in a quasi-mixed layer. The period of four time steps is clearly artificial, and likely has nothing to do with the actual period of any gravity waves in the real atmosphere. However, when averaged over the four time steps, the mean meteorological and turbulent state of the model approximates that of the real atmosphere averaged over one wave period.

18.2.3 Convective overturning and subgrid turbulence

The next steps refine the first guess for all of the elements of **Y** by preventing convective overturning, but allowing, at most, perfect mixing. To achieve this, the elements are forced to increase monotonically from the corners of the matrix toward the main diagonal. Starting from the corner element, if the next element closer to the diagonal (via row or column) is smaller, then it is reset to equal the corner element. If it is larger than the corner element, then it is left unchanged. The same procedure is then used starting from these elements (second from the corner), and continues until the diagonal is reached. At this stage, the new value Y'_{ii} at the main diagonal equals that of the largest neighbor.

Each element in the main diagonal now has a value equal to that of the largest element that was adjacent by either row or column. To this second guess is added a reference mixing potential, which represents the unresolved subgrid effects of smaller eddies:

$$Y_{ii} = Y'_{ii} + Y_{ref} \qquad (18.4)$$

where $Y_{ref} = Y_{den} \cdot \Delta z_1$, and Y_{den} is a constant parameter. When the bottom grid cell thickness is reduced, less mixing remains unresolved, therefore Y_{ref} becomes smaller in the parameterization.

Finally, the matrix is made symmetric.

$$Y_{ji} = Y_{ij} \tag{18.5}$$

Although Ebert et al (1989) show that transilient matrices, and consequently mixing potential matrices, can be asymmetric for free convection, the symmetric matrix is relatively fast to compute, and yields very reasonable forecasts of boundary-layer evolution and pollutant transport even for convective conditions, as will be illustrated in section 6. Although not shown in this paper, new asymmetric parameterizations are being developed based on convective available potential energy.

18.2.4   Unequal grid spacing

For unequally-spaced grid points (Raymond and Stull, 1990), let m_j represent the mass of air in source grid cell j (i.e., the thickness of grid cell j), and define an adjusted mixing potential, A_{ij}, for all pairs of i and j by

$$A_{ij} = m_j \cdot Y_{ij} \tag{18.6}$$

This new matrix is asymmetric for unequally spaced grids, even though $\mathbf{Y}$ is symmetric by definition.

18.2.5   Transilient matrices

An L-infinity type of matrix norm $||\mathbf{A}||$ is computed by finding the sum of each row (sum over j, at each i) in the $\mathbf{A}$ matrix, and selecting the maximum sum as the norm. The off-diagonal elements of the $\mathbf{A}$ matrix are then divided by this norm, which gives the off-diagonal transilient elements

$$c_{ij} = A_{ij}/||A|| \tag{18.7}$$

Finally, elements along the main diagonal are found from the constraint that elements in each row of $\mathbf{c}$ must sum to one (18.8). The c_{ij} matrix is also asymmetric, unless the grid points are equally spaced. For equally-spaced grid points (18.5) and (18.6) are still valid, but the m_j in (18.6) cancels the resulting m_j in the matrix norm.

The result is a matrix that: (a) satisfies all constraints such as non-negative elements; (b) is physically reasonable as a responsive-type parameterization that depends on the amount of instability in the mean flow; and (c) requires no matrix-inversion computations. As discussed by Raymond and Stull (1990; and Stull, 1993) for unequally-spaced grid points, the conservation of air mass constraint is given by

$$1 = \sum_j c_{ij} \tag{18.8}$$

while the conservation of tracer is given by

$$1 = \sum_i \frac{m_i}{m_j} c_{ij} \tag{18.9}$$

These constraints are also automatically satisfied by the parameterization, and are worthwhile computations to include in any computer model to verify that no programming errors were introduced.

18.2.6   Use of transilient matrices

Regardless of whether the grid points are equally spaced or not, the following equation applies for each forward time step:

$$S_i(t + \Delta t) = \sum_j c_{ij} \cdot S_j(t') \tag{18.10}$$

where S is the variable being forecast. Variable S must be defined per unit mass of air, in order for (18.10) to work for unequally-spaced grid points. Variables such as θ, U, V or W are already in this form, as they represent heat and momentum per unit mass of air. Moisture and pollutant concentration must also be given as a mixing ratio; e.g., grams of moisture or tracer per gram of air. The t' variable is during the portion of a split time step just before (18.10) was applied, but after all the other boundary conditions and dynamics have been incorporated.

18.2.7   Turbulent flux and mixed-layer depth

The recursive relation for vertical turbulent kinematic flux, F, presented by Stull (1984) is applicable to unequally-spaced grid points by using the actual thickness Δz_k of each grid cell k:

$$F_{k\ top} = F_{k\ bottom} + (\Delta z_k / \Delta t) \sum_{j=1}^{n} c_{ij} \cdot [S_k(t') - S_j(t')] \tag{18.11}$$

where $F_{k\ top}$ and $F_{k\ bottom}$ are the vertical turbulent kinematic fluxes at the top and bottom of grid cell k. By starting at the surface where $F_{1bottom} = 0$ by definition (no turbulence across a solid boundary), then (18.11) can be solved upward knowing that the flux at the top of a cell equals the flux at the bottom of the adjacent cell above it. At the top of the model, the flux must again be zero, which can serve as another check of the algorithms.

Although the transilient turbulence scheme does not explicitly forecast the mixed-layer depth h as a variable, this depth can be diagnosed from the flux calculation as the height where the heat flux is most negative. Although neither the turbulent flux nor the mixed-layer depth are required to make the forecast, they are often useful as auxiliary products of the forecast.

18.3   Split time step and the destabilization problem

Because the amount of turbulent mixing described by c_{ij} depends on the amount of instability in the mean flow, each time step must be split into at least two parts. During the first part, boundary conditions and body forcings such as dynamics, thermodynamics, cloud physics (not in this prototype model), advection, and radiation (not in this prototype model) are applied, which might destabilize the mean flow. During the second part of the time step, the value of the transilient matrix is calculated from the resulting instabilities, and then it is applied to mix the contents of grid cells. Thus, the destabilizing part is absolutely required in order for the transilient scheme to function.

By splitting the time step into destabilizing and mixing parts, however, an artificial dependence on discretization is introduced. In particular, boundary conditions such as conductive heating from the ground are applied as a flux $\overline{w'\theta'}_{sfc}$, and the influence of this flux on the potential temperature θ_1 of the lowest grid layer depends on the grid thickness and the time step:

$$\theta_1(t + \Delta t) = \theta_1(t) + \overline{w'\theta'}_{sfc} \cdot (\Delta t / \Delta z_1) \qquad (18.12)$$

This destabilization problem is associated only with flux boundary conditions and not with body-forcing terms.

As an example of the destabilization problem, suppose a surface heat flux of $0.2 \text{ K} \cdot \text{m/s}$ flows into a grid layer of $\Delta z = 100$ m, during a time interval of $\Delta t = 10$ min. From (18.11), one would expect the grid layer to warm $1.2°\text{C}$. Based on the temperature difference between this layer and the other layers above it for which there is no boundary condition applied, the transilient parameterization would produce a certain amount of mixing. Suppose that one changes the discretization so that the same flux would be applied to a bottom grid layer that is only 10 m thick. The layer would warm by $12.0°\text{C}$, which would cause greater temperature differences, greater instabilities, and lead to greater turbulent mixing to perhaps greater depths. Clearly, this is an artifact of the discretization that is not physically realistic.

In situations such as this, it is always helpful to look at the real atmosphere to better understand the physics as might be applied to improve the parameterization. On a sunny day when the ground is heated by the sun, we often observe a superadiabatic lapse rate in the surface layer. That such a layer exists tells us that there is some amount of destabilization caused by heat flux from the surface that is not quickly eliminated by turbulence, otherwise there would have been an adiabatic lapse rate. Thus, there is some relationship between the rate of destabilization and the rate of stabilization in nature that leads to an equilibrium temperature difference in the vertical associated with an applied heat flux. Such a relationship is the basis behind the many successful flux-profile relationships for the surface layer.

The vertical temperature difference or gradient between the lowest grid layer and the one above it is closely related to the change of temperature in the lowest layer during one time step. For example, start with an adiabatic temperature profile representing a statically neutral condition. Next, heat the lowest grid layer by an amount $\Delta\theta$. The vertical temperature difference between the bottom layer and the one above it is also $\Delta\theta$, which is directly related to the instability. From (18.13) we see that the amount of natural instability per amount of heat flux $(\Delta\theta / \overline{w'\theta'}_{sfc})$ translates to a natural value of the ratio $(\Delta t / \Delta z)$ when the domain is discretized. Such a natural or "best" value of $(\Delta t / \Delta z)$ allows a solution to the destabilization problem, as described below.

The obvious solution would be to always make the discretization of the model such that $(\Delta t / \Delta z)$ always equals the "best" value. This solution is rarely practical for a general model that could run with a range of discretizations.

A better approach is to further split the time steps of both the boundary conditions and the transilient turbulence parameterization as follows, where t' represents some intermediate time between the start and end of the full step:

- first apply all body forcings and emissions over whole time step
- BC1: apply boundary conditions, for $(\Delta t / \Delta z)$best:

$$\theta(t') = \theta(t) + (\overline{w'\theta'}_{sfc}) \cdot (\Delta t / \Delta z)_{best} \qquad (18.13)$$

- Turb1: calculate $\mathbf{Y}$, $\mathbf{A}$, and $\mathbf{c}$, but don't use them.
- BC2: apply the remaining boundary conditions:

$$\theta(t + \Delta t) = \theta(t') + (\overline{w'\theta'}_{sfc}) \cdot [(\Delta t / \Delta z) - (\Delta t / \Delta z)_{best}] \qquad (18.14)$$

• Turb2: now use **c** from Turb1 to calculate the mixing.

In this way, any Δt and Δz can be used, the full boundary conditions are eventually applied before the time step is finished, and the amount of turbulent mixing approximates some "best" value for the given conditions. This procedure will work regardless of whether $\Delta t/\Delta z$ is larger or smaller than the "best" value.

For surface drag in integrated form, the split into two parts becomes:

• BC1:

$$M(t') = \frac{M(t)}{[1 + M(t) \cdot C_d \cdot (\Delta t/\Delta z)_{best}]} \tag{18.15}$$

• BC2:

$$M(t + \Delta t) = \frac{M(t')}{\{1 + M(t') \cdot C_d \cdot [(\Delta t/\Delta z) - (\Delta t/\Delta z)_{best})]\}} \tag{18.16}$$

with the wind components calculated as shown at the end of section 5. The deposition need not be split into parts because it does not destabilize the flow.

The "destabilization problem" is eliminated, while the destabilization necessary for operation of the transient turbulence parameterization is retained. The method described above has been applied successfully for surface boundary conditions by finding an appropriate $(\Delta t/\Delta z)_{best}$ value as one of the model parameters. One might expect, however, that a different "best" value might exist for boundary conditions imposed in the interior of the domain, such as a radiative cooling flux at cloud top.

18.4   Calibration

Values of the three parameters for the transient turbulence method can be calibrated against idealized flow situations for which the solution is known in advance. Although idealized flow situations are affected by complex interactions from all of the parameters, cases can be chosen where one of the parameters dominates.

The value for $(\Delta t/\Delta z)_{best}$ is calibrated against a free convection solution of no mean wind, but a positive constant value of surface heat flux. For a mixed layer growing into a stably-stratified free atmosphere having constant lapse rate, it is known from field experiments and large-eddy simulation that the entrainment heat flux at the top of the mixed layer should be opposite in sign and 20 to 25% of the magnitude of the surface flux (Stull, 1976). A larger value of $(\Delta t/\Delta z)_{best}$ in the model allows the bottom grid cell to become warmer, thereby allowing deeper penetration of thermals into the capping inversion and greater entrainment. Thus, the magnitude of the entrainment flux ratio in the model increases as $(\Delta t/\Delta z)_{best}$ increases, allowing calibration to the observed flux ratio. A confounding factor in this calibration is Y_{den}, because larger Y_{den} values also cause the temperature of the lowest grid cell to become greater in convective conditions. We recommend $(\Delta t/\Delta z)_{best} = 10$ s/m.

For Y_{den}, the curvature and depth of the surface layer, and curvature within the boundary layer, provide a means of calibration. It is known from field experiments such as at Wangara (Clarke, et al, 1971) that the curved potential temperature profile in the unstable surface layer is a small fraction (2 to 5%) of the mixed-layer depth, while for a neutral boundary layer the curvature of the wind extends throughout the whole boundary-layer depth (Mason and Thompson, 1987; Andrén and Moeng, 1993). As Y_{den} (and thus Y_{ref}) increases, nonlocal mixing decreases in the model, and instabilities such as wind shear and superadiabatic lapse rates are reduced more slowly. Thus the modeled profiles become more curved as Y_{den} increases. We recommend $Y_{den} = 6 \times 10^{-5} \text{m}^{-1} \cdot \text{s}^{-2}$.

Another means to calibrate Y_{den} is via the time duration for initial instabilities to be eliminated in the absence of continued destabilization. Nieuwstadt and Brost (1986) have measured the amount of continued convection in the evening after daytime surface heating ceases. The model simulates continued weak convection in the interior of the mixed layer for a short time after the surface flux becomes non-positive. Calibration is possible because larger Y_{den} values cause slower decay rates of instabilities. However, this method is confounded by effects of the dissipation parameter.

Dissipation, Dis, can be calibrated by the decay of turbulence in the evening residual layer. In the residual layer the lapse rate is adiabatic, allowing even miniscule wind shears to generate turbulence in the absence of dissipation. Observations such as those by Mahrt (1981a) indicate the decay of turbulence in the evening. Larger values of Dis cause more rapid decay and termination of turbulence in the model. We recommend $Dis = 5 \times 10^{-6} s^{-2}$
.

18.5   Illustrative model

As a framework within which transilient turbulence theory can be illustrated and tested, a simplified 3-D model is described here. It has horizontal uniformity of the meteorology, but allows horizontal advection and vertical mixing of a heterogeneous field of pollutants. Instead of utilizing a classical advection- diffusion equation, mean horizontal advection is modeled using a method due to Bott (described next), and vertical turbulent mixing is handled with T3. Horizontal turbulent dispersion is neglected. Grid cells are equally-spaced in the horizontal, but are expanding in the vertical using the algorithm of Stull (1996). Horizontal grid spacing is on the order of 50 km, while vertical grid spacing is on the order of 50 m. There are no clouds, subsidence, precipitation, nor radiative heating in this prototype model.

Advection is modeled using the area-preserving, flux-form algorithm of Bott (1989a,b; 1991; Smolarkiewicz, 1989). The scheme is positive definite and produces only very small numerical diffusion. Although limited to advective speeds of Courant number less than one, it is computationally efficient. A nonstaggered-grid version is used here to be consistent with the transilient turbulence scheme.

Initial and boundary conditions are specified as follows. Winds are initialized to be geostrophic everywhere. Temperature can have any arbitrary initial vertical profile. The surface is flat and horizontally uniform. Surface heat flux is imposed as a time-varying boundary condition on the bottom grid cell: $Q_{\theta 1} = \overline{w'\theta'}_{surface}/\Delta z_1$, where $Q_{\theta 1}$ is the heat source into the bottom grid cell associated with an effective surface heat flux of $\overline{w'\theta'}_{surface}$, and Δz_1 is the thickness of the bottom grid cell. The surface momentum flux is found using a simple drag formulation, which yields the following source terms into the bottom grid cell: $Q_{u1} = -C_D \cdot M \cdot U/\Delta z_1$ and $Q_{v1} = -C_D \cdot M \cdot V/\Delta z_1$, where C_D is the drag coefficient, M is wind speed ($M^2 = U^2 + V^2$), and U and V are the wind components moving toward the east and north, respectively. Tracers in the lowest grid cell are subject to dry deposition to the surface, giving a source (loss) term in the bottom grid cell of $Q_{s1} = -V_d \cdot S/\Delta z_1$, where V_d is the deposition velocity and S is tracer concentration. These sources in the bottom grid cell are in addition to any other body sources (e.g., radiative heating, chemical reactions) that apply throughout the whole domain. Pollutant tracers can be emitted at constant rates E (g/s) into arbitrary grid cells (any x, y, z location), specified as boundary conditions: $Q_s = E/(\rho \Delta x \Delta y \Delta z)$, where ρ is air density, and Δx and Δy are grid-cell horizontal dimensions.

Integrated bottom boundary conditions are used for tracer deposition and for drag, because both of these processes have a requirement for positive definiteness and both have feedback from the variable to which the flux is applied. For example, if a forward time difference were applied to forecast the dry-deposition influence on tracer concentration S in the bottom grid cell: $S(t + \Delta t)=S(t) - (\Delta t/\Delta z)V_d S(t)$, then negative (unphysical) tracer concentrations could occur for large time step (Δt), small vertical grid increment (Δz) or large deposition velocity (V_d). Instead, the following integrated formula is used: $S(t + \Delta t) = S(t) \cdot \exp[-(\Delta t/\Delta z) \cdot V_d]$, which avoids negative concentrations.

Although wind components can be either positive or negative, drag acts oppositely to the wind direction to force the wind toward zero, but cannot change the sign of the wind. Again, if a forward time difference is used, an erroneous sign change could occur: $U(t + \Delta t) = U(t) - (\Delta t/\Delta z) \cdot C_d \cdot M(t) \cdot U(t)$. The integrated form below avoids that problem: $M(t+\Delta t) = M(t)/[1 + M(t) \cdot C_d \cdot (\Delta t/\Delta z)]$, with $U(t+\Delta t) = U(t) \cdot [M(t+\Delta t)/M(t)]$ and $V(t + \Delta t) = V(t) \cdot [M(t + \Delta t)/M(t)]$.

18.6 Simulation results for idealized scenarios

Although the model is fully three-dimensional, the following results utilize a wind in only the x-direction. For the two day simulation at the end of this section, $x - z$ cross sections of pollutants are shown. For all the other cases, one-dimensional vertical profiles of the meteorology are shown, because the meteorology is horizontally homogeneous in this prototype model.

For the purpose of displaying pollutant advection and dispersion in an $x - z$ cross section, the wind equations are modified to keep all the emitted pollutants within the same one $x - z$ plane in which they were emitted. Coriolis effects on the wind are removed and replaced with the following momentum adjustment scheme (which of course would be removed for actual weather forecasts):

$$U(i, j, k) = U(i, j, k) + (2/\pi) \cdot \Delta t \cdot f_{cor} \cdot [U_g(k) - U(i, j, k)] \qquad (18.17)$$

$$V(i, j, k) = V(i, j, k) + (2/\pi) \cdot \Delta t \cdot f_{cor} \cdot [V_g(k) - V(i, j, k)] \qquad (18.18)$$

Essentially, these equations relax the actual wind U toward geostrophic U_g with an e-folding time of $\pi/(2 \cdot f_{cor})$, where f_{cor} is the Coriolis parameter. This time period is $1/4$ of the inertial-oscillation period of the correct equations, which is approximatly the time span for subgeostrophic winds to accelerate to the point where $U = U_g$.

Table 1 indicates the initial and boundary conditions imposed to generate the test scenarios. For all these cases, an expanding grid is used in the vertical, with $\Delta z = 50$ m at the surface, smoothly expanding to $\Delta z = 500$ m at 3000 m (Stull 1996). Horizontal grid spacing is 50 km in x and y, with a y- domain of 150 km. The x-domain is given in Table 1. For these case studies, snapshots of the atmospheric state are shown at selected times in the forecast to illustrate transilient matrix results.

Winds are initialized to their geostrophic values (see Table 1), with $V_g = 0$ for these runs. The surface drag coefficient is constant at $C_d = 0.005248$. Emission of 30 g/s of lead oxides are from a continuous point surface source at $x = 150$ km and $y = 100$ km. This RpointS emission is immediately dispersed throughout the whole volume of the appropriate single grid cell. Dry deposition velocity is $V_d = 0.002$ m/s at the surface.

Case Name	BC1 $\overline{w'\theta'}$ $(K \cdot m/s)$	BC2 U_g (m/s)	IC $\partial T/\partial z$ (K/km)	Results shown at time (h)	x domain (km)
a neutral BL	0	10	0	1,6	500
b unstable ML	0.1	0	5	1,6	500
c mechanical ML	0	10	5	1,6	500
d both ML	0.1	10	5	1,6	500
e stable BL	-0.02	10	0	0-6	500
f diurnal	*	10	5	0-48	850

Table 18.1: Case-study conditions including boundary conditions (BCs) and initial conditions (ICs). See text for explanation of symbols. $*$ $\overline{w'\theta'}_{surface} = -0.02 + 0.12 \cdot \sin[\pi \cdot (t-6h)/(12h)]$ (K·m/s)

18.6.1 Neutral boundary layer

With no static stability to suppress turbulent transport and no heat flux at the surface, the turbulence in the neutral boundary layer quickly fills the whole model domain as wind shears develop in response to the forcings of surface friction and a geostrophic wind. Within 2 h a steady-state wind profile is reached (Fig. 18.1a) that is everywhere subgeostrophic, but which asymptotically approaches the geostrophic wind at high altitudes.

A neutral-boundary-layer similarity relationship that gives a reasonable fit to this asymptotic shape over the whole domain (not just in the surface layer) is:

$$U(z) = U_g \cdot \left[1 - \left(1 + \frac{z - z_o}{z_N}\right)^{-1}\right] \tag{18.19}$$

where z_N is a length scale for the neutral boundary layer, and z_o is an aerodynamic roughness length. This expression has the desirable characteristics that $U = 0$ at $z = z_o$, and $U \to U_g$ as $z \to \infty$. It also gives a much better fit to the simulated wind profile than an exponentially-decaying geostrophic departure.

For the steady-state wind profile in Fig. 18.1b, the best-fit length scale is $z_N = 35$ m and $z_o = 2$ m. A future project will be to relate such length scales to external forcings, boundary conditions, and to other scales used in past to describe neutral boundary layers.

Because of the split time step, a variety of different values of z_o and friction velocity, u_* can be calculated at different stages during each step. For example, when a logarithmic wind profile $U(z) = (u_*/k) \cdot \ln(z/z_o)$ is fit to the bottom several data points (Fig. 18.1b), the best fit parameters are $u_* = 0.75$ m/s and $z_o = 2$ m, where k is von Kármán constant of about 0.4. However, the steady-state momentum flux profile shown in Fig. 18.1c has a surface value corresponding to $u_* = 0.3$ m/s. The RimposedS drag coefficient (corresponding to $z_o = 0.1$ m) gives $u_* = 0.35$ m/s when applied to the final wind speed at the lowest grid point. Although each of the u_* values is consistent with the split time step approach, it is often desirable to calculate a value of u_* that is consistent with traditional flux-profile methods based on the final wind profile at the end of any time step. Intermediate values of u_* and z_o during the course of any one time step is usually of no interest to the user. Thus, one could treat the surface drag coefficient C_D as an intermediate model parameter, which can be tuned to produce a final wind profile with the desired z_o. By testing the best fit log-wind-profile value of z_o against the desired z_o, perhaps the drag-coefficient parameter can be relaxed from time step to time step toward a value that gives the desired result. This hypothesis will be tested in future versions of the model.

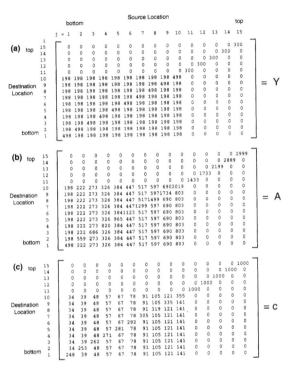

Figure 18.2: Transilient turbulence matrices for the neutral boundary layer at t = 6 h. All matrices are shown in atmospheric orientation (with destination grid index increasing upward from bottom to top), resulting in a main diagonal that runs from lower left to upper right (i.e., flipped upside-down from the usual mathematical orientation). Also, all matrices are multiplied by the scaling factors indicated below to produce these displays, and the resulting matrix elements are truncated to integer values to save space in this figure. (a) Initial symmetric mixing-potential matrix, Y, multiplied by 10^5. (b) Final asymmetric mixing-potential matrix, A, adjusted to incorporate variable grid spacing (multiplied by 10^5). (c) Transilient turbulence matrix, c, multiplied by 10^3, and rounded to the nearest integer. Note that each column and row of the actual c matrix indeed sums to exactly 1.0, even though the scaled and rounded numbers displayed here do not sum to 1000.

Figure 18.2 illustrates how mixing potential matrices are used to form transilient matrices. The initial mixing-potential matrix, $\mathbf{Y}$, is shown in Fig. 18.2a. This matrix is symmetric, with the largest elements along the main diagonal. The adjusted mixing-potential matrix, $\mathbf{A}$, is asymmetric because it includes the effects of variable grid spacing, as shown in Fig. 18.2b. Finally, Fig. 18.2c shows the transilient matrix, $\mathbf{c}$. The matrices in these figures have been multiplied by scaling factors and rounded to the nearest integer for display purposes. The actual $\mathbf{c}$ matrix does indeed satisfy the conservation constraints that each unscaled row and column sums to exactly 1.0.

18.6.2 Unstable (free-convective) mixed layer

Starting with no winds and a constant lapse rate that is statically stable, a convective mixed layer grows when surface heating is applied to the bottom grid cell. Figure 18.3a shows the evolution of the potential temperature profile during 6 h. A shallow surface layer with superadiabatic lapse rate forms under a well-mixed layer of constant potential temperature. Entrainment causes the top of the well mixed layer to be higher than would have occurred by thermodynamic encroachment alone. Note that Fig. 18.3 shows only the bottom half of the whole 3000 m vertical domain, because there was no turbulence in the top half.

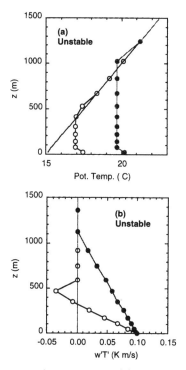

Figure 18.3: (a) Initial potential temperature (θ) profile (solid line), and simulated profiles after 1 h (open circles) and 6 h (solid circles) for a convective mixed layer. (b) Kinematic vertical turbulent heat flux profiles.

Heat fluxes exhibit the typical linear decrease with height, in Fig. 18.3b. Because of the imposed oscillation of the critical Richardson number, the amount of entrainment at the top of the mixed layer also oscillates. The individual profile at $t = 1$ h shows a reasonable negative heat flux at the top of the mixed layer associated with entrainment, while the profile at $t = 6$ h shows no entrainment. However, when averaged over four time steps (corresponding to the period of the critical-Richardson-number oscillation), there is a net negative heat flux at the top of the mixed layer of magnitude 20% of the surface flux (not shown).

The top of the mixed layer, h, measured as the height of most negative heat flux, smoothly rises with time as air is entrained into the mixed layer (Fig. 18.4). The turbulent entrainment zone can extend above the average value of h, and can contain intermittent turbulence

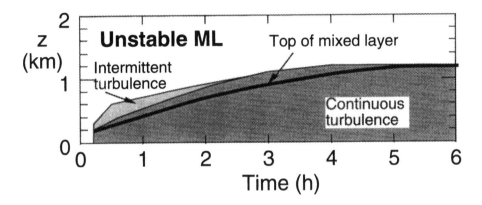

Figure 18.4: Evolution of the depth of the simulated convective mixed layer. Turbulent regions are identified as those grid cells for which the transilient coefficients along the main diagonal of the transilient matrix have $c_{ii} < 1$. The top of the mixed layer, h, is the height of most negative heat flux. Intermittent turbulence is caused in stable regions by the imposed oscillating critical Richardson number to simulate unresolved gravity waves.

because of the oscillating critical Richardson number. Figure 18.5 shows the transilient matrices for this free-convection case, at $t = 12$ h.

18.6.3 Mechanically mixed layer

In this simulation, wind shears generate turbulence against the consumption of a statically-stable initial lapse rate. Log-linear-shaped wind profiles form (Fig. 18.6a), while the potential temperature profile becomes well mixed (Fig. 18.6b). A fairly strong stable layer forms at the top of the mixed layer, which greatly limits the vertical spread of turbulence. Note that only the bottom 800 m of the 3000 m domain are shown in Fig. 18.6.

Momentum fluxes quickly decrease with height in Fig. 18.6c, exhibiting much less curvature than in the neutral boundary layer. The heat fluxes show the expected linear relationship with height, with negative fluxes aloft associated with mechanical entrainment, and zero flux at the surface as the imposed boundary condition. The boundary layer remains relatively shallow (Fig. 18.7), with a layer of intermittent turbulence associated with the parameterization of subgrid gravity waves.

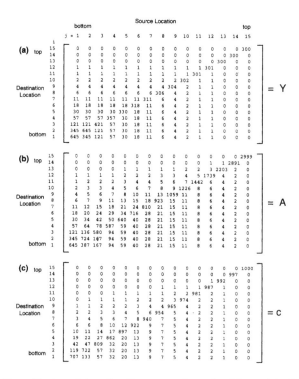

Figure 18.5: Transilient matrices for the convective mixed layer (similar to Fig. 18.2).

18.6.4   Both buoyant and mechanically mixed layer

Although growing through an initially statically stable environment, both wind shear and surface heating combine to generate strong turbulence and rapid growth of a deep mixed layer (Fig. 18.8). Winds shown in Fig. 18.8a become subgeostrophic but constant with height over most of the mixed layer, with evidence of a logarithmic wind shear in the surface layer. A deeper superadiabatic layer forms in the surface layer for this situation compared with that for free convection. For both this case and that of free convection, the problems of an unrealistically-deep surface layer are not encountered with this parameterization. Turbulent fluxes become linear with height for both heat and momentum, as required for a shape-preserving well-mixed layer. A deep entrainment zone is shown in Fig. 18.9 as the region where the top of turbulence extends above the "top of the mixed layer" as defined by the most negative heat flux. This is associated with wind shear and enhanced entrainment at the top of the mixed layer.

18.6.5   Stable boundary layer

Starting with an adiabatic temperature profile as often occurs at the end of the day, the stable boundary layer is simulated by applying a negative surface heat flux to cool the bottom grid cells. Turbulence is generated by wind shears associated with a constant geostrophic wind speed of 10 m/s.

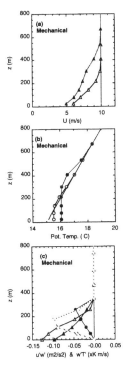

Figure 18.6: Profiles of (a) wind speed (triangles); (b) potential temperature (circles); and (c) vertical fluxes of momentum (triangles) and heat (circles), for a mechanically-mixed boundary layer in an initially stable environment. Solid lines without data points represent initial conditions, open data points show simulated results after 1 h, solid points give 6 h results.

A statically stable layer quickly forms in the simulated boundary layer, and strengthens and deepens during the night (Fig. 18.10b). The strong winds and wind shears (Fig. 18.10a) in this case cause the nocturnal inversion depth to grow to about 800 m by 6 h after surface cooling was initiated.

Figure 18.11 shows a large region of intermittent turbulence associated with the imposed oscillation of the critical Richardson number. This parameterization, designed to simulate the effects of unresolved gravity waves, appears highly successful. This procedure prevents the bottom grid cells from becoming prematurely decoupled from the rest of the boundary layer by excessive cooling. Yet, under lighter wind conditions (not plotted here), decoupling is simulated as desired. For the scenario simulated here, the overall average turbulence depth (found by applying a four-time-step running average) gradually decreases with time as the static stability increases during the night.

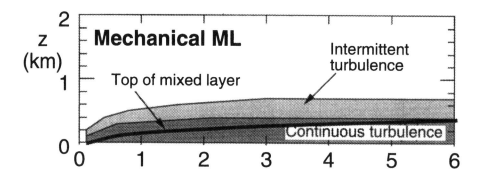

Figure 18.7: Evolution of the depth of the simulated mechanically-mixed boundary layer (like Fig. 18.4).

18.6.6 Diurnal cycles of boundary layer forcings (including pollution dispersion)

This two-day simulation starts at 0600 SLT (simulated local time). The results from the simulation are categorized into two daytime and two nighttime periods, where each day is from 0600 to 1800 SLT, and each night from 1800 to 0600 SLT.

Starting at 0600 SLT with a statically stable lapse rate and constant geostrophic wind with height, a sinusoidally increasing heat flux is applied at the surface during the first day. In response a mixed layer forms and deepens during the first day (Fig. 18.12a), and is accompanied by a well-mixed but subgeostrophic wind profile (Fig. 18.12b). At the bottom of the mixed layer is a shallow surface layer with superadiabatic lapse rate and approximately logarithmic wind profile. By 1800 SLT the mixed layer is about 1 km deep, and cooling has just started in the surface layer.

During the first night, when constant cooling is applied to the lowest grid point, a stable boundary layer forms with θ profiles that are roughly exponential with height, but gradually change to a more linear profile by morning (Fig. 18.12c). Above the stable boundary layer, winds gradually accelerate to their geostrophic value (Fig. 18.12d), while winds closer to the ground decelerate.

On the second day, the process repeats, except that θ is initially adiabatic in the residual layer (Fig. 18.12e). As a result, there is a rapid rise of the mixed-layer depth between 1000

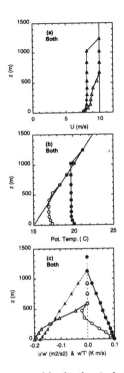

Figure 18.8: Mixed-layer profiles caused by both wind shear and buoyant convection (similar to Fig. 18.6).

SLT and noon after the nocturnal stable boundary layer is finally destroyed by heating. The winds in the shallow mixed layer at 1000 SLT are substantially slower than geostrophic (Fig. 18.12f) but accelerate to a faster but steady value for the remainder of the afternoon. Results from the second night are similar to those of the first, and are not plotted.

Figure 18.13 shows a time history of depths of the mixed layer and turbulence. Day 1 is clearly different from Day 2, where the latter exhibits a rapid rise of the mixed layer between 1030 and 1100 SLT. During this second morning there is a period of time between about 930 and 1030 SLT when turbulence has already sporadically reached the higher capping inversion, but when the average mixed layer has not. This corresponds to the time when there is only a weak and shallow stable layer aloft, as the last remains of the previous nocturnal stable layer.

Figure 18.14 shows isopleths of lead-oxide concentration after being advected and turbulently mixed from the surface continuous point source. Tracers from this source are input into the nearest grid cell and instantly dispersed over the whole volume of that single grid cell. These tracers later disperse by turbulence into a growing mixed layer during Day 1, and advect downwind.

During Night 1 a residual layer forms holding the pollutants from the previous day, and these pollutants advect rapidly downwind. Those pollutants at the base of this residual cloud are partially depleted by dry deposition at the surface, and are advected downwind at a slower speed than the pollutants aloft. Meanwhile, the new surface emissions are trapped

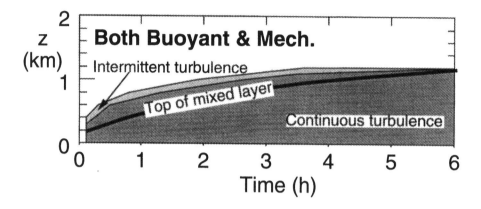

Figure 18.9: Evolution of the turbulent depth of a mixed layer forced by both shear and buoyancy (similar to Fig. 18.4).

near the ground in the developing stable nocturnal boundary layer.

By the second day, the cloud of pollutants aloft have advected so rapidly downwind that they are nearly separated from the new surface emissions. As a result, when the mixed layer grows high enough at noon to fumigate the residual pollutants back to the surface, there are plugs of high pollutant concentration both close to the source and near the downwind edge of the model, with a region of low concentration in between. During the second night the cloud of pollutants in the residual layer again separate from the new surface emissions.

18.6.7 Discussion

The idealized scenarios above were selected to illustrate some of the wide-ranging meteorological conditions for which the T3 model can be applied. The last case, in particular, illustrates how complex combinations of simultaneous forcings, such as varying buoyancy and shear, pose no exceptional difficulties. Because the transilient method is a general turbulence parameterization rather than specifically a boundary layer parameterization, it is not restricted to operation in archetypical boundary-layer scenarios such as a mixed layer in free convection or a neutral boundary layer.

Also, the expanding vertical grid was chosen to demonstrate the capabilities in situations such as long-range pollutant-transport models or global climate models where computer limitations restrict the number of grid points in the vertical. As the top of the boundary

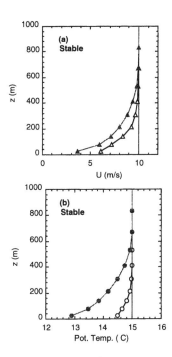

Figure 18.10: Profiles of (a) wind speed and (b) potential temperature for a stable nocturnal boundary layer, where open and solid data points show results at 1 h and 6 h into the simulation, respectively. Solid line in (a) indicates both the initial wind profile, and the imposed geostrophic winds. Solid line in (b) indicates the initial adiabatic temperature profile, to simulate the evening transition from a convective mixed layer to a stable boundary layer.

layer grows into grid layers of greater thickness, resolution is obviously lost but the nature of turbulence and dispersion within the boundary layer is modeled properly.

The problem of excessive mixing of the bottom couple grid cells remains (Chrobok, 1988), especially for grid thickness less than 30 m. During the calibration of the transilient parameterization, it was found that this problem could easily be eliminated by increasing the reference potential (Y_{ref}) that is added to the elements of $\mathbf{Y}$ along the main diagonal. However, a side effect of this cure was to cause excessive entrainment and excessively negative heat flux at the top of the mixed layer. For an evenly-spaced vertical grid arrangement, this is not usually a problem because a "best" reference potential can usually be found. However, for unequally-spaced vertical grids, the nonlocal interaction between many grids of different thicknesses made it impossible to adequately resolve this problem. The values of parameters recommended earlier in this paper are tuned to a bottom grid thickness of 40 km or greater, in anticipation of use of the model where coarse grids are required in order to make forecasts for large-horizontal domains.

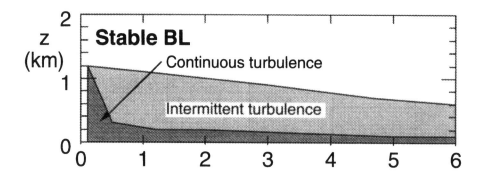

Figure 18.11: Evolution of turbulent depth of the stable boundary layer under strong wind shear. The intermittency in turbulence by unresolved gravity waves is simulated by the oscillating critical Richardson number.

18.7 Conclusions

The nonlocal nature of turbulence, as captured with the transilient parameterization given above, has been shown to simulate a range of atmospheric stabilities and forcings. Intrinsic in the transilient model is the concept that pollutants move with the air in nonlocal mixing. Thus, a forecast of turbulence and boundary-layer evolution is a natural component of a forecast of pollutant dispersion.

The parameterization scheme above yields symmetric matrices, even though large eddy simulations suggest that the matrices should be asymmetric for free convection. New parameterizations have recently been proposed using convective available potential energy – an asymmetric measure that yields an asymmetric transilient matrix. Applications of the transilient matrix, published elsewhere, have been in 3-D mesoscale weather forecast models, 1-D atmospheric and oceanic boundary layer models, pollutant transport models, models for pollutant transport across forest canopies (Inclan et al 1996), cumulus convection, and clear air turbulence.

Acknowledgements

Both authors appreciated the opportunity to work at the IBM Environmental Sciences & Solutions Centre in Bergen, Norway, where this research was started. The first author

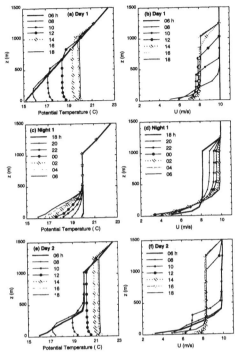

Figure 18.12: Profiles of potential temperature (a, c, e) and wind speed (b, d, f) every two hours during the first 1.5 days of a two-day simulation of a diurnally varying boundary layer. The second night was very similar to the first night, and is not plotted here. Initial conditions are shown as the solid lines without data points in (a & b). A geostrophic wind of 10 m/s is applied at all heights and times.

thanks Prof. Roberto San José for the initial interest in transient turbulence theory, which resulted in the invitation to visit IBM-Bergen. We appreciate Henryk Modzelewski's aid in using the computers at IBM. We also thank Andreas Bott at the University of Mainz for the non-staggered version of his advection algorithm. The United States National Science Foundation provided partial support under grant ATM-9411467, as did the Canadian Natural Sciences and Engineering Research Council.

Bibliography

Alapaty, K., J.E. Pleim, S. Raman, D.S. Niyogi, and D.W. Byun, 1997: Simulation of atmospheric boundary layer processes using local and nonlocal closure schemes. J. Appl. Meteor., 36, 214-233.

Andrén, A. and C.-H. Moeng, 1993: Single-point closures in a neutrally stratified boundary layer. J. Atmos. Sci., 50, 3366-3379.

Bott, A., 1989: A positive definite advection scheme obtained by nonlinear renormalization of the advective fluxes. Mon. Wea. Rev., 117, 1006-1015.

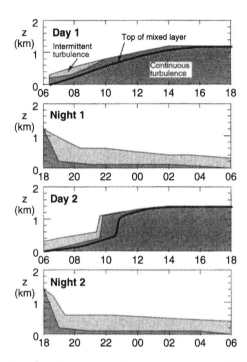

Figure 18.13: Evolution of turbulence depth and average top of the mixed layer, h , during the 48 h boundary-layer simulation.

Bott, A., 1989: Reply. Mon. Wea. Rev., 117, 2633-2636.

Bott, A., 1992: A positive definite advection scheme for use in long range transport models: extension to monotonicity. Air Pollution Modeling and Its Application IX, H. van Dop (Ed.), NATO Challenges of Modern Society, Vol 16. Plenum Press, NY.

Chrobok, G., 1988: Zur numerischen Simulation konvektiver Grenzschichten mit Intergralen Schliessungsansatzen. Diplomarbeit im Fach Meteorology, November 1988, Inst. Fur Meteorologie und Klimatologie der Universitat Hannover. 92 pp. (copies available from Stull)

Clarke, R.H., A.J. Dyer, R.R. Brook, D.G. Reid and A.J. Troup, 1971: The Wangara Experiment: Boundary Layer Data. Div. of Meteor. Phys. Tech. Paper No. 19. CSIRO, Melbourne. 350pp.

Crum, T.D. and R.B. Stull, 1987: Field measurements of the amount of surface layer air versus height in the entrainment zone. J. Atmos. Sci., 44, 2743-2753.

Ebert, E.E., U. Schumann and R.B. Stull, 1989: Nonlocal turbulent mixing in the convective boundary layer evaluated from large eddy simulation. J. Atmos. Sci., 46, 2178-2207.

Finnigan, J.J., F.Einaudi, and D. Fua, 1984: The interaction between an internal gravity wave and turbulence in the stably-stratified nocturnal boundary layer. J. Atmos. Sci., 41, 2409-2436.

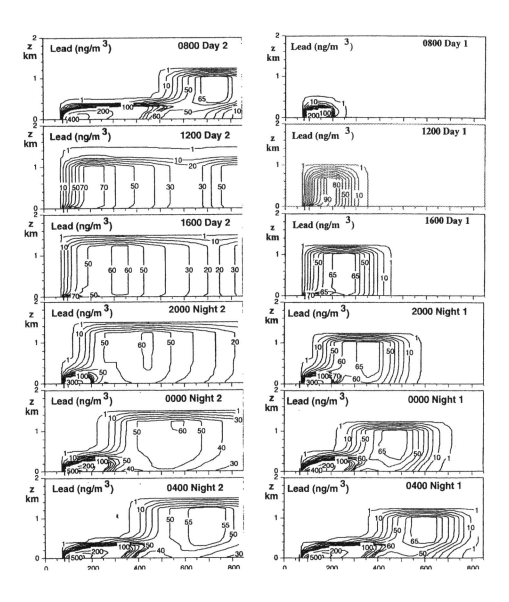

Figure 18.14: Simulated spread of neutrally-buoyant lead pollutants continuously released from a surface point source (solid black circle indicates source location), shown every 4 h during a two-day simulation. Times are simulated local time (SLT). Contour interval is 10 ng/m^3 for concentrations between 10 - 100 ng/m^3, and 100 ng/m^3 for greater concentrations. In addition, the contour for 1 ng/m^3 is plotted, and other special contours are shown where appropriate. (a) Day 1 and night 1. (b) Day 2 and night 2.

Inclan, M.G., R. Forkel, R. Dlugi, and R.B. Stull, 1996: Application of transilient turbulence theory to study interactions between the atmospheric boundary layer and forest canopies. Bound.-Layer Meteor., 79, 315-344.

Mahrt, L., 1981a: The early evening boundary layer transition. Quart. J. Roy. Meteor. Soc., 107, 329-343.

Mahrt, L., 1981b: Modelling the depth of the stable boundary layer. Bound.-Layer Meteor., 21, 3-19.

Mason, P.J. and D.J. Thompson, 1987: Large eddy simulations of the neutral-static-stability planetary boundary layer. Quart. J. Roy. Meteor. Soc., 113, 413-443.

Nappo, C.J., 1992: Gravity-wave-generated turbulence in the stable boundary layer over complex terrain. Preprints from the Tenth Symposium on Turbulence and Diffusion, 29 Sept P 2 Oct 1992, Portland, Oregon. Published by the American Meteorological Society, 45 Beacon St., Boston, MA02108-3693. J64-J67.

Nieuwstadt, R.T.M. and R.A.Brost, 1986: The decay of convective turbulence. J. Atmos. Sci., 43, 532-546.

Raymond, W.H. and R.B. Stull, 1990: Application of transilient turbulence theory to mesoscale numerical weather forecasting. Mon. Wea. Rev., 118, 2471-2499.

Smolarkiewicz, P.K., 1989: Comment on 'A positive definite advection scheme obtained by nonlinear renormalization of the advective fluxes', Mon. Wea. Rev., 117, 2626-2632.

Stull, R.B., 1976: The energetics of entrainment across a density interface. J. Atmos. Sci., 33, 1260-1267.

Stull, R.B., 1986: Transilient turbulence theory, Part III: Bulk dispersion rate and numerical stability. J. Atmos. Sci., 42, 2070-2072.

Stull, R.B., 1988: An Introduction to Boundary Layer Meteorology. Kluwer Academic Press. Dordrecht. 666 pp.

Stull, R.B., 1993: Review of nonlocal mixing in turbulent atmospheres: transilient turbulence theory. Bound.-Layer Meteor., 62, 21-96.

Stull, R.B., 1996: A procedure to determine smoothly-varying vertical grid spacings. Atmosphere-Ocean., 34, 581-588.

Stull, R.B. and A.G.M. Driedonks, 1987: Applications of the transilient turbulence parameterization to atmospheric boundary layer simulations. Bound.-Layer Meteor., 40, 209-239.

Stull, R.B. and T. Hasagawa, 1984: Transilient turbulence theory, Part II: Turbulent adjustment. J. Atmos. Sci., 41, 3368-3379.

Zhang, Q. and R. Stull, 1992: Alternative nonlocal descriptions of boundary layer evolution. J. Atmos. Sci. , 49, 2267-2281.

Index

Milton Keynes UK
Ingram Content Group UK Ltd.
UKHW051945071024
449327UK00026B/2169